Cisco Voice over Frame Relay, ATM, and IP

Steve McQuerry,
Kelly McGrew,
Stephen Foy

CISCO SYSTEMS

CISCO PRESS

Cisco Press
201 W 103rd Street
Indianapolis, IN 46290 USA

Cisco Voice over Frame Relay, ATM, and IP

Editors: Steve McQuerry, Kelly McGrew, Stephen Foy

Copyright© 2001 Cisco Systems, Inc.

Cisco Press logo is a trademark of Cisco Systems, Inc.

Published by:
Cisco Press
201 West 103rd Street
Indianapolis, IN 46290 USA

Printed in the United States of America

1 2 3 4 5 6 7 8 9 0 04 03 02 01

First printing April 2001

Library of Congress Cataloging-in-Publication Number: 99-67923

ISBN: 1-57870-227-5

Warning and Disclaimer

This book is designed to provide information about Cisco Voice over Frame Relay, ATM, and IP networks. Every effort has been made to make this book as complete and as accurate as possible, but no warranty or fitness is implied.

The information is provided on an "as is" basis. The authors, Cisco Press, and Cisco Systems, Inc., shall have neither liability nor responsibility to any person or entity with respect to any loss or damages arising from the information contained in this book or from the use of the discs or programs that may accompany it.

The opinions expressed in this book belong to the author and are not necessarily those of Cisco Systems, Inc.

Trademark Acknowledgments

All terms mentioned in this book that are known to be trademarks or service marks have been appropriately capitalized. Cisco Press or Cisco Systems, Inc., cannot attest to the accuracy of this information. Use of a term in this book should not be regarded as affecting the validity of any trademark or service mark.

Feedback Information

At Cisco Press, our goal is to create in-depth technical books of the highest quality and value. Each book is crafted with care and precision, undergoing rigorous development that involves the unique expertise of members from the professional technical community.

Readers' feedback is a natural continuation of this process. If you have any comments regarding how we could improve the quality of this book, or otherwise alter it to better suit your needs, you can contact us through e-mail at feedback@ciscopress.com. Please make sure to include the book title and ISBN in your message.

We greatly appreciate your assistance.

Publisher	John Wait
Editor-in-Chief	John Kane
Executive Editor	Brett Bartow
Cisco Systems Program Manager	Bob Anstey
Managing Editor	Patrick Kanouse
Development Editor	Kitty Jarrett
Production Editors	Argosy
	Marc Fowler
Lead Course Developer	Joe Drotter
Contributing Course Developers	Allison Humphries
	Kelly McGrew
	Priscilla Oppenheimer
Technical Editors	Christina Hattingh
	John Livengood
	Martin Walshaw
Team Coordinator	Tammi Ross
Cover Designer	Louisa Klucznik
Production Team	Argosy
Indexer	Tim Wright
Proofreader	Bob LaRoche

CISCO SYSTEMS

Corporate Headquarters
Cisco Systems, Inc.
170 West Tasman Drive
San Jose, CA 95134-1706
USA
http://www.cisco.com
Tel: 408 526-4000
      800 553-NETS (6387)
Fax: 408 526-4100

European Headquarters
Cisco Systems Europe
11 Rue Camille Desmoulins
92782 Issy-les-Moulineaux
Cedex 9
France
http://www-europe.cisco.com
Tel: 33 1 58 04 60 00
Fax: 33 1 58 04 61 00

Americas Headquarters
Cisco Systems, Inc.
170 West Tasman Drive
San Jose, CA 95134-1706
USA
http://www.cisco.com
Tel: 408 526-7660
Fax: 408 527-0883

Asia Pacific Headquarters
Cisco Systems Australia,
Pty., Ltd
Level 17, 99 Walker Street
North Sydney
NSW 2059 Australia
http://www.cisco.com
Tel: +61 2 8448 7100
Fax: +61 2 9957 4350

Cisco Systems has more than 200 offices in the following countries. Addresses, phone numbers, and fax numbers are listed on the Cisco Web site at www.cisco.com/go/offices

Argentina • Australia • Austria • Belgium • Brazil • Bulgaria • Canada • Chile • China • Colombia • Costa Rica • Croatia • Czech Republic • Denmark • Dubai, UAE • Finland • France • Germany • Greece • Hong Kong • Hungary • India • Indonesia • Ireland • Israel • Italy • Japan • Korea • Luxembourg • Malaysia • Mexico • The Netherlands • New Zealand • Norway • Peru • Philippines • Poland • Portugal • Puerto Rico • Romania • Russia • Saudi Arabia • Scotland • Singapore • Slovakia • Slovenia • South Africa • Spain Sweden • Switzerland • Taiwan • Thailand • Turkey • Ukraine • United Kingdom • United States • Venezuela • Vietnam • Zimbabwe

About the Editors

Steve McQuerry, CCIE #6108, is an instructor and consultant with more than ten years of networking industry experience. He is a Certified Cisco Systems Instructor (CCSI) teaching routing and switching concepts for Global Knowledge. In addition to teaching the CVoice course, Steve regularly delivers the ICND, BCMSN, BSCN, CIT, CID, and OSPF/BGP courses. Additionally, Steve has developed and taught custom Cisco switching courses for large corporate customers. In his consulting capacity, he provides design, integration, and troubleshooting services to enterprise customers. Previous to founding his own network consulting and education firm, Steve was a Senior Consultant at the Information Connection, where he worked on various network implementation projects including responsibilities for network design, project management, security planning, and LAN/WAN troubleshooting. As a member of the Network Planning Team at the University of Kentucky Hospital, Steve designed and implemented networking solutions for the hospital's LAN and WAN. Steve holds a B.S. degree in engineering physics from the Eastern Kentucky University.

Kelly McGrew is a Certified Cisco Systems Instructor (CCSI) and vice-president of mcgrew.net inc., a network training and course development firm. He has worked as a trainer throughout the world. Kelly holds the CCNP-Voice Access Specialist and CCDA certifications. He has more than 15 years of experience in the networking industry, including experience with a variety of LAN and WAN protocols seldom seen in today's IP-centric world. Kelly has held a variety of positions for leaders in the networking industry. These include positions as a network systems engineer for CompuServe Network Services and MCI/WorldCom, an instructor/consultant for Chesapeake Computer Consultants, Inc., a program manager for Microsoft Corporation, and an instructor/consultant under a leased-employee relationship for Cisco Systems, Inc. He is a graduate of The Evergreen State College (B.A.) and obtained an M.B.A. from City University. Kelly is an associate member of the IEEE and member of the ASTD. Kelly currently focuses on teaching and course development in the Voice over Layer 2/IP arena. He currently resides in Olympia, Washington, with his wife, Tammy (also a CCSI and the president of mcgrew.net inc.), their son, Duncan, and the world's best doggie, Lady Buttons.

Stephen Foy is an internetworking instructor for Global Knowledge, Cisco's largest worldwide training partner, with more than 18 years of experience in the networking industry. Stephen is a Cisco Certified Network Professional (CCNP) and Certified Cisco Systems Instructor (CCSI) conducting classes for various vendors including Cisco Systems, Digital Equipment, and Motorola. He teaches the ICND, ACRC, BCRAN, CATM, MCCM, and CVoice classes on a regular basis.

About the Lead Course Developer

Joe Drotter is a course developer specializing in various voice technologies for Cisco Systems. Currently he works for the Cisco Interactive Mentor (CIM) group, developing a CD-based interactive course on quality of service (QoS) for VoIP. He has more than 28 years of experience in the IT world, holding a variety of positions with Motorola (instructor/consultant), DEC (network designer/course developer), Stone and Webster Engineering (programmer/analyst), and as a graduate school associate professor of data communications for Southern NH University. He is a graduate of Merrimack College (B.A.) and obtained an M.B.A. at New Hampshire College. Joe currently resides in the "Live Free or Die" state of New Hampshire, or as it has been renamed, the "Live, Freeze, and Die" state.

About the Technical Reviewers

This book's reviewers contributed their considerable practical, hands-on expertise to the entire development process for *Cisco Voice over Frame Relay, ATM, and IP.*

Christina Hattingh is a member of Technical Marketing organization at Cisco Systems. In this role she works closely with product management and engineering, and focuses on assisting Cisco Sales Engineers, partners, and customers to design and tune enterprise and service-provider Voice over X networks. Prior to this she developed PBX Call Center products at Nortel Networks. Her development experience in X.25 and network management provides background to the issues involved in migrating customers' traditional data and voice networks to packet-based technologies.

John Livengood, Certified Cisco Systems Instructor (CCSI), is the general manager of voice class for Advance Network Information, formerly known as Information Innovation, Inc. He was a contract instructor for his own company, netLessons, delivering Cisco University seminars and Cisco certified classes. The classes John teaches include ICND and CIPT. Before starting his own company, John worked for Boeing, supporting large-scale network services. He played a major role in rebuilding over 50 percent of the 200 network distribution closets on the Wichita Boeing campus. John started in the computer industry in 1971. In the U.S. Navy, he worked on a network of CDC processors responsible for the database tracking all vessels in the Atlantic and Mediterranean theaters. He spent 22 years in aviation supporting engineering and factory systems. John was the regional coordinator for DECUS (Digital Equipment Computer Users Society) in the late 1980s.

Martin Walshaw, CCIE #5629, CCNP, CCDP, is a systems engineer working for Cisco Systems in the Enterprise Line of Business in South Africa. His areas of specialty are multiservice (voice and video) as well as security, which keeps him busy both night and day. During the last 12 years or so Martin has dabbled in many aspects of the IT industry, ranging from programming in RPG III and COBOL to PC sales. When Martin is not working, he likes to spend all of his available time with his wife, Val, and his son, Joshua. Without their patience, understanding, and support, projects such as this one would not be possible.

Dedications

I would like to dedicate this work to my beautiful wife, Becky, and my wonderful children, Katie, Logan, and Cameron. Your love and support give me the strength to spend the extra hours needed to complete a project like this. I love you all very much.

—Steve McQuerry

This work is dedicated to my family—Mom and Dad, Mom and Dad Lee, Tammy, and Duncan—and to the Nikolsburg Rebbe, Rabbi Yechiel Michel Liebvitz, shlita.

In the spirit of NetAid, 20% of my royalties from this work are being donated to charities that help those in need. Many of us in the computer network field have been blessed with meaningful work, enjoyable travel opportunities, and a greater-than-average income. It only seems fitting to recognize that there are those who have less than we do and to reach out to help them. Regardless of your station in life, I encourage you to do the same.

—Kelly McGrew

Acknowledgments

I would like to take this time to acknowledge all of those who have been instrumental in making my contribution to this book possible:

- First of all, let me say that working with the personnel at Cisco Press has been a real treat. A special thanks to Brett Bartow for convincing me to do another book and for having the faith in me that allowed me to get into this part of the business. Thanks to Amy Lewis for answering my endless barrage of questions and for getting me all the support material I could ever ask for. Thanks to Kitty Jarrett for putting up with a techie who thinks he knows how to write. Kitty, this project is truly better because of the effort that you have put into it. Also Thanks to John Kane and Tammi Ross for the guidance and support.

- Cisco Internet Learning Solutions Group and the course developers: Joe Drotter, who developed the concept for the course as well as the majority of the content for the initial release and the latest release; Allison Humphries, Kelly McGrew, and Priscilla Oppenheimer, who all contributed to the course development; Christina Hattingh and other Cisco employees, who created presentations on the subject of voice and data integration used to enhance some of the topics; and Stephen Foy, for his contributions in reviewing the latest version of the course in the beta stage.

- The technical editors: Christina Hattingh, John Livengood, and Martin Walshaw.

- The co-authors: Steve Foy and Kelly McGrew. Thanks for putting up with me on this project.

- My students and fellow instructors.

- My family, for their never-ending patience and understanding during this project and all of my projects.

- God, for giving me the skills, talents, and opportunity to work in such a challenging and exciting profession.

—Steve McQuerry

Firstly and mostly, I must thank Brett Bartow of Cisco Press for his undying patience during my work on this book. Over a six-month period during which I was supposed to be diligently working on this book, I spent only four weeks at home and the rest of the time working long days on the road, which made it difficult for me to meet my schedule with Brett. Thanks, Brett, for having the patience of Job. I also owe a great debt to Kitty Jarrett, without whose deft editing hand this work would have been much less coherent. To Steve McQuerry and Steve Foy: a big thanks for doing their share on this project.

Secondly, having done much of the work on this during the brief periods I spent with my family, I must thank my wife, Tammy, and our son, Duncan, for their understanding when I was "heads-down" on this work.

Thirdly, this work would not have been possible without the assistance of several people who helped me along the way. They are: my mother and father, Mom and Dad Lee, Dan and Dottie Lehuta, Peggy O'Neall, Ron Seymour, Tom Martin, Brian Bailey, Maureen O'Brien, Scott Bulger, Theran Lee, Doug Longstreth, Sean Stinson, Andrea Lindsley, Gurdeep Singh Pall, Bernard Aboba, Glenn Tapley, Terry Slattery, Terry's wife, partner, and a most charming lady—Peggy Slattery, Bruce Enders, Paul Simoneau, David Eitelbach, Jawad Khaki, Ken Crocker, and Gary Rubin.

Those who helped me with content include Joe Drotter—who is the real author of this work, Christina Hattingh—one of the finest engineers I've had the honor to know and work with, Alison Humphries of The Garnett Group, and numerous engineers and reviewers both inside and outside of Cisco Systems, Inc., most of whose names I do not know. Anthony Wolfenden, Andy Espinosa, Jean Altman, and James Del Buono of Cisco Systems, Inc. World Wide Training were of great help in providing access to equipment, people, and facilities. This book would not have been possible without their assistance.

I was also privileged to have the continued friendship, irregular e-mail, or occasional companionship of David Eitelbach, Wendy Huff, Tom Mehlhaff, John Odegaard, Larry Perrin, and Tony Varela during the few periods I was home. Their laughter, smiles, and friendship have helped keep me together. Finally, I must thank Reb Yaakov Sholem Schwartz for his support, encouragement, and assistance over a period much longer than the gestation period of this title. Any errors, despite the best efforts of others, are solely mine.

—Kelly McGrew

I would like to thank the love of my life, Charlene, for keeping me focused during this project. Also, thanks to my children, Shannon and Sara, for their patience while Daddy was busy. Finally, I'd like to thank all of my co-workers at Global Knowledge for their support and encouragement.

—Steve Foy

Contents at a Glance

Contents

Foreword

Cisco Voice over Frame Relay, ATM, and IP presents in book format all the topics covered in the challenging instructor-led and e-learning certification preparation courses of the same name. CVoice teaches you the knowledge and skills needed to install and configure Cisco voice and data network devices. You will learn how to analyze and select voice hardware and software, design and configure Voice over Frame Relay, ATM, and IP networks, integrate and optimize enterprise networks in remote branch offices using integrated access technology, develop a process for integrating Cisco equipment with PBXs, and appraise existing branch and regional office voice network services and choose the optimum transmission method for voice traffic. If you are seeking to gain a practical understanding of Cisco's implementation of voice technology (VoIP, VoFR, VoATM), you will benefit from the information presented in this book.

Cisco and Cisco Press present this material in text-based format to provide another learning vehicle for our customers and the broader user community in general. Although a publication does not duplicate the instructor-led environment, we acknowledge that not everyone responds in the same way to the same delivery mechanism. It is our intent that presenting this material via a Cisco Press publication will enhance the transfer of knowledge to a broad audience of networking professionals.

Cisco Press will present existing and future courses through these Coursebooks to help achieve the principal objectives of the Cisco Internet Learning Solutions Group: to educate the Cisco community of networking professionals and to enable that community to build and maintain reliable, scalable networks. The Cisco Career Certifications and classes that support these certifications are directed at meeting these objectives through a disciplined approach to progressive learning. The books Cisco Press creates in partnership with Cisco Systems will meet the same standards for content quality demanded of our courses and certifications. It is our intent that you will find this and subsequent Cisco Press certification and training publications of value as you build your networking knowledge base.

Thomas M. Kelly
Vice-President, Internet Learning Solutions Group
Cisco Systems, Inc.
November 2000

Introduction

The text of this book is derived from the instructor-led customer course presented by Cisco Systems, Inc., training partners. The course focuses on introducing students to the latest version of how Cisco handles voice transmission over its latest equipment. This book is designed as an alternative to the course in preparing readers for voice and data integration using Cisco equipment. The overall focus of this material is to teach fundamental voice/telephony skills and knowledge in order to build intelligent, scalable Cisco voice networks.

This Book's Audience

This text targets many people in the internetworking industry. The main focus is for technicians and engineers who know data technologies and wish to learn voice technologies and how to integrate voice and data in their networks. It is also designed for those who know voice technologies and want to learn how to use Cisco equipment to merge the voice and data technologies together. While this is not an introductory-level course, it does provide some basic voice technologies overviews for those who are uninitiated in this realm of the communications industry.

As mentioned, this is not an introductory-level book. The authors assume that you have experience and knowledge with Cisco routers, routing technologies, WAN technologies, and the Cisco IOS software. At a minimum, you should have experience configuring Cisco IOS software. You should have been exposed to the information from the Cisco Interconnecting Cisco Network Devices (ICND) or Introduction to Cisco Router Configuration (ICRC) course or book. Also, you should have exposure to and familiarity with the topics in Building Scalable Cisco Networks (BSCN) or Advanced Cisco Router Configuration (ACRC).

It would also be beneficial if you have knowledge of voice essentials like those learned in the Telecommunications Research Associates (TRA) "Voice Essentials" self-study CD. This CD is available from Cisco Systems and from Telecommunications Research Associates (TRA), St. Mary's, Kansas (800-872-4736).

Readers need to have hands-on experience installing and configuring Cisco routers for Frame Relay, ISDN, or ATM networks, and being familiar with these technologies' terms is also beneficial to the understanding of the text. Finally, experience installing voice technology and basic understanding of voice technology terms is assumed.

Objectives

The objectives of this book are to teach the reader how to install and configure Cisco voice and data network routers; how to configure Cisco voice-enabled equipment for Voice over Frame Relay, ATM, and IP; how to configure voice ports, dial peers, and special commands to enable voice transmission over a data network; and how to perform voice traffic analysis to determine how to improve the quality of services (QoS) for delay-sensitive voice traffic.

Conventions Used in This Book

This book uses the following conventions:

- Important or new terms are *italicized*.

- All configuration examples appear in `monospace` type, and command syntax uses the following conventions:

  — Commands and keywords are in **bold** type.

  — Arguments, which are placeholders for values the user inputs, appear in *italics*.

  — Square brackets ([ ]) indicate optional keywords or arguments.

  — Braces ({ }) indicate required choices.

  — Vertical bars (|) are used to separate choices.

This Book's Organization

In order to facilitate the learning process, the book is broken up into five parts. The text was designed so that it can be read as a whole or each part can be independently read. For example, a reader who has experience with voice technologies and understands how voice is carried over communications facilities may choose to skip over Part I. If you have no interest in the hardware you may choose to skip over Part II, and if you have no interest in the changing technologies you may choose to skip over Part IV. The book was designed, however, to work as a whole unit, by taking the reader through the basic understanding of voice technologies, Cisco hardware options, configurations, and, finally, the changing voice technologies. The section layout that follows provides greater detail of each section and chapter.

Part I, "Voice Technologies," reinforces the reader's understanding of the trend of voice and data integration in Chapter 1, "Merging Voice and Data Networks." Chapter 2, "Introduction to Analog Technology," provides a review of basic analog voice technology. Chapter 3, "Introduction to Digital Voice Technology," describes how voice is carried over digital services. These chapters lay the foundation for integrating voice over existing data networks. By understanding how voice technologies work, readers can better understand the goals of Cisco Systems to provide voice and data integration.

Part II, "Cisco Products and Solutions," introduces the reader to Cisco platforms that support voice applications. Chapter 4, "Cisco Voice Hardware," describes the available routers and options that support voice-over applications. Chapter 5, "Applications for Cisco Voice-over Routers," shows the roles that each device takes in the data/voice integration role. Chapter 6, "Setting Up Cisco Routers," explains how the hardware is installed into the routers and how it interfaces with voice technologies.

Part III, "Configuring Cisco Voice Solutions," teaches the reader how to configure Cisco voice products for operation. Chapter 7, "Configuring Voice Ports and Dial Peers for Voice," provides the basics for voice-port configuration and how voice is mapped to dial peers for voice-over applications. Chapter 8, "Configuring Cisco Routers for VoFR," helps the reader understand how to configure the voice-enabled routers to transport voice over Frame Relay networks. Chapter 9, "Configuring Cisco Routers for VoATM," describes the steps required to provide voice over an ATM network. Chapter 10, "Configuring Cisco Access Routers for VoIP," shows how to provide voice traffic over a standard IP network.

Part IV, "Changing Voice Technologies," shows how the data and voice networks are changing and why there is a growing trend to integrate voice and data networks. Chapter 11, "Cisco IP Telephony Solutions: New-World Telephony," shows how the face of telephony communications is changing and how voice-enabled routers are taking an active role in this new world. Chapter 12, "Old-World Technology: Introduction to PBXs," reviews the technologies for current PBX operation. Chapter 13, "Network Design Guide," describes the six-step process involved in integrating voice and data networks and provides some case studies to demonstrate these steps.

Part V, "Appendixes," provides Appendix A, "Answers to Review Questions," and Appendix B, "Further Reading."

This Book's Features

This book features actual router output and configuration examples to aid in the discussion of the configuration of these technologies. There are also many pointers, tips, and cautions spread throughout the text. You will also find many references to standards, documents, books, and Web sites that will be helpful in understanding networking concepts. At the end of each chapter your comprehension and knowledge will be tested by review questions prepared by a Cisco Certified Systems Instructor.

NOTE   The configuration options discussed in this text are referenced from Cisco IOS software version 12.0.

Voice Technologies

After reading this chapter, you should be able to perform the following tasks:

- Describe the traditional use of networks for both voice and data traffic.

- Describe the new-world network model that integrates both voice and data into a single network.

- Describe the Cisco network design model in general terms.

- Briefly describe Cisco's AVVID technology and the standards used for implementing AVVID.

CHAPTER 1

Merging Voice and Data Networks

Tremendous changes have rocked the computer networking industry over the past decade. Those changes have included widespread networking in businesses, such frequent networking in homes. As a result, Microsoft Windows 98 Second Edition now comes with built-in support for small networks (DHCP server and NAT), the World Wide Web, and the phenomenon of integrated voice and data. There is also a whole host of wireless technologies such as Cisco Systems Aironet, infrared links between Personal Digital Assistants (PDAs) and PCs, cell phone modems, and the JetCell product from Cisco Systems that permits cell phones to register with a campus network and uses the "in-house" voice over IP network to service those cell phone calls. This chapter provides a brief history of how we got where we are today, some of the tools we use for voice-over services, a brief explanation of network design, and where Cisco's products fit in the network, both from a network design perspective and from a customer perspective.

Traditional Networks

Let's look at a very brief history of networks in general. First we'll look at telephone, or voice, networks, and then we'll look at data networks. *Traditional networks* are networks that use the voice network to transmit voice, video streams, and data. Frequently a carrier's network is delivered to the end customer as if it were three distinct networks, one for each of the services listed above. More recently it has become common to share the service provider's network but still segregate the network into three discrete networks at the point of ingress to the customer's site. After we briefly look at traditional networks, we'll see how new ideas and technologies are combining these networks into a single network—from desktop through the network to the desktop at the opposite end of a call, video conference, or PC application. This makes a single network infrastructure, protocol management, and addressing structure possible.

Voice Networks

The phone system began as a system based on copper wires from Point A to Point B. For several decades the current on the wires was used to transmit only analog voice calls. In the mid-20th century, technologies were developed to transmit calls via digital networks. Today nearly all calls are handled within the network cloud by digital transmission techniques. Most home and small-business service is still analog, from the home handset to the central office (CO). For many readers this is probably the system with which you grew up. In Europe and other areas of the world, the Integrated Services Digital Network (ISDN)—a digital service that provides support for integrated voice, video, and data—is the predominant service for all but the smallest businesses and homes.

Analog Voice Networks

We'll discuss analog technology in much greater detail in Chapter 2, "Introduction to Analog Technology." For now, think of the old black-and-white movies where the operator used to pull cables from the switchboard and plug them into the proper jack to complete a call. What the operator was doing was providing a continuous copper circuit from one end of the call to the other. The entire circuit transmitted the voice as an analog wave, similar to the transmission of sound (as electricity) from a modern stereo receiver or audio/video receiver to the speaker. A dedicated pair of copper wires was required for each call because of the nature of an analog call in the old phone network. The advent of frequency division multiplexing (FDM) permitted a single analog circuit to carry several voice calls, still using analog technology, and typically over a coaxial cable. Eventually techniques were devised to permit the voice stream to be digitized and sent as a digital signal.

Digital Voice Networks

The gradual replacement of the analog voice network with a digital voice network began in the United States and other more developed countries in the 1960s. Many of these changes were required in order to carry not only greater volumes of voice traffic, but also data from the then-new computer industry. In the digital telephone network, calls from homes are typically digitized when they reach the local exchange carrier's (LEC's) CO. In a modern business the digital line can extend all the way to the private branch exchange (PBX), a kind of miniature CO that serves a private organization, and on to the phone handset itself. In these networks, the only analog voice stream is from the mouth to the microphone of the handset. We'll cover digital voice networks in detail in Chapter 3, "Introduction to Digital Voice Technology," and PBXs in Chapter 12, "Old-World Technology: Introduction to PBXs."

Data Networks

The widespread use of the phone network for personal communications led to the use of the same infrastructure for data networks. In the early days of data communications, the protocols used to communicate between machines were all proprietary. As the computer communications industry grew, the need for different machines to interwork with each other required common communications protocols.

In this section, we discuss traditional data networks so that you can compare them to new-world data networks, which are covered later in this chapter. Even though we categorize these networks as traditional as if they are no longer common, there are several places in the world where they are still the rule. One of the greatest challenges for those converting from traditional networks to new-world networks is making sure that the current network is well understood. Using a U.S.-centric model for a global network will not provide an optimal solution based on providers' local offerings and may lead to suggesting solutions that simply are not available in some parts of the world.

Network Cloud Terminology

The terms LEC, IXC, "tail circuit" or "last mile," and "local loop" are all U.S.-centric. In many countries the local telecommunications services have not been deregulated and there is only one carrier—the government. In that case, the terms are irrelevant and may be confusing to readers from outside the United States. It is not intended to be confusing—far from it. Because of the size and familiarity many readers have with the U.S. market, these terms are used to help distinguish the differences where they exist. In those countries that do not have a deregulated telecommunications market, all services are obtained through the local Post, Telephone, and Telegraph (PTT) agency.

Leased-Line Networks

In a leased-line network, the customer typically leases all of the services from one or more service providers. Leased-line networks may be simple point-to-point networks connecting two points, running a simple Layer 2 protocol such as HDLC and a single Layer 3 protocol such as IP. A leased-line network may also consist of a network running multiple protocols and connecting multiple sites, some in point-to-point configurations and some in a series topology called multipoint. A leased-line network typically consists of three components:

- The local circuit, frequently called the "tail-circuit" or the "last mile," from the customer's local site to the inter-exchange carrier's (IXC's) local point-of-presence (POP) via the LEC's CO.

- A circuit through the IXC's network to the remote site.

- Another local circuit at the remote site. See Figure 1-1 for an illustration of these network components.

In a site that does not cross tariff boundaries or in an area where competitive local exchange carriers (CLECs) provide service, all of the traffic may stay on the same network throughout. Leased-line networks typically come in two flavors, point-to-point and multi-drop.

Figure 1-1   *A Typical Leased-Line Network*

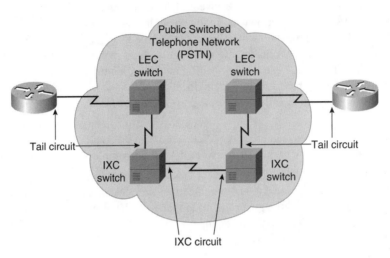

Point-to-Point Networks

A point-to-point network is just what it sounds like: a network with only two connections, one at each end. Point-to-point networks provide a specified amount of bandwidth, which is available whether it is used or not. An example of a common implementation of point-to-point networks that offer great error recovery and flexibility is the X.25 public data networks. These are discussed in more detail in the sidebar "X.25 Networks" later in this chapter. Point-to-point connections are frequently replaced with Frame Relay networks, as described in the "Frame Relay Data Networks" section of this chapter. The earliest days of point-to-point networks required the same kind of equipment on each end of the link. For the past few years, however, Point-to-Point Protocol (PPP) has provided a mechanism for heterogeneous networking. The PPP suite provides a rich set of features, discussion of which is beyond the scope of this book.

Multi-Drop Networks

Multi-drop networks were a common implementation using proprietary protocols such as IBM's Synchronous Data Link Control (SDLC) protocol and Digital Equipment Corporation's Digital Data Communications Message Protocol (DDCMP), as well as standardized protocols such as the High-Level Data Link Control (HDLC) protocol's Multilink Procedures (MLP) extension.

These networks permitted the phone company to bridge connections within their CO or POP so that a single frame sent from a primary station (typically the host or the host's front-end processor, or FEP) would be transmitted to all secondary stations. The frame included an address that one of the secondary stations recognized as its own, and that station would process the frame; all other stations on the multi-drop link would discard the frame.

Multi-drop networks were characteristic primarily of the proprietary mainframe and minicomputer worlds. Many enterprises have replaced these multi-drop networks with Frame Relay networks where Frame Relay is available. In many areas, the lack of Frame Relay circuits requires continued use of multi-drop networks.

Network Design

This book is not solely or primarily about network design, but we do discuss network design as it relates to converged voice and data networks in Chapter 13, "Network Design Guide." This section is meant to serve as a general review for those who have studied network design and a gentle introduction for those who have not studied this subject.

Cisco has a well-developed and highly respected design model that comprises three layers:

- Access layer
- Distribution layer
- Core layer

Each layer has specific characteristics and is used to perform specific tasks within the overall network architecture. The model is illustrated in Figure 1-2. We briefly discuss these characteristics and tasks in the following sections.

The Access Layer

The access layer is the layer that is closest to the users and provides them access to the resources on the network, hence the name. The primary task of the access layer is to provide a point of ingress/egress to users of the network. Characteristics of the access layer are:

- It uses switched or shared LAN technology in a campus environment.
- It uses remote access technologies (leased line, Frame Relay, ISDN, and asynchronous dial) in a WAN environment.
- It may provide for deterministic fail-over in mission-critical environments.

Figure 1-2   *Hierarchical Wide-Area Design*

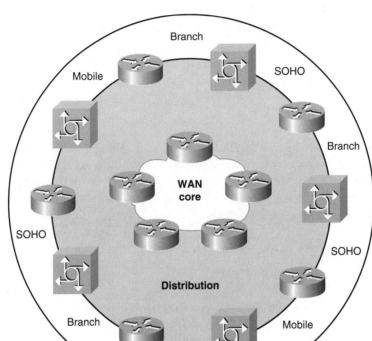

The Distribution Layer

The distribution layer is the layer of the design model that provides for the aggregation of traffic from the access layer and serves as the point of ingress to the core layer. The distribution layer may have several characteristics depending upon the specific features implemented. Among these are:

- Policies for routing, quality of service (QoS), or security
- Address summarization
- Broadcast or multicast domain boundaries
- Media translations (for example, between 10BaseT Ethernet and 100BaseT or Gigabit Ethernet)
- Routing redistribution between different routing protocols
- Bandwidth aggregation from the low- and medium-speed access links into higher-speed backbone links

This list is just a subset of characteristics that might be considered to apply to the distribution layer. Frequently the foundation of a distribution layer area is a multilayer switch, and for that reason they are called "distribution blocks" or "switch blocks." Figure 1-3 illustrates the use of distribution blocks in a network and the layers to which each series of devices belongs. See the Cisco Press book titled *Cisco LAN Switching* for more information on the concept of distribution blocks.

Figure 1-3   *Enterprise Campus Network Design Model*

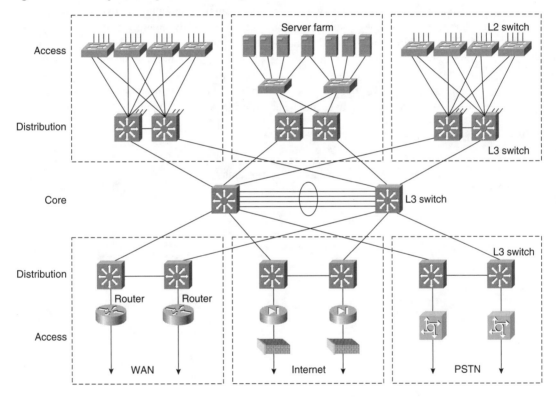

The Core Layer

The purpose of the core layer is to perform high-speed switching between one distribution area and another distribution area. Typically no filtering or route summarization is done here; this layer simply switches frames or cells quickly from one port to another port.

Sharing Circuits with Drop-and-Insert Devices

From the 1970s through the 1990s it was common to have both a voice network and a data network on entirely separate infrastructures in an enterprise. Many enterprises still use this kind of network, since the total bandwidth of their data and voice traffic will fit within one T1 or E1 circuit (see Chapter 3 for more information on T1 and E1 circuits).

To better use the bandwidth, they insert a channel bank between the telephone company's (telco's) circuit termination point and their telephone and computer equipment. The channel bank takes the input of a single trunk from the telephone company and sends some channels to the phone equipment and others to the data equipment. This type of deployment is illustrated in Figure 1-4. The function that channel banks provide is called drop-and-insert because the user can drop channels from, say, the voice stream and insert them into the data stream. Even with this kind of network, though, when telephone calls aren't being made there is unused (but paid-for) bandwidth. Likewise, unused bandwidth that is allocated to the data stream cannot be used during periods of heavy telephone circuit usage.

Figure 1-4   *A Shared Network Using Drop-and-Insert Devices*

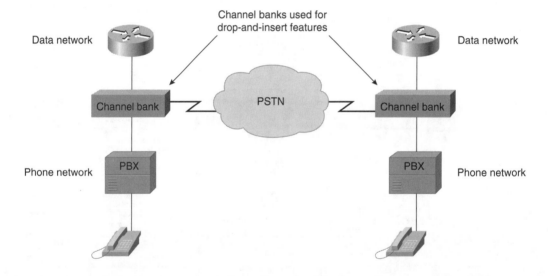

New-World Networks

The solution to the problems posed by traditional networks was to develop a new way to network voice and data, as described in the following sections.

Circuit Switching, Packet Switching, and Virtual Circuits

The traditional telephone system uses switches throughout the network to connect the two endpoints together. Once a call—irrespective of whether it is a voice call or a data call—is established, there is a continuous path from one end to the other that is dedicated solely to that call. This is the meaning of the term "circuit switching"; it refers to switching a shared infrastructure from one dedicated use to another dedicated use. This term is applied equally to either voice circuits or data circuits. Since many voice calls consist of "dead time" when neither party is speaking, there is a tremendous waste of resources in a circuit-switched network. On the other hand, since the circuit is entirely dedicated to that one call there is no need for any of the techniques we will discuss later to provide what we call *quality of service (QoS)*. In a packet-switched network the network is shared between users and between applications, so a single user may have several applications running that use the local network but are switched on an individual packet-by-packet basis to their ultimate—and frequently different—destinations.

In the paragraph above we discussed circuit switching versus packet switching. Another concept we build on extensively in this book is that of "virtual circuits." We have probably all heard of virtual reality, but what is a virtual circuit? Remember that circuit switching allocates an entire circuit to either a voice or data call, while packet switching switches packets over a shared infrastructure. Using a shared infrastructure to build a circuit that appears to be a permanently allocated or dedicated circuit between two points is what we are referring to when we speak—or write, as the actual case may be—about "virtual circuits." Virtual circuits are of two types, either permanent or switched, and are encountered in X.25, Frame Relay, and ATM environments. A permanent virtual circuit (PVC) is one that is established between two points and left in place in perpetuity—or until there is a technical problem with the infrastructure. A switched virtual circuit (SVC) is one that is built on demand and destroyed or torn down when the need for the circuit is over. All of the transports we have encountered—X.25, Frame Relay, and ATM—can make use of SVCs. It is most common to see them in X.25 and ATM environments, although they are becoming more common in some service providers' networks.

This chapter opened by noting that tremendous changes have rocked the computer networking industry recently. This is best illustrated by the technology reviewed in this section. In "old world" communications, an enterprise typically had a network dedicated to voice, another to data, and perhaps even a third dedicated solely to video links, as shown in Figure 1-5.

Figure 1-5   *Separate Data, Voice, and Video Networking*

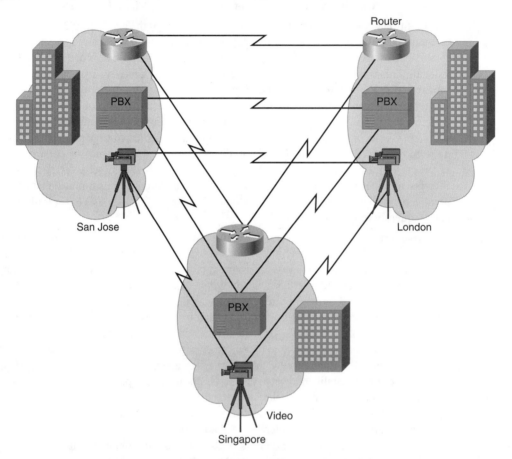

Now voice, video, and integrated data are coming together into a single network. Cisco refers to its vision for this as the Architecture for Voice, Video, and Integrated Data (AVVID). An enterprise that deploys an AVVID network can reduce costs and increase productivity through the use of a single network for all of its communications needs. This network is standards-based, using technologies such as Frame Relay, IP, and other IETF-based standard protocols, and International Telecommunications Union (ITU) standards for signaling and voice compression. An implementation of this type of network is illustrated in Figure 1-6. We'll look at each of these subjects briefly now, and in detail in later chapters of this book.

Figure 1-6   *A Converged Network for Voice, Video, and Integrated Data*

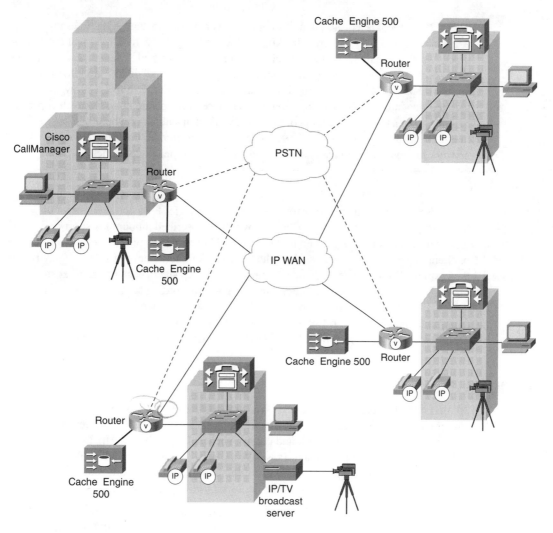

Frame Relay Data Networks

Frame Relay networks were first deployed in the early 1990s, but did not really become deployed in large numbers until the mid-1990s. Prior to the advent of Frame Relay, most large switched networks were based on the X.25 family of standards. The X.25 protocol is an extremely flexible technology, but each link is of a fixed speed. The X.25 protocol is a high-overhead protocol (see sidebar), where Frame Relay has a very low overhead. Frame Relay fixed the overhead issue and the fixed-speed issue.

X.25 Networks

The X.25 protocol is a protocol built to operate over circuits that are sometimes unreliable; X.25 has multiple layers of error checking and sequencing. These reliability features require processes that take time and induce delay into the end-to-end path. Many implementations of X.25 were limited to what today would be called "low-speed" links—56 kilobits per second (kbps) or 64 kbps. Nothing in the X.25 standard limits X.25 to such low speeds. In fact, the standard specifically states that the interface may operate up to 2 megabits per second (Mbps). One of the editors of this book worked in a network where X.25 was carried internally at speeds of 8 Mbps! The X.25 protocol is unsuitable for an integrated service network because of its typically slow speed and high overhead requirements.

Frame Relay permits a customer to subscribe with a carrier for a certain amount of bandwidth, but with the ability to "burst" over the subscribed amount for short periods of time. This behavior is well suited to the short and bursty nature of much of the traffic seen in client/server environments, such as the Internet. Another benefit of Frame Relay is that the customer does not need a dedicated circuit from each site to each of the other sites. The Frame Relay cloud takes care of interconnecting sites according to the customer's preferred design. Frame Relay can be deployed in full-mesh networks, partial-mesh networks, or star configurations. For a fuller discussion of the important elements of Frame Relay network design, see *Cisco Internetwork Design* or *Building Cisco Remote Access Networks*.

AVVID

AVVID includes four building blocks: basic infrastructure, such as routers and switches; call processing, used to resolve dialed numbers to gateways or other end-devices; applications, such as call control; and clients, such as IP telephones, H.323 videoconferencing equipment, and PCs. AVVID uses Internet Protocol as the common transport for the converged network, permitting common QoS, policy, routing, management and security services to be deployed throughout a network.

One of the key aspects of AVVID is the unified messaging component that permits e-mail, voice mail, and faxes to be routed to the appropriate device. Voice mail, for example, may be forwarded as a WAV file as an attachment to an e-mail. AVVID leverages efficient network design with appropriate technology to provide higher availability at a lower total cost of ownership.

Techniques for Combining Voice and Data

Earlier in this chapter we discussed how voice and data have been carried in the past. We looked at separate networks and a single network with dedicated portions of that network carrying either voice or data. New-world telephony treats voice and video just like any other data. The data stream is converted to a digital signal, compressed, and packetized for transport by the IP suite of protocols. We can establish quality of service for the IP packets that carry voice, video, or other time-sensitive traffic (such as whiteboarding), and in this case the network differentiates between that time-sensitive data and traditional application data. The old-world phone system has distinct steps required to establish and tear down a call. Those signaling functions must be replicated in new-world telephony. AVVID supports open standards to provide those signaling functions.

Voice Compression and Packetization

The old-world telephone system dedicates 64 kbps for each phone call, even during periods of silence. The new-world telephone system uses low-bandwidth voice-compression standards such as the ITU-T's G.729 and G.729a for phone calls requiring as little as 11.2 kbps during periods of burst and minimal bandwidth during periods of silence. As new standards are developed and implemented, that bandwidth will fall. The voice input is first digitized using open standards, then packetized into small data packets to avoid delay and jitter.

Delay and Jitter

Delay is when the packet takes so long to arrive at the other end that the speaker at that end may interpret the delay as silence on the other person's part. *Jitter* occurs when packets arrive with varying amounts of delay. Delay is normal, but long delay is annoying. Varying amounts of delay create jitter. Delay and jitter are common in traditional networks, but several steps are taken to reduce their impact. De-jitter buffers are built into switches in the network to ensure there is enough of the bit-stream buffered to resynchronize the bit-stream if necessary. The network is engineered to make sure that delay does not become a degrading issue (like using multiple satellite hops from end to end). Delay and jitter are discussed more in Chapter 3 and are covered in detail in *Voice over IP Fundamentals* from Cisco Press. You can use various QoS features to avoid or compensate for delay and jitter; these will be addressed in Chapter 10, "Configuring Cisco Access Routers for VoIP."

It is also possible to transmit voice directly over OSI Layer 2 without the benefit of IP. The compression and encapsulation procedures are the same as when the data is transmitted over IP; the only changes are the encapsulation layer and the specific mechanisms used to provide QoS. Irrespective of the layer, the principles of encapsulation, fragmentation, and queuing are used in providing QoS.

Signaling Protocols

There are several kinds of signaling in the old-world telephone system. The new-world telephone system must support the old system for interoperation with legacy systems. The new-world paradigm must also provide signaling systems that are standards-based and that are designed to operate over IP networks. There are several protocols designed to do this, including the ITU-T protocol H.323, and the IETF protocols Media Gateway Control Protocol (MGCP) and Session Initiation Protocol (SIP). These protocols are discussed in more detail in Chapter 10. Traditional protocols such as ITU-T Q.931 and ITU-T Q.700 (Common Channel Signaling System 7, or just SS7) are also supported. Connectivity between traditional PBXs was provided via tie-lines that used E&M signaling, which is still supported in several Cisco platforms. There are also protocols that permit signaling between proprietary old-world systems, such as the European Computer Manufacturers Association (ECMA) QSIG, based on the ITU-T's Q.931. QSIG is discussed briefly in Chapter 12; for a more thorough treatment of this protocol, see *Voice over IP Fundamentals*. British Telecommunications (BT) developed a proprietary protocol in the 1980s to connect PBXs known as the Digital Private Network Signaling System (DPNSS), which is being replaced by QSIG in many installations.

Cisco's Role in New-World Networking

Cisco Systems plays several roles in the arena of new-world telephony. It is a manufacturer and is active in many standards bodies that define the standards for the new world. Cisco's equipment fits into all three of the design layers discussed earlier in this chapter. In this section we briefly look at the technologies Cisco is deploying and where they fit, both in the design model and from a customer perspective.

Cisco's Role from a Design Viewpoint

Recall the three-layer design model discussed earlier: access, distribution, and core. Each of these layers plays a role in new-world networking.

The Access Layer

The access layer is the layer that provides entry to the network for users. At the access layer, one would see technology that either integrates traditional telephony—for example, connecting to a legacy PBX or handset—or that is based on open standards—for example, Cisco's IP Phone or SoftPhone or another standards-based telephony device. There are also collaboration tools like Microsoft's NetMeeting and similar products that permit simultaneous voice, video, and whiteboarding.

SWITCH + Phones

Integrating Traditional Telephony

Cisco equipment that integrates traditional telephony has interfaces for devices that use T1/ E1, ISDN, traditional FXO/FXS connections, or tie-lines between PBXs using an E&M interface. These are the legacy interfaces we have inherited from nearly 100 years of analog telephony. The access device, such as a Cisco 2610, would have handset(s) connected directly to it to provide the ability to call from one location to another. There may be a line to the local CO to provide access to the PSTN.

New-World Telephony

New-world networking devices such as the IP Phone act as a DHCP client to request an IP address or be statically configured locally with a permanent IP address. They use IETF and ITU-T standards throughout to provide quick and easily deployed solutions to problems. These devices connect directly to a switch and may even take their DC power from the switch. This pure IP environment permits great flexibility and granularity of service, management, policy, QoS, and security services throughout the network.

The Distribution Layer

ROUTER OR CATALYST SWITCH

The distribution layer is the point at which some of the policy, QoS, and other services mentioned above may be implemented. At this layer, one will find either Cisco routers or Catalyst switches with an embedded router to forward packets to their destination. If necessary, the class of service (COS) may be changed to a more appropriate class to provide the QoS necessary for real-time services. Features such as Random Early Detection (RED), Weighted RED (WRED), or other "flavors" of RED may be used to "throttle back" reliable connections such as HTTP to provide bandwidth for voice and video.

The Core Layer

The core layer is designed to forward packets as quickly as possible. In this environment, one will probably find Catalyst switches, GSRs, and LS1010s—depending on the network size, complexity, and distribution-layer speeds—to quickly forward frames to the appropriate distribution layer. In a service provider's core or a very large enterprise core, one may find such switches as IGX, MGX, and/or BPXs.

Cisco's Role from a Customer Viewpoint

The customer will not only need to look at the basic design of the network as outlined above, but will also need to determine specific applications. The customer type will determine the Cisco equipment deployed in the network.

Enterprise Telephony

Enterprise telephony can be thought of as a purely internal system. Say the Widget Company, for example, decides to cut down on its long-distance bills for interoffice calls. It also wants a unified messaging system and the ability to conduct videoconferencing or online training. This customer will want to implement the appropriate access technology, depending on whether they will coexist with legacy equipment or prefer to switch to new-world technology. They may or may not need to reconfigure their distribution and core layers, depending upon their design strategy in their old-world network.

Service Provider Telephony

Many service providers are adopting Cisco's AVVID technology to replace the legacy, proprietary equipment that limits their business flexibility. Cisco has a wide range of equipment that is only briefly covered or not covered by this text. These include all the equipment previously mentioned as well as routers that implement interactive-voice response (IVR) and permit authentication via TACACS+ or RADIUS for prepaid long-distance service. Also included in this expanded view of Cisco's product offerings are the large-scale switches such as the IGX, BPX, and MGX. These switches provide the ability to send voice, video, and data using ATM as the transport, at interface speeds of up to OC-12 (622 Mbps) with backplane switching speeds in the several gigabits per second range.

Summary

In this chapter we have looked briefly at the development of both voice and data networks. From completely separate networks they evolved through shared networks with dedicated bandwidth and now into fully converged networks. Different technologies and design strategies have been deployed throughout the years, leading us to the scalable and easily replicated three-layer hierarchical design model. The old-world network was built on dedicated lines and proprietary protocols; the new-world network is built on a shared or converged network using open standards. These standards replicate the functions of the old-world proprietary standards, with the benefit of interoperability with legacy equipment during periods of coexistence and migration. Cisco Systems provides a wide range of products to meet the needs of enterprises and service providers, from access to core.

Review Questions

The following questions should help you gauge your understanding of this chapter. You can find the answers in Appendix A, "Answers to Review Questions."

1 Name the three components of a leased-line network.

   _____

   _____

   _____

2 Name the two types of leased-line networks.

   _____

   _____

   _____

3 Name the three layers of the Cisco network design model and briefly describe their functions.

   _____

   _____

   _____

4 What are the chief benefits of implementing an AVVID network?

   _____

   _____

   _____

5 What is a commonly implemented ITU-T standard used for voice compression?

   _____

   _____

   _____

6 What are two business markets that can benefit from implementing Cisco's AVVID technology?

   _____

   _____

   _____

After reading this chapter, you should be able to perform the following tasks:

- Describe the components of a telephone call.
- Describe the components of a telephone handset.
- Describe analog signaling types and instances when they would be used.
- Describe switch types, their purpose, and the locations where each type would be deployed.
- Describe echo in the telephone system, the cause of echo, and how to deal with echo that is excessive.

CHAPTER 2

Introduction to Analog Technology

Voice can be transmitted across a traditional telephony network in two forms: as analog or as digital waveforms. This chapter covers the basics of analog telephony. It includes the following topics:

- Telephone call components
- Telephone set components
- Telephone signaling
- The local loop, switches, and trunks
- 2- to 4-wire conversion and echo

This chapter also provides an introductory overview of the basic telephone handset and its internal components. Analog phone line types, switch functions, trunk types, and trunk signaling are also described. The evolution of digital voice technology is covered in Chapter 3, "Introduction to Digital Voice Technology."

In this chapter, you will learn about where the various components fit into the telephone network, and you will learn about the signaling protocols the devices use to communicate.

Telephone Call Setup and Completion

To understand the components of a telephone call, let's look step-by-step at what happens when you make a phone call over a typical telephone network. The following events occur when you place a call to another party (see Figure 2-1 for a visual representation):

Figure 2-1   *Telephone Call Setup and Completion*

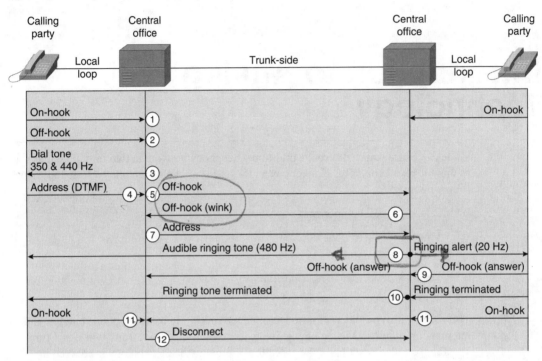

1   Initially the phone is hung up, or in the idle state. The wire circuit between the local handset and the central office (CO) switch is open or incomplete. Another term for this is "on-hook," which refers to the earpiece of older ringer-style phones being hung on the "hook" when not in use.

2   Lifting the handset off the cradle to place a call causes the circuit to be closed or completed (or to "go off-hook," in traditional telephony jargon). Current now flows through the telephone circuit.

3   The switch will detect the current and return dial tone. Dial tone is informational signaling and is not necessary. In specialized business applications (like credit card verification), to reduce the call setup time the local CO may not provide dial tone.

4 The caller requests a specific destination address by entering the digits of the phone number. Dialing or pressing buttons, depending on the type of phone, typically does this. These digits are entered into the dial plan of the originating switch to find the path to the destination or terminating switch. The International Telecommunication Union Telecommunication Standardization Sector (ITU-T) E.164 telephone number is the equivalent of the IP address in that it provides end-to-end addressing.

5 The originating switch then seizes a trunk to the appropriate destination switch. This is performed within the telephone network. The originating switch—in the analog world—transmits a voltage over an analog line to indicate it is requesting that the trunk be placed in service. In the digital world, this voltage is simulated through the use of signaling bits. In the modern, digital, world using ITU-T Signaling System 7 (SS7) protocols, the signaling dialogue is handled in a different—and more complex—manner.

6 The destination switch in some cases signals back to the originating switch with a "wink" to indicate that it is ready to place the trunk in service and has allocated the trunk. In other cases, no return signaling is used to indicate that the trunk is available. The originating switch will be programmed to wait a short period of time (delay-start) or transmit immediately (immediate-start). If there are no trunks available, the switch will signal back to the caller a special kind of informational signaling that we cover later in this chapter, in the section "Signaling." In other cases the signaling is—as mentioned in the previous step—completed using SS7.

7 The originating switch transmits the destination address (phone number) to the destination switch.

8 The terminating switch signals the called party by sending ringing voltage to the called party. The terminating switch also signals back to the originating switch that the local loop has been seized for call delivery by transmitting a "ring back" signal to provide informational signaling to the user that the phone is ringing. In an SS7 network, a signal to the originating switch will cause that originating switch to send ring-back tone to the caller. There is no need to provide this ring-back signal other than to provide a certain amount of comfort to the user. Some lines are configured such that no ring-back is sent in order to decrease the call setup and delivery time. This is common in credit card transaction access lines. If the destination line is in use, the terminating switch will signal that condition back to the originating switch, which in turn will send a normal busy to the caller.

9 The receiving party forces the line off-hook when the handset is lifted and current flows through the loop. There is now an end-to-end connection. The terminating switch will signal this back to the originating switch.

10 The destination switch will then terminate the ring-back signaling so that the call can proceed with no interference.

11 At the conclusion of the call both ends will hang up, putting the local loop in the on-hook state.

12 The originating switch will drop the voltage on the trunk in the analog world causing the terminating switch to drop the trunk.

As you can see, there are several steps involved in even a simple phone call within the same metropolitan area. The example shown in Figure 2-1 includes three separate instances of signaling:

- The caller's local-loop signaling
- The switch-to-switch signaling
- The called party's local-loop signaling

In this book you will learn how to configure both the local-loop signaling and the trunk-side signaling.

Telephone Set Components

There are several components in even the tiniest handset used for making calls. These are illustrated in Figure 2-2 and described below:

Figure 2-2 *Telephone Set Components*

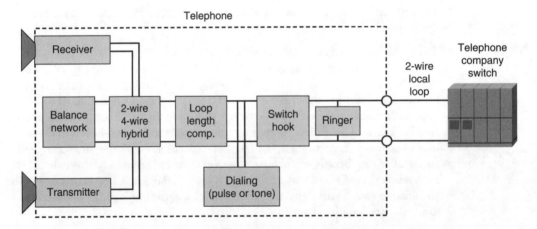

- Handset—This is the part of the telephone you hold in your hand to speak and listen to a conversation. It is also common to have the "handset" built into a headphone and microphone set for those people who spend a lot of time on the phone or need their hands free while speaking and listening on the phone. Inside the handset are a transmitter and a receiver. You speak into the transmitter and listen from the receiver.

- Switch hook—This is the switch that is pushed down when the handset rests on its cradle (on-hook). When you lift the handset to place a call, you release the switch hook and it pops up. The circuit is now off-hook and current flows through the telephone. When the telephone is placed back on-hook, current flow ceases.

- Hybrid 2- to 4-wire converter—Four wires, organized in two pairs, run from the handset, one pair from the transmitter and another pair from the receiver, to the *hybrid*, which provides the conversion between the 4-wire handset and the 2-wire local loop. The converter is the communications bridge between the handset equipment and the 2 wires to the telephone company.

- Sidetone—This is a planned, audible result emanating from the hybrid in the phone, through which a portion of speech is allowed to bleed over into the earpiece during a conversation so that users can judge how loudly they are speaking.

- Dialer—This is the touch pad or rotary dial that signals the telephone company that you are placing a call. When you push the buttons on a touch pad or spin the dial on a rotary telephone that is in the off-hook state, you send a signal to the telephone company, specifying the location you are calling. Keep in mind that for flexibility many push-button phones have a setting that permits them to send either tones or pulses for signaling.

- Ringer—When someone is trying to call you, the telephone company notifies you by sending alternating current (AC) voltage through the wires to your telephone set. The voltage triggers a device, the ringer, that makes a ringing sound. An electrical component called a capacitor prevents direct current (DC) from flowing through this circuit when the phone goes off-hook and dial tone DC voltage is received.

Signaling

There are several types of signaling used in the telephone network. In this section we will describe the common signaling types that you will encounter when configuring Cisco IOS to perform voice services. Typically, signaling can be viewed as occurring in one of two locations:

- Line side—End-device to switch signaling
- Trunk side—Switch-to-switch signaling

The types of signaling we discuss are:

- Supervisory, used to provide line-seizure signaling
- Address, used to encode the address of a call
- Informational, used to provide information to the users of the telephone system

Supervisory Signaling

Supervisory signaling is used to indicate the state changes in the telephone network. This is the signaling that is used to set up and tear down a connection. Supervisory signaling is used to indicate on-hook and off-hook states and line or trunk seizure. In the following sections we will discuss different instances of supervisory signaling.

On-hook and Off-hook States

The section "Telephone Call Setup and Completion" describes the phone being placed off-hook or on-hook. These are states that the local loop is in; the actual supervisory signaling has to do with the manner in which the circuit is wired and with the electrical signaling that takes place when these state changes occur.

Local-Loop Seizure

The circuit between the CO and the handset was previously defined as the *local loop*. There are two basic types of signaling used to signal state changes in this loop: loop start and ground start.

Loop-Start Signaling

The loop-start signaling process can be summarized as follows:

1   When the line is in an idle state, it is said to be on-hook, and the telephone or PBX opens the 2-wire loop. The CO has battery on ring and ground on tip, although polarity is not generally an issue in analog telephony.

2   If you lift the handset off the cradle to place a call, you cause the switch hook to go off-hook and close the loop. Current will now flow through the telephone circuit. The CO will detect the current and return dial tone.

3   If your telephone is ringing to notify you of an incoming call, the CO applies AC ring voltage superimposed over the –48 VDC battery, causing the ring generator to notify you of a telephone call. When the telephone or PBX answers the call, closing the loop, the CO will remove the ring voltage.

For example, Figure 2-3 shows four examples of a loop start circuit. In Figure 2-3a you see the normal idle condition. A –48 VDC battery on the CO, or switch, side provides the current for dial tone. A ring generator provides the AC current for ringing the phone. On the telephone or PBX side, there is an open circuit indicating the phone being in the on-hook state.

Figure 2-3   *Loop-Start Call Progression*

a. Idle condition (on-hook)

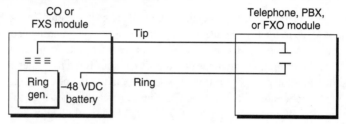

b. Telephone or PBX seizure (off-hook)

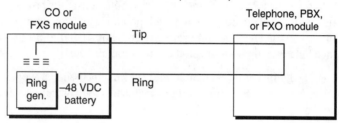

c. CO seizure (on-hook)

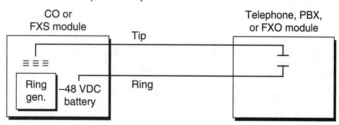

d. Call completion (off-hook)

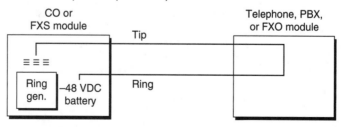

FXO—foreign exchange office
FXS—foreign exchange station

In Figure 2-3b the telephone or PBX side of the connection has been closed: The phone has gone off-hook. This permits the current to flow from the CO to the phone, providing the dial tone.

Figure 2-3c illustrates the CO "seizing" the line. The CO signals line seizure by applying a ringing voltage to the line, hence the ring line is of heavier shading. The local handset then rings and will be answered. Figure 2-3d illustrates the phone being answered and going off-hook. It is important to remember that the CO cannot force the other end of the circuit closed in a loop-start configuration. The CO can make the phone ring, but it can't force someone to answer it. Likewise, the CO cannot force the phone set to disconnect the call. This causes problems when using FXO to emulate a trunk-side protocol connecting to a switching device like a PBX, key system, or router.

Loop-start signaling is a poor signaling solution for high-volume trunks because it is possible to seize the trunk simultaneously from both ends. This problem is known as *glare*. Glare may periodically occur on a home telephone. If you are at home and you pick up the telephone to call out and a person is already at the other end of the connection, you have experienced glare—you both seized the loop simultaneously.

Glare is not a significant problem at home, but imagine being at work with ten (or more) times the phone usage. Signaling methods that detect loop or trunk seizure at both ends will solve the problem.

Ground-Start Signaling

Ground-start signaling is a modification of loop-start signaling that corrects for the probability of glare. It solves the problem by providing current detection at both ends. Loop-start signaling works when you use a telephone at home, but ground-start is preferable when high-volume trunks are involved.

A telephone connection can be in one of the following states:

- Idle (on-hook)
- PBX or telephone seizure (off-hook)
- CO seizure (ringing)

A summary of the ground-start signaling process is as follows (see Figure 2-4):

Figure 2-4   *Ground-Start Call Progression*

a. Ground-start idle condition

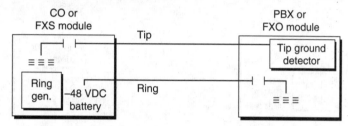

b. PBX seizure, step 1

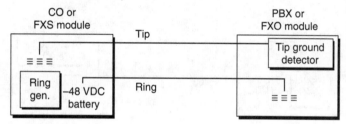

c. PBX seizure, step 2

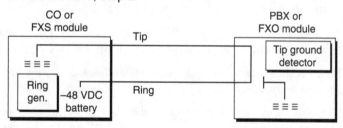

d. CO seizure, step 1

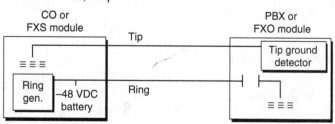

Figure 2-4   *Ground-Start Call Progression (Continued)*

e. CO seizure, step 2

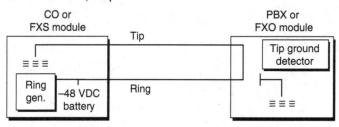

f. CO seizure, step 3

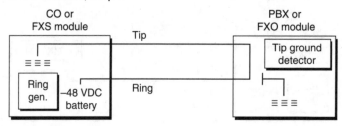

1   When the line is in idle state (see Figure 2-4a), the PBX monitors for ground on the tip lead. Battery from the CO appears on the ring lead.

2   If you lift the handset off the cradle to place a call, your PBX grounds ring lead. The CO senses the ring ground and then grounds tip lead (see Figure 2-4b).

3   The PBX senses the tip ground from the CO, closes the 2-wire loop, and removes ring ground (see Figure 2-4c).

4   If someone is calling you, the CO grounds tip lead and superimposes ringing voltage over ring lead battery (see Figure 2-4d). The PBX must recognize the incoming seizure within 100 ms. The tip ground and ringing conditions are sensed and the PBX closes the loop and removes the ring ground (see Figure 2-4e). The last step is the CO sensing DC current and removing the ring ground (see Figure 2-4f).

E&M Seizure (E&M Interface Callout)

E&M, which stands for any of "Ear and Mouth," "Earth and Magneto," or "receive and transmit," is a signaling system developed to interconnect lines between PBXs via the Public Switched Telephone Network (PSTN) or a leased line. E&M provides three different signaling scenarios, each of which is discussed below. E&M lines are frequently referred to as "tie-lines" because they tie together two PBXs.

The E&M Interface

The E&M interface is an 8-wire interface, of which a minimum of 6 wires must be used. Up to 4 wires may be used to provide signaling and there is either a 2-wire or a 4-wire talk path. Different combinations of signaling characteristics and the number of signaling path wires yield several different E&M "types," which are discussed in detail below. The basic operation of the E&M interface is as follows (see Figure 2-5, which depicts a Cisco voice-capable router connected to a PBX):

Figure 2-5   *E&M Interface and Connection*

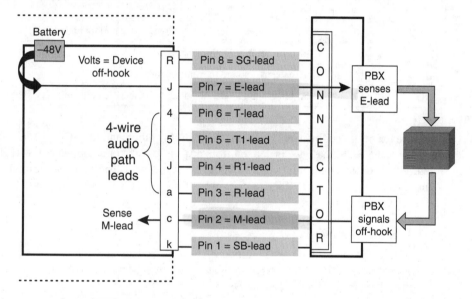

1   The router receives a call destined for the PBX. To seize the E&M lead to the PBX, the router will raise voltage on the E-lead.

2   The PBX, sensing the voltage on the E-lead, may check to make sure it has an address register to store the incoming digits, then signal back on the M-lead to indicate it is ready to accept the call and allocate the trunk. In some cases the receiving device does not signal back to indicate that the transmitter may begin transmitting.

The same operation is used when the PBX needs to forward a call to the router; however, the PBX will raise the M-lead to signal the router of its request for a line and the router will respond back on the E-lead. In this regard some people compare the E&M interface to a DTE/DCE-like relationship.

The interface has the types of leads listed in Table 2-1.

Table 2-1   *E&M Leads*

Lead	Description
SG	Signal ground, the ground lead for the E-lead in some E&M types
E	"Ear" lead, used to initiate a trunk seizure
T	Tip lead, used as an input lead in a 4-wire talk path
T_1	Tip 1, used as the output lead for the talk path (PBX to router) in either a 2-wire or a 4-wire talk path
R_1	Ring 1, used as the output in either a 2-wire or a 4-wire talk path
R	Ring, used as the input in a 2-wire talk path
M	"Mouth" lead, used to respond to trunk seizure
SB	Signal battery, the ground lead associated with the M-lead in some E&M Types

Be sure to check your manuals when installing E&M equipment. Like the EIA/TIA-232-F interface, the E&M interface is defined from a specific piece of equipment view (PBX-centric). If you are connecting a non-PBX device, some of the leads may not be on the same pin numbers on both sides. If the circuits are not completed no traffic will pass.

Wink-Start Signaling

Wink-start signaling is the most common E&M trunk seizure signal type. It essentially provides a hardware handshake to ensure that the receiving device has located and allocated a digit register for the incoming address.

The following scenario provides a summary of the wink-start protocol event sequence (see Figure 2-6):

1   The calling device initiates line seizure by going off-hook.

2   The called end does not immediately return an off-hook acknowledgment once detecting the seizure of the line by the calling office.

3   Instead of returning an off-hook acknowledgment, the on-hook state is maintained until the receive digit register is attached.

4   The called office toggles the off-hook lead for a specific time. This is the "wink" in wink-start signaling.

5   The calling office receives the wink and forwards the digits to the remote end.

Figure 2-6   *E&M Wink Signaling*

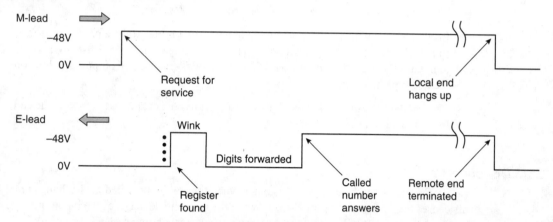

6 The called end makes the appropriate forwarding decision based on the incoming address.

7 The called party answers the phone.

8 The remote PBX raises the M-lead during the call.

Delay-Start Signaling

Delay start does not provide any feedback from the called device to the calling device. Basically, the calling device raises the E-lead, waits a specified amount of time, then begins forwarding digits. The following scenario provides a step-by-step overview of delay-start signaling in a router-to-PBX configuration:

1 When you place a call, the originating switch raises the E-lead.

2 The originating switch looks at the status of the remote switch's signal.

3 The originating switch waits until the remote switch supervision is on-hook—that is, there is no current on the M-lead.

4 Once the remote is on-hook, the originating switch will output digits.

Immediate-Start Signaling

The following scenario provides a summary of the immediate-start protocol event sequence:

1 Your company's router seizes the E-lead by going off-hook.

2 Instead of receiving a double acknowledgment, your router waits a predetermined time and forwards the digits "blindly." The originating switch goes off-hook and maintains the condition for at least 150 milliseconds (ms) before outputting digits on the audio path.

3 The remote PBX only acknowledges your PBX after the called party answers the call by raising current on the M-lead.

E&M Disconnects

In all E&M signaling, the disconnect is made when the calling or called device hangs up. This causes the device to drop its E-lead. When the opposite device sees the incoming E-lead drop current, it will remove current from the corresponding M-lead. The trunk is now available for use by another call.

Address Signaling

A telephone number is the destination address and the formats for telephone numbers are specified in the ITU-T E.164 standard. There are two common means of transmitting this address from the telephone to the switch (CO, PBX, or Cisco voice-capable router): by using pulsed digits or by using tone, specifically dual-tone multi-frequency (DTMF), as described in the following sections.

Pulsed Digits

Although somewhat outdated, rotary dial telephones are still in use and are easily recognized by their big numeric dial-wheel, which is spun to send digits when placing a call. Some push-button phones are capable of sending either pulsed digits or DTMF. When you dial digits to place a call, the digits must be produced at a specific rate and be within a certain level of tolerance. Each pulse consists of a make and a break. The *break* segment is the time that the circuit is open. The *make* is the period during which the circuit is closed. The cycle corresponds to the following ratio in the United States: 60 percent break, 40 percent make. Figure 2-7 illustrates this cycle. There are typically about 10 pulses per second, but this is dependent upon the pulse duration.

Figure 2-7 *Address Signaling—Pulsed Digits*

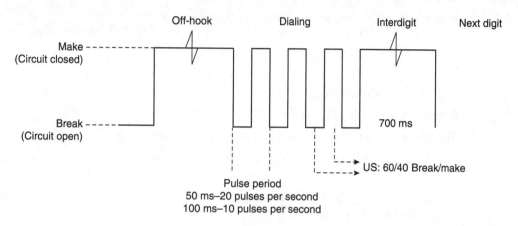

A governor inside the dial controls the rate at which the digits are pulsed. In older-style phones the governor is mechanical, and in newer phones it is electronic.

DTMF

With a touch pad or push-button phone, you push buttons on a keypad when placing a call. Each button on the keypad is associated with a set of high and low frequencies. When you look at the keypad, each row of keys is identified by a low frequency tone and each column is associated with a high frequency. The combination of both tones notifies the telephone company of the number you are calling, hence the term DTMF.

Figure 2-8 illustrates the combination of tones you can generate for each button on the keypad.

Figure 2-8 *DTMF Frequency Matrix*

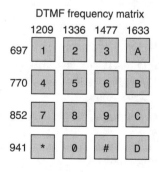

Informational Signaling

Informational signaling is the signaling that is used to provide information to the user of the telephone system. It is not necessary for the electrical working of the phone system, but if you didn't hear the phone ring, you wouldn't know to answer it! Informational signaling can be divided into two subclasses: call progress indicators and alerting indicators. *Call progress indicators* are used to notify you of your call's status. Each combination of tones notifies you of different events in the call's progress. *Alerting indicators* are designed to alert you to the state of the telephone call or line conditions.

The following are the call progress indicators and what they tell you:

- Dial—A dial tone is given when the telephone company is ready to receive digits from the user's telephone.

- Busy—A busy signal is given when a call could not be completed because the phone at the remote end was already in use.

- Ring-back (normal or PBX)—Ring-back is the sound you hear when you are calling someone else. It is a confirmation that the telephone company is attempting to complete your call.

- Congestion—A fast busy signal is given when there is congestion in the long-distance telephone network that prevents your telephone call from being processed.

- Reorder—A reorder or fast busy signal—also called engaged or overflow outside of the USA—is given when all the local telephone circuits are busy, which prevents your telephone call from being processed. This tone is frequently intercepted and a recorded announcement played to the user to explain the problem.

The following are the alerting signals and what they tell you:

- Ringing—The phone rings to notify you that there is an incoming call.

- Receiver off-hook—When you leave your receiver off-hook for an extended period without placing a call, the telephone company will notify you to place it back on-hook.

- No such number—When you place a call to a nonexistent number, the telephone company will notify you with a "no such number" signal.

- Confirmation tone—Some telephones provide a confirmation tone. It lets you know that it is working on completing the call. Users usually like the tone better than dead air while the call is connecting.

The tones for some of these signals—in North America—are listed in Table 2-2.

Table 2-2   *Call Progress and Indicator Tones*

Tone	Frequency (Hz)	On Time (sec)	Off Time (sec)
Dial	350 + 440	Continuous	Continuous
Busy	480 + 620	0.5	0.5

Table 2-2   *Call Progress and Indicator Tones (Continued)*

Tone	Frequency (Hz)	On Time (sec)	Off Time (sec)
Ring-back, line	440 + 480	2	4
Ring-back, PBX	440 + 480	1	3
Congestion (toll)	480 + 620	0.2	0.3
Reorder (local)	480 + 620	0.3	0.2
Receiver off-hook	1400 + 2060 + 2450 +2600	0.1	0.1
No such number	200 to 400	Continuous Frequency	Continuous Frequency

The Local Loop, Switches, and Trunks

In this section we discuss the infrastructure that makes up the circuit from the CO to the handset, the switches that compose the PSTN, and the trunks.

The Local Loop

The local loop is the circuit—generally copper wires—that runs from the CO to the customer premises. The term *loop* is derived from the fact that the wire actually forms a loop when the telephone is off-hook. On the customer-premises side there is typically a handset, which we have already discussed, or a PBX, which we will discuss in detail in Chapter 12, "Old-World Technology: Introduction to PBXs." On the CO side there are switches and various trunks, which we discuss here.

Switches

Switches are used in the telephone industry much like switches are used in the data networking industry. They provide a scalable architecture for building large and fast networks. This section discusses switches in general and the three types of voice switches found in the PSTN:

- CO switches
- Intermediate switches
- PBXs

Voice Switches

A voice switch is an analog or digital device that directs your voice call to the proper destination. A voice switch can be at your site or at the telephone company. If you are placing calls from home, you are most likely relying on the telephone company's switch to complete your call. Many businesses rely on PBXs to do some of their switching.

Without a switch in your telephone network, you would need a telephone line to each destination you wish to call. Obviously this would be impossible! Switches make it possible to set up a dedicated line for the duration of your telephone conversation and tear it down again when the conversation ends.

NOTE   A switched telephone network is analogous to switched virtual circuits. The connection exists only for the duration of the call and is then broken down.

Various vendors (including Lucent Technologies and Nortel in North America; Ericsson, Siemens, and Alcatel in Europe; and NEC and Fujitsu in Japan) manufacture traditional telephone company switches.

The switch selectively establishes and releases connections between transmission facilities to provide dedicated paths for the exchange of messages between two users. Paths are established before the information exchanges begin and, until the users terminate the sessions, these paths are maintained for the switch's exclusive use.

CO Switches

Whenever you place or receive a call, the line you use is ultimately connected to the telephone company's CO switch. That switch will route the call to the proper destination. Private phones are almost always directly connected to the CO.

When you place a telephone call, the CO forwards the call to one of the following:

- Another CO switch
- Another end user's telephone, if it's connected to the same CO
- An intermediate switch

The CO provides the following components to make your telephone work:

- Battery—This is the source of power to both the circuit and the phone and is also used to determine the status of the circuit. When you lift the handset to let current flow, your telephone company is providing the source that powers the circuit and your telephone. Because your telephone company is powering your telephone from the CO, electrical power outages should not affect your basic analog telephone service.

NOTE   Some telephones on the market offer additional features that may require an additional power source that you supply. Cordless telephones are one such example. You may not be able to use a cordless telephone in the event of a power outage.

- Current detector—The current detector monitors the status of the circuit by detecting whether the circuit is open or closed. When the handset rests in the cradle, the circuit is on-hook and there is no current flow. When the handset is raised from the cradle, the circuit is off-hook and current flows in the circuit.

- Dial tone generator—Once the digit register is ready, the dial tone generator generates a dial tone to acknowledge the request for service.

- Tone detectors—These detect and/or recognize DTMF from the phone.

- Dial register—When the PBX detects current flow on the interface, the dial register receives the dialed digit.

- Ring generator—Once the PBX detects a call for a specific subscriber, the ring generator alerts the called party by sending a ring signal to the subscriber.

Intermediate Switches

An intermediate switch is a switch that forwards a call to other switches in the network. It connects switches via trunks. These are also called tandem or toll switches. Intermediate switches may connect a local CO to another switch within the service area of the LEC, to an IXC's toll center, or to another intermediate switch. The classic problem of connecting every device to every other device in a network is the N^2 problem. To connect each device to every other device in the network, one needs N^2 number of links, where N is the number of devices. Switches eliminate the N^2 problem.

PBXs

Telephone companies generally charge for each phone number and the line associated with it. Most large organizations have no need for everyone in the building to be on the telephone at once. In fact, it would probably be highly unproductive in most organizations! To save money, an organization can bring several lines from the telephone company's CO switch to their own private branch exchange, or PBX, but far fewer lines than the number of workers at the site. When you need an outside line, the PBX will switch your call to a line to the telephone company. If you are calling someone within the organization, the call is switched by the PBX. PBXs and key systems are covered in detail in Chapter 12.

NOTE	A PBX is sometimes called a private automatic branch exchange (PABX) in Europe.

PBXs are easily identified by the following characteristics:

- A PBX is located at the customer site.
- Purchase and maintenance are the responsibility of the PBX owner, not of the telephone company.
- A separate battery backup to the system is generally required.
- The PBX is used as the connection medium to other customer switches and to the outside world.
- Voice and data switching capabilities are enabled through digital technology.
- There are several interface options with other equipment (for example, voice mail).

E&M Line Types

In this section we take a closer look at exactly which leads are used and the signaling for each of the E&M types.

Type I E&M Lines

Type I is a 2-wire E&M signaling type common in North America; approximately 75 percent of North American PBXs are Type I. Only the E-lead and M-lead are used for the signaling path, which requires that the PBX and tie-line equipment share a common signaling ground reference.

NOTE	It is always a good idea to ground the Cisco voice-enabled router to any other equipment to which it will be attached. This will ensure that there is no voltage differential that may harm the equipment. Most PBXs do not fit in standard equipment racks and therefore need a separate grounding or earthing wire.

With the Type I interface the tie-line equipment generates the E signal to the PBX by grounding the E-lead. The PBX detects the E signal by sensing the increase in current through a resistive load. Similarly, the PBX generates the M signal by sourcing a current to the tie-line equipment, which detects it via a resistive load.

The Type I interface is illustrated in Figure 2-9.

Figure 2-9   *E&M Type I Interface Signaling*

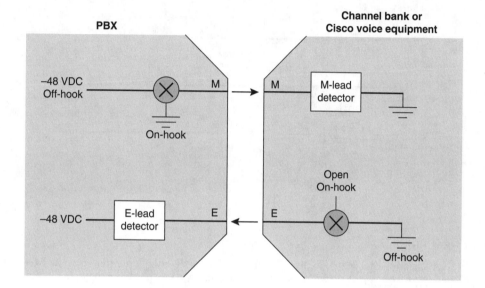

Type II E&M Lines

E&M types II, III, and IV are 4-wire interfaces. One wire is the E-lead and the second wire is the M-lead. The remaining two wires are SG and SB. In type II, SB and SG are the return paths for the M-Lead and E-lead, respectively.

The Type II interface requires no common ground; instead, each of the two signals has its own return. For the E signal, the tie-line equipment permits current to flow from the PBX; the current returns to the PBX's SG-lead or reference. Similarly, the PBX closes a path for current to generate the M signal to the tie-line equipment on the SB-lead.

The Type II interface is illustrated in Figure 2-10.

Figure 2-10 *E&M Type II Interface Signaling*

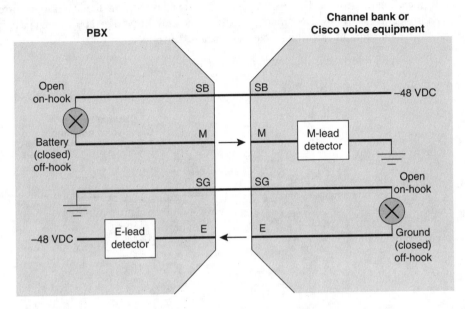

Type III E&M Lines

A variation of Type II, Type III is an uncommon signaling type that uses the SG-lead to provide common ground. The E-lead operates similarly to Type I. With this configuration, the PBX drops the M signal by grounding it, rather than by opening a current loop.

The Type III interface is illustrated in Figure 2-11.

Type IV E&M Lines

Type IV is symmetric and requires no common ground. Each side closes a current loop to signal; the flow of current is detected via a resistive load to indicate the presence of the signal.

The Type IV interface is illustrated in Figure 2-12.

Figure 2-11 *E&M Type III Interface Signaling*

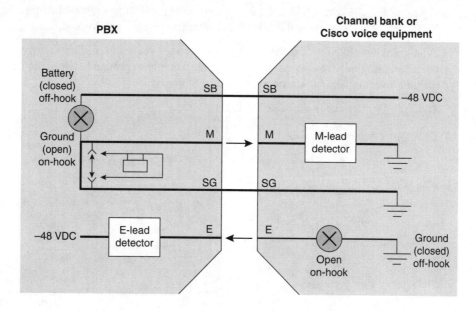

Figure 2-12 *E&M Type IV Interface Signaling*

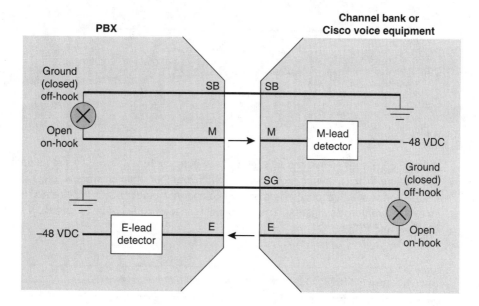

Type V E&M Lines

Type V is also a 2-wire E&M signaling type; it is the most common signaling type outside North America. One wire is the E-lead and the second wire is the M-lead.

Type V is a simplified version of the Type IV interface you will see. This is a symmetric interface, using only two wires. Type V requires a common ground between the PBX and the tie-line equipment; this is provided via the SG-leads.

The Type IV interface is illustrated in Figure 2-13.

Figure 2-13 *E&M Type V Interface Signaling*

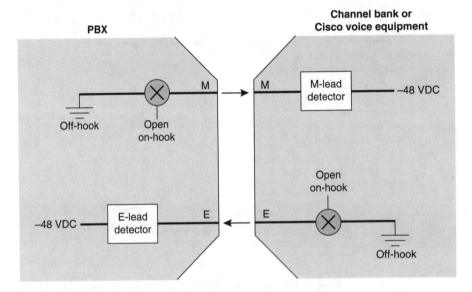

Trunks

A telephone call is routed through multiple switches before it terminates at its final destination. When a switch receives a telephone number, it determines whether the destination is within a local switch or if the telephone number needs to be routed to another switch to a remote destination. Trunks connect the telephone company and PBX switches, or PBXs and PBXs.

NOTE   Switches provide logic and trunks provide the path between switches.

The trunk's primary function is to provide the path between switches. The switch must route the call to the correct trunk or telephone line. A trunk is shared by many different subscribers, although only one uses it at any given time. As telephone calls complete, trunks are released and made available to the switch for subsequent calls. Between two switches, there may be many trunks.

A few of the more common trunk types are described in this section. They are:

- Private trunk lines
- CO trunks
- Foreign exchange trunks
- Direct inward/direct outward dialing (DID/DOD) trunks

Private Trunk Lines

An organization that has multiple PBXs can connect them with private trunk lines. Generally, private trunk lines serve as dedicated circuits that connect PBXs to each other. Subscribers who have PBXs that they want linked will lease trunks from the telephone companies on a monthly basis and reduce their cost by avoiding paying for the use of telephone lines, the other option, on a per-call basis. As mentioned earlier in this chapter, these are known as tie-lines. E&M and QSIG are examples of signaling used on these trunks.

CO Trunks

A CO trunk is a direct connection between a local CO and the PBX that enables calls to be routed from the private network at your work site, for example, to the public telephone network. CO trunks also interconnect COs. Examples include ISDN BRI/PRI and T1 CAS trunks.

Foreign Exchange Trunks

Foreign exchanges are trunk interfaces that allow you to connect the phone directly to the switch. These connect between the CO and local customer equipment.

Foreign exchange office (FXO) and foreign exchange station (FXS) can be thought of as different sides of the connection to a switch. The FXS provides battery and dial tone; the FXO expects to receive battery and dial tone from the adjacent device. We will discuss this in more detail in Chapter 5, "Applications for Cisco Voice-over Routers."

FXO Trunks

The FXO sits on the switch end of the connection. It plugs directly into the line side of the switch so the switch thinks the FXO interface is a telephone. The switch notifies the FXO of an incoming call by sending ringing voltage to the FXO. Likewise, the FXO answers a call by closing the loop to let current flow. Once current is flowing, the FXO interface uses any current technology to transport the signal to the FXS.

FXS Trunks

The FXS sits at the remote site and is the switch interface, albeit a remote one, to which the telephone is connected. It provides the dial tone and other signaling to the telephone. The telephone thinks it is the switch.

DID/DOD Trunks

DID trunks are implemented as one-way trunks that allow you to dial into a PBX without operator intervention. The outside subscriber will dial the extension digits of the desired destination that the connecting CO passes to the PBX. The CO will know which calls to pass through the DID trunk because it associates a block of numbers with each DID trunk.

NOTE   The term DID is a North American term. The European equivalent is called a "DDI."

DOD trunks are also one-way trunks that allow you to connect directly to the CO. They are outbound trunks. If you are at the office, for example, and want to place a call outside your company's network, you dial an access code such as 9, and the PBX forwards your call out to the CO. At that time the CO will provide a second dial tone and use the remaining digits you dialed to forward the call to its destination.

In high-volume situations, the use of separate DID and DOD trunks guarantees there will not be glare, since only one side of the connection will ever attempt to seize the line.

Telco Trunking

The telephone network has a variety of trunks based on the old Bell-system five-level hierarchy. These are trunks that connect strictly between telephone company equipment and do not interface with end-user customer equipment. Competing telephone companies will interconnect in order to provide global service.

Inter-Office Trunks (IOTs)

IOTs are trunks that connect one CO to another CO. This class of trunk is very common in metropolitan areas where a single LEC controls all access. These trunks are of T1 and greater bandwidth.

Intertandem Trunks (ITTs)

ITTs are trunks used solely for switching from one tandem switch to another. These trunks connect switching centers to each other to eliminate the N^2 problem. These trunks are of T1 or T3 bandwidth.

High-utilization Trunks (HUTs)

HUTs are similar to ITTs. These trunks are T3 or greater and connect switching centers that have large traffic loads.

2- to 4-Wire Conversion and Echo

A local loop is made up of 2 wires. Once it reaches the switch, the connection is changed to 4 wires with a 2- to 4-wire hybrid converter so your signal can be transported across the trunks in the network.

If there is a good impedance match between the lines, the hybrid is said to be *balanced*, with little or no reflected energy. However, if the hybrid is inadequately balanced, and a portion of the transmit voice is reflected back toward the receive side, echo results.

Echo in Voice Networks

Two types of echo exist:

- Talker echo—If you talk and hear your voice reflected back to you, you are experiencing talker echo. In effect, you hear yourself twice.
- Listener echo—If you are listening to another speak and hear the speaker's voice twice, you are experiencing listener echo.

Some form of echo is always present. However, echo is a problem if the magnitude or loudness of the echo is high. Echo is also a problem if the delay time between when you speak and when you hear your voice reflected is significant. If you are the listener, echo is a problem if you hear the speaker twice.

NOTE   Note that echo is an analog phenomenon. If the end-to-end connection is digital (digital phone behind a PBX to another digital phone [PBX or BRI], there is no echo—unless, of course, there's an analog trunk in the middle somewhere.

NOTE   Everyone's echo tolerance is different. However, echo delay over 50 ms is generally problematic for most people.

If you sense a problem with echo in a telephone network, there are two ways to solve the problem:

- Echo suppression
- Echo cancellation

Echo Suppression

Voice is often a strong signal and the echo is an attenuated signal. The echo suppressor determines which signals match to you and which signals match to the person you are speaking to, or both. If the echo suppressor determines that the echo is on the return path, the echo suppressor either attenuates or breaks the transmission path. If the echo suppressor determines that both speech and echo are present at the same time from a combination of both parties on the phone, the echo cannot be attenuated without affecting the voice level.

Echo Cancellation

Due to echo suppression's shortcomings in addressing certain echo conflict situations, such as the one described previously, a more sophisticated method of eliminating echo is echo cancellation.

Rather than break or attenuate the transmit path, as is the case in echo suppression, echo cancellation uses an echo canceller to build a mathematical model of the speech pattern and subtracts it from the transmit path.

NOTE   The echo canceller only removes the echo from one end of the circuit. If the echo is an issue at both ends of the circuit, another separate echo canceller would need to be applied at the other end.

Summary

In this chapter we have covered the basics of analog telephony and some information about digital telephony. If you are an individual with an engineering interest in analog telephony, there is a great deal of engineering literature for you to use in further research.

Keep in mind that signaling principles in the telephone network are no different than the signaling in a data communications network. If you are using a dial-in ISP, your local PC must perform some kind of signaling with the modem and the modem at your ISP must perform some signaling to the dial-in server. These are two separate instances of signaling. The signaling within the telephone network for that dial-in ISP call is the same as that used for a voice call and is a third (caller's local loop), fourth (telephone network), and fifth (called party's local loop) instance of signaling.

Review Questions

The following questions should help you gauge your understanding of this chapter. You can find the answers in Appendix A, "Answers to Review Questions."

1 What is the name for the normal state of a telephone that is hung up?

2 What component in the telephone set converts from 2-wire to 4-wire, and what other function does it serve?

3 Name the three types of signaling used to establish a call and their general purpose.

4 List the three types of E&M start signaling.

5   What is the most common E&M tie-line type used in North America and what are its distinguishing characteristics?

_____

_____

_____

6   What is the most common E&M tie-line type used outside of North America and what are its distinguishing characteristics?

_____

_____

_____

7   What are the two types of address signaling covered in this chapter?

_____

_____

_____

8   What device solves N^2 problem?

_____

_____

_____

9   What are the two trunk types one would order to provide dedicated lines for incoming calls and outgoing calls?

_____

_____

_____

10   What are two methods of dealing with echo?

_____

_____

_____

After reading this chapter, you should be able to perform the following tasks:

- Review, identify, and define digital telephony fundamentals and basics.

- Contrast digital and analog signaling.

- Identify and contrast the different digital frame formats, signaling formats, and coding types.

- Compare the various levels of voice quality.

- Categorize the various types of digital voice compression.

- Examine the ISDN digital architecture and identify its basic components.

Introduction to Digital Voice Technology

As you learned in Chapter 1, "Merging Voice and Data Networks," voice can be transmitted across a traditional network in either analog format or digital format. This chapter covers the basics of voice in a digital network.

As shown in Chapter 2, "Introduction to Analog Technology," analog transmission requires one set of wires for each telephone call. Each FXS or FXO requires two wires, and variations of E&M can use up to eight wires for each call. If you are provisioning multiple voice channels between a PBX and a router using traditional analog interfaces, port density becomes a problem. The number of individual port connections on both the router and the PBX can be cost prohibitive and difficult to maintain.

The digitized alternative to this problem, commonly called T1 in the U.S. (and E1 in most other countries), assumes a fixed number of digitized voice channels that consume a fixed amount of bandwidth across a serial, time-division multiplexed set of wires. Because the interface is digital and not prone to the distortion effects that are common to most analog transmission methods, it provides greater reliability than analog transmission. Maladies such as crosstalk and radio frequency interference (RFI) are eliminated because only ones and zeros are transmitted. Repeaters may be placed in the transmission path for longer-distance T1 transmission lines to reshape the signal in much the same way that an Ethernet repeater does. The combination of the aggregated channels makes the transmission method very high speed (1.544 megabits per second [Mbps] for T1 and 2.048 Mbps for E1). Originally developed for voice transmission, T1 is a common replacement for individual analog interfaces.

In the following sections, we will examine the method by which analog signals are digitized, including some common compression schemes and other innovative ways to save bandwidth. Next, we'll cover standard ways in which the compression schemes are graded for quality. Then we'll move on to a discussion of the signal formats used on the transmission path. Finally, we'll delve into examination of ISDN and Signaling System 7 and the Q.SIG standard.

Digitizing Analog Signals

Expressing an analog signal as a digital one is a daunting task. Since an analog signal, by its very nature, has an infinite amount of values that can be expressed in terms of amplitude, frequency, and phase, converting those values to a comprehensive one and zero expression scheme is very difficult. A mathematical means by which the conversion could be accomplished needed to be developed, and the result of that research (explained in the next four sections) led to the development of a device called a *codec* (*coder–decoder*). The analog telephony signal (human voice) is applied at the input to the codec, and a coded one and zero digital bit stream is developed at the output. Conversely, the process can be inverted to convert the digital bit stream back to analog at the other end of the communications path, with the same codec working in reverse.

There are four steps involved in digitizing an analog signal:

1 Sampling

2 Quantization

3 Encoding

4 Compression (optional)

The following sections explain the details of each step, and focus on digitizing voice-grade telephony channels in preparation for transmission over T1 or E1 wires.

Sampling

In 1924, while working for American Telephone & Telegraph, Henry Nyquist searched for a means by which analog signals could be converted to digital. His research was rooted in military applications, where a minimal amount of transmission wires could be used to multiplex a large number of analog circuits. He developed what is now known as the Nyquist Theorem. It states that in order to digitize an analog signal, the signal must be sampled at a rate equal to that of twice the highest frequency of the signal to be digitized. A voice channel's highest frequency, therefore, will be sampled only twice during one sine wave cycle in order to reproduce it as analog again. Although sampling the signal less often at the highest frequencies does not sound like very good representation, this method assumes that most of the intelligible information in an analog voice channel is nearer the middle of the bandwidth, and sampling will take place there much more frequently. Put another way, the human voice contains sounds that are more often middle-pitched than mostly high-pitched or low-pitched. The significance of the sampling will therefore be more valuable at the middle frequencies than at the high or low frequencies.

So what is the bandwidth of a voice-grade analog channel? The answer depends on the fidelity and capabilities of the analog transmission equipment. We can hear sounds from about 200 hertz (Hz) to about 20,000 Hz. Human speech is in the range from about 250 Hz to 10,000 Hz. A typical telephone channel only carries from about 300 Hz to about 3000

Hz, depending on age and other factors. Voice-grade telephony channels, then, have a limited bandwidth of frequencies compared to the natural human voice (see Figure 3-1). This limitation is imposed for the following reasons:

- The goal of a telephony channel is to transmit intelligible speech.
- The speaker must be recognizable by the listener.
- The speaker's emotions must be discernable by the listener.
- Costs and technology constraints in the telephony transmission system must be considered.

Figure 3-1   *Audible Frequencies Comparison: Human-Audible Spectrum Compared to Speech Spectrum and Telephone Channel*

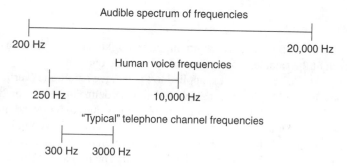

Voice-grade telephony channels exhibit varying frequency responses because of physical differences in wires used, amplifier condition and age, and other electrical factors. In the end, all analog voice telephony channels must meet the four criteria noted above. Frequency responses of 300 Hz to 3000 Hz are common in average telephone channels, as are 270 Hz to 3300 Hz and other widely assorted variations. Because of the variations (and the lack of a standard to define the limits), Nyquist decided to drive a stake in the ground and considered all voice-grade analog channels to have a bandwidth of 0 to 4000 Hz. Our sampling rate, then, is twice 4000 Hz, or 8000 times per second.

The process of sampling an analog signal is simple. As shown in Figure 3-2, the analog waveshape is examined for its vertical amplitude at a rate equal to 8000 times per second (thus satisfying Nyquist's Theorem). Each sample thus represents a snapshot of the signal's amplitude (or vertical height) for 1/8000 second. It will take 8000 of these samples (the vertical lines in the figure) to re-create 1 second's worth of sound at the other end of the T1 line. The electronic component that accomplishes the sampling is called a codec.

Figure 3-2   *An Analog Sine Wave Sampled at 1/8000-Second Intervals*

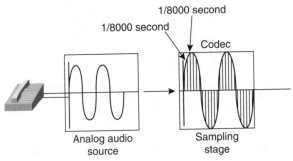

Quantization

The next step in the process of digitizing an analog signal is to express a mathematical value for each of the samples taken. Each sample can be quantified, or assigned a numerical value, if a scale of some sort is applied to the relative amplitude (vertical height) of the samples. Figure 3-3 shows a scale with definitive delineations to which we can assign values. In the case of digitizing analog voice, the values assigned are in the form of 8-bit binary words. Although an 8-bit binary word will produce 256 combinations, only 255 are used. There are 127 delineations above the zero reference line, 127 below the zero reference line, and 1 to mark the zero reference line itself. The bit pattern consisting of all zeros is never used, and the reason for eliminating it will be explained later. Note that the zero reference line will be represented by the all-ones bit pattern. Because of this, any of the 24 T1 channels that are idle will transmit the all-ones pattern in their respective time slots.

Figure 3-3   *Assigning Measurable Quantities to the Samples*

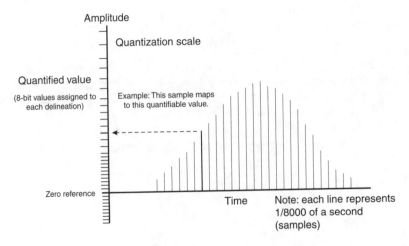

As you can see in Figure 3-3, the scale used is nonlinear in nature. This means that the delineations on the scale are set up to be spaced unevenly. Notice that the marks are close together nearer to the zero reference line, and they move further apart away from the zero reference line. This design serves two purposes. First, the human ear is designed to hear in a nonlinear fashion, and a person can discern sounds of a lower volume more clearly than loud sounds. That is, we get better granularity of the samples at lower volumes than at higher volumes as a result of the nonlinear scale. Second, the original analog signal is compressed in amplitude before it is sampled and quantified so as to reduce the amount of noise present with the signal. After sampling and quantization, less noise is transmitted to the other end as a result, so when the signal is re-created, the signal-to-noise ratio is greatly improved.

There are two standard methods today for quantifying the signal and setting up the nonlinear scale. Since the technology was first invented and deployed in the United States, the original quantifying method used by AT&T is still in use today in North America. This quantifying (or, as it is sometimes called, *companding*) method is called μ-law (pronounced "mu-law"). In the late 1960s, this digitizing technology was presented for international standardization to what was then the CCITT (Consultative Committee for International Telephone and Telegraph) and now known as the ITU-T (International Telecommunication Union Telecommunication Standardization Sector). The study group assigned to the project modified the original companding method. As a result, the international standard for companding, A-law, was released. A consequence of this ruling is that North America, which uses μ-law, is incompatible with the rest of the world, which uses a-law. (Another consequence of this is that North America uses T1 while the rest of the world uses E1.) International agreement states that while interfacing to any a-law country, the μ-law country must convert to a-law before interfacing.

NOTE   In some parts of Japan, μ-law is used to better interface with North American countries.

The quantization process is not always accurate, however. Consider what would happen if a sample's amplitude fell exactly between two marks on the scale. Figure 3-4 shows the third sample to be between two valid scale delineations. How do we quantify this particular sample? For a definitive quantifiable measure, should the codec guess in the up direction or the down direction? No matter which way it guesses, the sample taken will never be exactly accurate at this 1/8000-second instance. When we re-create the sample at the far end, it will be inaccurate because we cannot represent a value between two scale delineations. This mistake in representing value is called a *quantization error*, and it results in quantization noise when the analog signal is re-created. *Quantization noise* is noise that the listener may not necessarily hear (since it's a mistake for only 1/8000 second), but if enough quantization errors occur together, they may create noise that can be heard. For this reason, the codec applies a low-pass filter, or *smoother circuit*, to the re-created analog signal to filter noise created by quantization errors.

Figure 3-4   *Quantization Error: One of the Samples Shown Is Exactly Between Two Sampling Points*

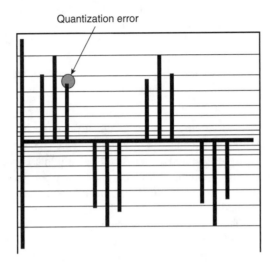

Quantization error

Encoding

Once each sample is taken and then quantified against the quantization scale, the codec assigns a numerical value to the quantified sample. As noted earlier, we use 8-bit binary representations for each quantified sample. The 8-bit binary word that represents each of the marks on the quantization scale (see Figure 3-3) has a predefined format. Figure 3-5 depicts the meaning of each bit of the 8-bit word. The first bit is used to represent polarity. It simply relates whether this sample is above the zero reference line or below it.

The next 3 bits represent Segment, sometimes called Chord. These bits represent the area of the scale in which this sample is found. Each Segment has an equal number of steps, which are represented by the remaining 4 bits. For example, an 8-bit value of 10110101 decodes as follows: The quantified sample is above the zero reference line (since the leftmost bit is a 1), the sample is situated in the third segment (the next 3 bits, 011, have a decimal value of 3), and the sample is on the fifth delineation within segment three (the last 4 bits, 0101, have a decimal value of 5).

Figure 3-5   *Format of a Word Sample: Sample Is 8-Bits Long, Specific Bits Have Individual Meaning*

P	Se	Se	Se	St	St	St	St

P = Polarity
Se = Segment
St = Step

Now that three of the four steps of digitization are complete, it is helpful to review what's happened:

1 Sampling: Samples are taken at a rate of 8000 times per second.

2 Quantization: Each sample is quantified in comparison to a scale that has delineations grouped in segments.

3 Encoding: Each quantified sample will produce an encoded 8-bit word that represents the sample's amplitude.

Altogether then, 8000 samples per second times 8 bits per sample yields 64000 bits per second to represent one second of sound or speech. This particular coding scheme is designated as ITU-T standard G.711, or pulse code modulation (PCM). That's a very large quantity of bandwidth to consume across expensive WAN networks for one telephone call. But it was a very innovative way to multiplex multiple channels across a digital transmission medium (T1).

Compression

Voice compression schemes were developed as an effort to save bandwidth on the WAN. As an attempt at maintaining the quality of PCM at 64000 bits per second (bps), many different approaches were taken. The results were variations on the same theme: lower bit rates, but with quite a loss in quality. Some attempted to cut the bandwidth and follow a similar approach to PCM, while others blazed new trails with bold approaches like continuously variable slope delta (CSVD). We will present some of the more common approaches in the following sections, including those chosen by standards bodies and selected by Cisco Systems. It is important to note that two of the terms used in this chapter, coding and compression, have two different meanings. Coding is a means by which the analog signal is represented. Compression means we're trying to improve on the original 64000-bps PCM method.

NOTE   Note that compressing the original PCM-coded signal is completely optional, and is done specifically to save WAN bandwidth. Any compression of the original PCM will affect voice quality. You need to balance the cost of WAN bandwidth against the cost to compress with the loss of quality of the original digitized PCM signal. Rating the resulting quality is described later in the section titled "Voice Compression Techniques Compared."

The following sections examine three technologies that can be used for voice compression:

- Wave form compression: Follows the approach used for PCM encoding
- Vocoder compression: Synthesized voice with processing intelligence

- Hybrid compression: A combination of wave form and vocoder compression

An in-depth examination of each compression approach follows, with pros and cons listed for each.

Wave Form Compression

Wave form compression is a subset of compression schemes, which include PCM and its related derivations. This family of compression schemes is known as *wave form coding* because quantization and encoding (that is, sampling for quantity and assigning a binary value to the quantity) tracks and follows the actual analog wave form as it develops in real time. Adaptive differential pulse code modulation (ADPCM) includes a variety of wave form coding methods, including 40000-bps, 32000-bps, 24000-bps, and 16000-bps varieties. Collectively, these compression schemes are designated as ITU-T standards G.726 and G.727. The 32000-bps version was developed first, and was planned to allow the capacity of T1 to double, from 24 voice channels to 48.

As an example of how wave form compression schemes work, the 32000-bps version uses a 4-bit word, instead of an 8-bit word as PCM uses (thus resulting in 32000 bps instead of 64000 bps). An individual quantization represents the comparative difference between the quantization itself and the last sample (that is, the differential). The 4-bit coded word uses 1 bit to represent the amplitude change since the last sample (that is, an increase or a decrease in amplitude from the last sample) and the remaining 3 bits to represent how much to increase or decrease the amplitude. The relative value of the 3 bits can change based on the last few samples seen (that is, it is adaptive).

The 24000-bps version uses a 3-bit word, and the 16000-bps version uses a 2-bit word. As you may have guessed, the quality of ADPCM suffers somewhat in comparison to PCM and degrades significantly with each version. The 16000-bps version is almost unintelligible. The 40000-bps version uses a 5-bit word, and was actually developed to improve on the poor quality of the 32000-bps version.

Pros of wave form compression are:

- Reduced bandwidth consumption
- Simple and inexpensive to process

Cons of wave form compression are:

- Poorer audio quality at lower bandwidths

Vocoder Compression

A second method of speech coding is called *vocoder* (for voice coder). This method can produce low bit rates with very intelligible speech, but it sounds machine-like. It is based on the premise that the human voice is made up of a base frequency or sound, produced by the vocal cords, which is varied using a sound chamber (the mouth and throat). The variations, like the tongue against the teeth, lips formed in a circle, and the rolling of r as in the Spanish language, are called *fricatives*. The vocoder method assumes that the base frequency can be played continuously for each word spoken at the far end of the channel, with the transmitter sending only the fricatives.

During the vocoder processing, the coder must have the ability to recognize and encode each different fricative produced by human speech. This ability requires some measure of intelligent processing power and some preprogrammed knowledge of human speech patterns. Special processing electronics are required to accomplish this task. An example of a vocoder is the speech created by the computer used by the famous physicist Stephen Hawking. Although this form of synthesized voice is very practical for this type of use, vocoders are not practical for voice telephony since we must be able to identify the speaker and sense his or her emotional state.

Pros of vocoder compression are:

- Reduced bandwidth consumption

Cons of vocoder compression are:

- Expensive to process
- Requires specialized electronics
- Sounds synthetic
- Speaker is not recognizable

Hybrid Compression

The third category of voice coding is *hybrid coding*. It combines the best of the wave form and vocoder compression techniques to create high-quality voice at low bit rates. Used extensively in the digital cellular telephone industry, hybrid coding schemes can save significant bandwidth across a WAN channel. One of the more recent developments in this category is called *code excited linear predictive (CELP)* coding. This method maintains a "codebook" of waveshapes that are representative of sounds that the human voice can produce. Each waveshape is assigned a binary code. When the speaker talks, entire waveshapes are sampled and compared to the codebook, and the closest waveshape in the codebook to the original is sent across the channel in the form of its assigned binary pattern. Although of high quality, CELP requires powerful processing circuitry and lots of memory in order to perform its task. It also causes a significant amount of delay due to the coding process.

A second variation of CELP, called *low-delay CELP (LD-CELP)*, builds the codebook directly from the speaker's voice rather than from fixed waveshapes. This results in a shorter processing delay and possibly a more accurate voice representation. LD-CELP has been designated as ITU-T standard G.728 and operates at 16000 bps.

A third variation of CELP has been made possible by specialized microprocessor chips called *digital signal processors (DSPs)*. Conjugate structure algebraic CELP (CS-ACELP) can encode very high-quality speech at a rate of 8000 bps. The original LD-CELP algorithm was modified somewhat to make it more efficient and sample more exactly to produce this complex scheme. The codebook is even more adaptive with CS-ACELP, utilizing far more complex mathematics to evaluate and encode the signal. Because the codebook is more adaptive, it can react to different languages more readily. The original CELP codebook was designed using sounds produced with American English and was therefore limited. CS-ACELP can adapt its codebook's waveshapes to many variations of human speech and can therefore adapt itself to the language being spoken. It has been designated ITU-T standard G.729 and has some basic variations.

The ITU-T Standard G.729a is also an 8000-bps CS-ACELP coding method, but its algorithm has been simplified to make it more efficient. Although it produces very high-quality voice reproduction, its fidelity is slightly less than that of the original G.729. Two other variations of G.729 are G.729B and G.729AB. These two variants are also 8 k but include a built-in VAD (Voice Activity Detection) algorithm, which serves to save even more bandwidth. VAD is described in a later section titled "Digital Speech Interpolation."

Pros of hybrid compression are:

- Excellent audio quality
- Very low bit rates
- Adapts to speaker

Cons of hybrid compression are:

- Requires specialized processing chips (DSPs)
- Requires memory
- Induces processing delay

Voice Compression Techniques Compared

As the demand for voice-over-X technologies continues to increase, the IT industry will continue to develop new coding and compression techniques to conserve bandwidth and align with the particulars of the lower layer services available. When bandwidth is affordable and high quality is a must, or if connecting equipment is of traditional de facto standard, PCM coding at 64000 bps stands at the ready. Where corners can be cut and variations of ADPCM fit the need (such as in hotels and call centers), less bandwidth for less cost and quality can be satisfactory. More recent users, whose needs may not have to align with older equipment, will opt for the CELP compressions. Your choice will depend on many factors. The newer CELP standards offer very low bit rate transmission with excellent quality, and carriers and private customers will migrate toward these standards without hesitation.

Digital Speech Interpolation

In addition to compression, there are other innovative ways to save bandwidth. If we were to examine the mechanics of a human conversation, we would find that during portions of the conversation, the speaker goes silent to listen. As the listener is silent, the transmission of voice from the listener to the speaker is almost non-existent, with the exception of background noise. An opportunity exists here to close the communications path from the listener to the speaker and save bandwidth only in that direction. In addition, as the conversation develops, the speaker intentionally and unintentionally makes pauses in the conversation. These pauses—to make a point, to end a sentence, to allow the listener to digest what's been said—provide more opportunities to save expensive WAN bandwidth. The Cisco Voice Port option used to accomplish the savings is called VAD, for voice activity detection. Figure 3-6 shows how this saving can be applied.

Figure 3-6   *Voice Activity Detection: Saving Bandwidth Due to Listener's Silence*

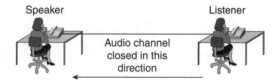

Digital speech interpolation is a DSP function that examines voice to determine power (that is, volume) and change of power, and to determine frequency and change of frequency. When periods of silence are present, the sending entity will send an "enter silent period" notification, at which time the receiver will play a period of white noise (or *comfort noise*) to the listener. During the period of silence, the sender need not send empty packets or frames to the receiver, which means there is more room for data frames or packets to get through. When the speaker again begins to talk, the DSP again opens the audio channel, allowing speech to be transported.

Although it is valuable during a conversation, the digital speech interpolation approach can devastate some voice mail systems that end recording when speech is not present. Care should be taken when enabling this feature on Cisco voice-capable routers. Later releases of Cisco IOS allow the system administrator to control the timing of the VAD action when silence appears so as to not clip the speech between words and phrases. Also, keep in mind that the savings in bandwidth may be attractive, but because of the artificial comfort noise, it may not be aesthetically pleasing to the listener.

Telephone Voice Quality

As shown in Figure 3-7, AT&T has defined three levels of voice quality for comparing different compression schemes (see Table 3-1).

Figure 3-7 *Comparison of Voice Quality Compression Technologies*

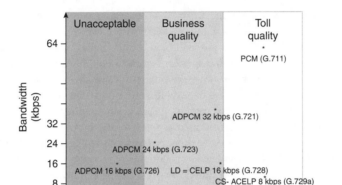

Table 3-1 *AT&T Voice Quality Levels*

Quality Level	Description	Examples
Toll	Indistinguishable from a straight "all analog" copper connection. Produces the best sound quality and has no distinguishable disturbances or distortions.	Calling within the continental United States usually results in toll quality.
Business	Noticeable distortions and poorer audio quality, but the speaker is still recognizable and an intelligent conversation can still take place.	Calling between two continents or using a poor transmission system as in third-world countries.
Unacceptable	Annoying and disturbing distortions that interfere with speech. Words and phrases are misinterpreted or not understood due to the severity of the distortions, and a conversation is difficult to maintain.	Calls to remote portable systems, as in military battlefield applications or remote oil drilling applications.

The ITU-T standard G.711 is best in the toll quality category—that is, it cannot be distinguished from straight copper wire end-to-end, and it is the digital coding standard by which all others are compared. The industry's second attempt at coding, ADPCM 32000, falls into the business quality category. ADPCM has many variations, as noted in previous chapters; as variations use less bandwidth, quality suffers. The 16000 variety of ADPCM falls into the lower category of quality called unacceptable, because the distortions and distractions are too numerous and frequent to allow an intelligible conversation.

As shown in Figure 3-7, the CS-ACELP coding methods have a quality that surpasses all of the ADPCM schemes and comes very close to PCM. These coding methods may be expensive in terms of processing power required, but their payoff in WAN bandwidth saved is undeniable.

The ITU-T, like AT&T, has defined parameters for measuring voice quality. A fair comparison of voice quality is critical to designing a comprehensive voice/data network because traditionally, the higher-bandwidth schemes tended to deliver better quality than the lower-bandwidth schemes. This can translate to higher costs for corporations deploying voice across data networks, and the solution becomes less cost-effective than was first thought.

The following sections explain the categories of quality as outlined by the ITU-T, the method of voice quality measurement used, and the quality scores for various voice compression methods.

ITU-T's Voice Quality Measurement

Unlike AT&T's three-category method, the ITU-T defines five categories of voice compression methods, as shown in Table 3-2. Each of these five categories is assigned a value, on a scale from 1 to 5, with 5 indicating toll quality and 1 indicating unsatisfactory. The following section relates how the ITU-T uses this scale to compare various voice compression methods.

Table 3-2   *ITU-T Voice Quality Levels*

Quality Level	Description	Examples
Toll quality (Grade 5)	Toll quality emulates and sounds like a copper wire. Exhibits similar quality to that of an analog end-to-end call with signal-to-noise ratios and harmonic distortions within acceptable limits.	Calls within a PBX from user to user or within a central office, such as calling a neighbor in the same geographic area.
Transparent quality (Grade 4)	This is very similar to toll quality, with some tolerable distortions and almost imperceptible distractions. The distortions may be discernable to the most critical user, but are not annoying.	Calling long distance, from state to state, or between neighboring countries.

continues

Table 3-2 *ITU-T Voice Quality Levels (Continued)*

Quality Level	Description	Examples
Conversational quality (Grade 3)	Conversational quality has perceptible distortions and annoying distractions. In this category, the user begins to noticeably hear the degraded quality of the channel and may need to ask the speaker to repeat portions of the conversation.	Intercontinental calls or calls to third-world countries.
Synthetic quality (Grade 2)	Synthetic quality is tolerable, but has very annoying distractions and poor reproduction of the speaker's voice fidelity. Because the reproduction is so poor, the listener hears what almost sounds like a machine-like re-creation.	Ship-to-shore telephony communications.
Unsatisfactory (Grade 1)	When the voice reproduction is this poor, the listener is sometimes forced to ask for a new channel. A simple exchange between the speaker and listener is strained to the point of the speaker repeating what's been said again and again.	Interference caused by malfunctioning equipment or induced from outside sources, such as radio interference.

Mean Opinion Score

The ITU-T recommends measurement of voice quality using the five categories shown in Table 3-2 by a subjective methodology called the Mean Opinion Score (MOS). Test subjects are gathered into a lab environment and asked to rate voice quality through varying methods of compression. The rating uses the five grades shown in Table 3-2. The participants listen to a recorded message that is chosen based on its varying fricatives. The message used to measure MOS for English speaking listeners is "Nowadays, a chicken leg is a rare dish." As they listen to the message, they realize that there are hard sounds (as in the "K" sound in *chicken*) and soft sounds (as in the "CH" in *chicken*), long vowels (as in the "A" sound at the end of *nowadays*) and short vowels (as in the "I" in *dish*). As the listeners rate the different compression methods, the average, or mean, of all the participants is calculated after the tests. The ITU-T considers a MOS of 4.0 as toll quality.

Although MOS is widely used, a newer method for measuring voice quality is being accepted by the industry. This method is called Perceptual Speech Quality Measurement, or PSQM, and is assigned the ITU-T standard P.861. PSQM was developed to measure voice quality in transmission systems originally developed for data, such as running voice-over IP networks. The PSQM quality measurement can sometimes be a more precise measurement tool than MOS since it is not subjective and is sensitive to impairments seen on voice-over data networks, such as delay and missing packets or frames. Some manufacturers of PSQM equipment include the capability to convert the PSQM result into MOS scores. Cisco Systems uses PSQM as a more precise means by which to measure voice quality over data networks.

You must also keep in mind that the ITU-T's recommendation for using MOS as a quality tool only specifies how to conduct the tests. The ITU itself does not publish individual MOS scores for individual codecs. Results will therefore vary from vendor to vendor and test to test.

MOS Rating of Digital Voice

Clear distinctions emerge when the MOSs of different compression schemes are listed side by side. Table 3-3 lists the MOSs of selected compression schemes as published by the ITU-T. From Table 3-3, we can see that G.711 PCM scores very high, has a short framing size or sampling interval (.125 seconds equals 8000 samples per second), and demands very little of the processor (requires .34 millions of instructions per second [MIPS]). G.726 ADPCM scores a bit lower than G.711, with more processor demand and the same sampling rate. G.729 and G.729a both require more processing power, but even though their bit rates are lower, they score significantly higher on MOS tests. This is where corporations can realize a savings in combining voice and data networks. For a little more WAN bandwidth, voice can be economically carried with data and still be delivered at a quality that approaches G.711 PCM.

Table 3-3 *Examples of ITU-T MOS Ratings of Different Compression Schemes*

Codec	Compression Method	Bit Rate	MIPS	Compression Delay (ms)	Frame Size	MOS
G.711	PCM	64	0.34	0.75	0.125	4.1
G.726	ADPCM	32	13	1	0.125	3.85
G.728	LD-CELP	16	33	3–5	0.625	3.61
G.729	CS-ACELP	8	20	10	10	3.92
G.729a	CS-ACELP	8	10.5	10	10	3.9

MOS Under Varying Conditions

Although it is a good starting point, the ITU's method of ratings of voice compression should be used only for comparative purposes. In the real world of deploying voice with data, we are faced with more varying conditions than simply a collection of people in a laboratory listening to quotes about chicken legs. In the real world, we must consider that we may have soft speakers, the network may exhibit bit errors (where data may be retransmitted, voice must be delivered even with errors because it's real-time), and/or the network may have framing errors. We must also consider tandem codings, where the compressed voice is decompressed at an intermediate site, then compressed again into the same or different format and forwarded to a third site. All of these conditions will cause the MOS to suffer and the voice quality to degrade quite rapidly, as in the practice of making a copy of a copy of a videotape.

MOS Scores Compared

Given the MOS ratings of different types of coding schemes, we can plot the MOS results on a comparative graphic. As shown in Figure 3-8, vocoders always tend to score very low, no matter what their bit rate. Wave form coders (such as PCM and ADPCM) score much better, but require higher bit rates. Their quality drops significantly at lower bit rates. However, the hybrid coders (such as LD-CELP and CS-ACELP) score very high even at lower bit rates. As Figure 3-8 illustrates, 8K CS-ACELP is very close in quality to 64K PCM.

Figure 3-8   *MOS Subjective Analysis*

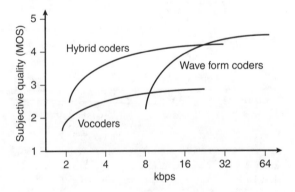

The results shown in Figure 3-8 are simply an example from the author's perspective and do not relate scientifically to any particular test. Most typically, wave form coders tend to score highest and vocoders score lowest. Table 3-4 lists the MOS grades used to illustrate the MOS scores in Figure 3-8. As an example, hybrid coders score about 3.9 at a data rate of 8 k, which is typically representative of results for G.729.

Table 3-4   *MOS Grades*

Grade	Quality	Level of Impairments
5	Excellent	Imperceptible
4	Good	Just perceptible, not annoying
3	Fair	Perceptible and slightly annoying
2	Poor	Annoying but not objectionable
1	Bad	Very annoying and objectionable

Channel Signaling Types and Frame Formats

When Cisco voice users connect to the Public Switched Telephone Network (PSTN) or to PBX systems via a digital T1 or E1 interface, most telephony vendors agree on the standard G.711 PCM coded format. This coding format is used by default in most digital telephony systems. In order to better understand the responsibilities of both the PBX vendor (or the PSTN provider) and the installed Cisco Voice router, we must take a look at the format of the digital interface. The following sections explore the workings of both T1 and E1 as they relate to interconnecting devices for the purpose of voice transmission. We will examine the physical properties of the connections as well as the logical or framing properties. Synchronization and timing is also explained, as are methods for conveying on-hook and off-hook conditions of each telephone.

T1/DS1

Most people in the IT business unknowingly throw around acronyms and abbreviations without regard for their original meaning or their intended purpose. Such is the case when we talk about the digital service available in North America that we call T1 or DS1. In reality, the two terms have separate meanings, and each should be taken in its own context.

The term T1 dates back to its original deployment in the 1960s, when all installed circuits at this speed (1.544 Mbps) had to run on terrestrial (hence the *T* in T1) facilities. Dedicating land-line copper to the circuit was important at the time T1 was first installed because regenerative repeaters had to be installed every 4000 to 6000 feet along the path of the facility; otherwise, the signal would degrade. Therefore, when speaking of the term T1, we use it in the context of its physical characteristics. Modulation of the ones and zeros along the wire is a physical characteristic of T1. Electrical termination impedance is a physical characteristic of T1.

The term DS1 describes the framing characteristics of this digital transmission method (for digital signaling level 1). DS1 describes the facility as capable of carrying 24 DS0s, or a

DS1 frame, and grouping the frames into contiguously structured Superframes (that is, 12 frames) or Extended Superframes (that is, 24 frames). The details and reasoning behind the DS1 framing structure are addressed in the next section. Subsequent sections will explain options for framing structures, how the individual telephone channels signal for on-hook and off-hook, and the critical issue of end-to-end timing.

DS1 Digital Signal Superframe Format

Since the original design behind DS1 was to carry voice (not data, as most believe), we must consider that the framing format is set up so as to carry one PCM sample from each of 24 simultaneous telephone conversations along the same wire. Figure 3-9 shows the format of DS1 frames, each carrying, in sequence, an 8-bit sample from the first telephone channel, followed by an 8-bit sample from the second telephone channel, and so on up to 24 channels. The samples, when added together, result in 192 bits (24 frames × 8 bits per frame), which are sent in a serial fashion. At the end of each frame is a special bit, the 193rd bit, which serves as a synchronization function. We will address it later, in the section "Robbed Bit Signaling."

Figure 3-9   *DS1 Framing Format*

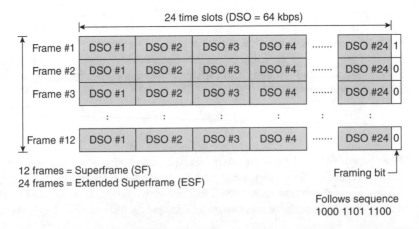

12 frames = Superframe (SF)
24 frames = Extended Superframe (ESF)

Framing bit

Follows sequence
1000 1101 1100

It is important to note that the transmitting device (that is, the PBX) will need to transfer 8000 frames per second to the receiving device (the Cisco Voice router) in order to re-create one second's worth of audio for all of the 24 separate telephone channels. (Recall that the sampling rate of PCM is 8000 times per second, with each sample issuing an 8-bit word at each interval.) Now for the math: If a frame consists of 193 bits, and 8000 frames are transmitted every second, then 1544000 bps (that is, 193 × 8000) is the resulting data rate.

Viewed another way, 24 telephone channels each use 64000 bps using PCM, so 24 × 64000 = 1536000 bps, plus the 8000 framing bits that must be sent (one with every frame), for a total of 1544000 bps.

Robbed Bit Signaling

As discussed earlier in the chapter, T1 was developed to *replace* individual analog interfaces. A common analog interface used between a PBX and a Cisco Voice router is the E&M interface. In Chapter 2 we discussed the fact that E&M interfaces carry one voice conversation at a time, so T1 *emulates* 24 E&M connections. But there's more to digitizing E&M interfaces than just transmitting 8-bit audio samples. We must also convey the status of the M lead from the PBX to the Cisco Voice router. This is accomplished through a method called *robbed bit signaling.*

In order to understand how robbed bit signaling works, we must go back and take a closer look at the 193rd bit used for frame synchronization in each frame. This synchronization function is accomplished by having the 193rd bit follow a unique pattern during transmission. The pattern that the 193rd bit follows is called the *D4 framing pattern.* The sequence followed, in the Superframe format, is 1000 1101 1100. Figure 3-9 shows that the first frame in a Superframe carries a 1 bit in its 193rd position. The second frame carries a zero. The third frame also carries a zero. This continues through the 12 frame sequence at which time a new Superframe begins with the 193rd bit following the same sequence again. Using this unique bit pattern in the 193rd position, the receiving T1 device can identify frame number 1, frame number 2, frame number 3, and so on up to frame number 12 in a given Superframe.

NOTE   In the 12-bit D4 framing pattern, a receiving T1 device need only receive 4 frames to declare synchronization. This is because any 4 consecutive framing bits in the sequence are unique from any other 4. If a T1 device receives 4 frames, for example, with the framing pattern 0110, it can establish that these are frame numbers 4, 5, 6, and 7 in a Superframe. If you study the framing pattern, you will see that any 4 consecutive bits are unique.

What does all this have to do with signaling? The robbed bit signaling method will identify frame number 6, and use the least significant bit in each of the 24 channels to convey the status of the M-lead from the PBX to the Cisco Voice router. (Remember that there are 24 separate E&M emulated connections, so there are 24 separate M-leads to convey.) As each of the 24 channels in frame number 6 are representing an 8-bit PCM word (coded from a 1/8000-second sample), the transmitting T1 device robs, or overwrites, the least significant bit and uses it to convey the M-lead. This happens in the 12th frame as well. The robbed bit from the 6th frame is called the A bit, and the robbed bit from the 12th frame is called the B bit. Figure 3-10 shows the format of the 6th and 12th frames. Taken together, the A and B bits can convey four different states: 00, 01, 10, and 11. This becomes important when telephones and telephone lines are capable of more than just on-hook and off-hook. A taken with B allows the PBX to convey additional signals, such as transfer or hold.

Figure 3-10 *Robbed Bit Signaling Format*

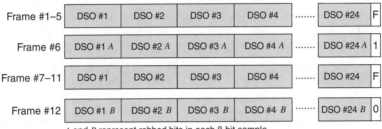

A and *B* represent robbed bits in each 8-bit sample.

Since robbed bit signaling overwrites and uses one of the bits from an 8-bit audio sample in each of the DS0s (that is, every 6th frame), the samples in the 6th and 12th frame lose some of their quantization reference. Instead of the sample being represented by 8 bits, 7 bits are effectively used for each channel every 6th frame (the 8-bit samples are left alone for all the other frames). As a result, quantization errors are introduced, thereby compromising the quality of the received and re-created analog signal. Smoother circuits, as discussed earlier in the chapter, are then even more important to get rid of the resulting quantization noise.

Extended Superframe

In the early 1980s, AT&T identified a need to include an in-band management channel for T1. This management channel would be used to command T1 devices to initiate loopbacks on certain DS0s to be tested, and to return the number of framing or other such errors to the central office for evaluation purposes. In addition, the format of T1 could be changed in such a way as to allow a CRC to be included to identify, but not correct, framing errors. It was decided that the 8000 "framing" bits would be repurposed for these uses. As this new framing method was designed, 4000 of the 8000 bits would be used for the management channel, 2000 of the bits would be used for CRC checks, and 2000 would still be used for frame synchronization.

From this idea, the Extended Superframe (ESF) format was born. While Superframe employs a format that consists of 12 contiguous frames, ESF uses 24 frames. Of the 24 frames, 12 frames carry management information in their 193rd bit position, 6 frames carry CRC information, and 6 frames carry synchronization bits. Table 3-5 illustrates the ESF format starting with the 1st frame and ending with the 24th. Each of the 24 frames in the ESF format has its 193rd bit used for a special purpose. Robbed bit signaling can now represent 16 different situations using ABCD bits, or, optionally, four situations using AB bits, or two situations using only the A bit. Notice that the first frame carries a management bit, called a data link control (DLC) bit. The second frame carries a CRC (or block check [BC]) bit. The 3rd frame carries a DLC bit again, and the 4th frame carries a synchronization, or F_e (framing extended) bit. This cycle continues until the 24th frame is transmitted, then it starts again.

Table 3-5   *Extended Superframe Format*

	Use of 193rd Bit			DS0 Bits		Signaling Bits		
Frame	**Fe**	**DL**	**BC**	**Traffic**	**Signaling**	**2**	**4**	**16**
1		M		Bits 1-8				
2			C1	Bits 1-8				
3		M		Bits 1-8				
4	0			Bits 1-8				
5		M		Bits 1-8				
6			C2	Bits 1-7	Bit 8	A	A	A
7		M		Bits 1-8				
8	0			Bits 1-8				
9		M		Bits 1-8				
10			C3	Bits 1-8				
11		M		Bits 1-8				
12	1			Bits 1-7	Bit 8	A	B	B
13		M		Bits 1-8				
14			C4	Bits 1-8				
15		M		Bits 1-8				
16	0			Bits 1-8				
17		M		Bits 1-8				
18			C5	Bits 1-7	Bit 8	A	A	C
19		M		Bits 1-8				
20	1			Bits 1-8				
21		M		Bits 1-8				
22			C6	Bits 1-8				
23		M		Bits 1-8				
24	1			Bits 1-7	Bit 8	A	B	D

Two standard protocols are used on the management (M), or DLC, channel: ANSI T1.403 and AT&T's proprietary standard. The Cisco router may be configured to use either. The decision to configure for either method is determined by the protocol being used by the PBX or PSTN provider.

You can also see from Table 3-5 that the ESF format provides for robbed bit signaling in the 6th, 12th, 18th, and 24th frames. This results in an ABCD robbed bit format, allowing for 16 different signaling representations. In the event that only four representations are necessary, the 6th frame carries an A bit, the 12th frame carries a B bit, the 18th frame repeats the A bit, and the 24th frame repeats the B bit. If only two representations are needed, each robbed bit position carries the A bit.

Clocking and Line Coding

Most T1 equipment has options for clocking reference for the purpose of bit synchronization. These options are as follows:

- Internal: Synchronization is derived from an independent internal oscillator in the T1 equipment.

- Clock from Network: Synchronization is derived from the bits arriving from the received signal. The synchronization is then tied to the internal oscillator and used for transmitting all other signals.

- Loop Timed: Synchronization is derived from the received signal but is not referenced to the internal oscillator. It is, however, used to transmit signals on this interface only.

Most of the time, T1 equipment is optioned for Clock from Line, also known as Clock from Network. This clocking method ensures that the received bits will follow the clocking rate of the network, which sources its timing from a very accurate clock at the heart of the digital network. In order for the receiving T1 equipment to develop the clock, the receiving T1 device watches the pulses applied to the circuit and measures the timing between pulses to develop a timing source. The timing source develops by applying the electrical pulses on the wire to a crystal oscillator, which vibrates at a relatively constant rate. This type of clocking circuit regenerates itself by applying the crystal's output pulses back to the input of the circuit, along with the pulses received from the T1 signal for reference. An absence of pulses on the circuit causes the reference clock to drift. Transmitted ones pulse the circuit, and transmitted zeros do not. Because of the lack of pulses when transmitting zeros, carriers require that the customer follow certain rules to ensure that enough pulses are present on the circuit to develop a proper clock reference. These rules are referred to as the Ones Density rules and are stated as follows:

- No more than 15 consecutive zeros are allowed.

- At least 12.5 percent of the user traffic must be ones.

NOTE   A T1 circuit is said to use a modulation scheme called Bipolar Alternate Mark Inversion Return-to-zero (or AMI). This means that when ones are to be transmitted, they will pulse the circuit with a square wave in an alternating bipolar fashion. Successive ones will pulse in opposite directions, positive and negative, and successive pulses of the same polarity are not allowed. When successive pulses do modulate in the same polarity by mistake, this error is called a bipolar violation (BPV). Zeros do not pulse the circuit at all.

When a T1 is used to carry voice traffic in the form of PCM, the Ones Density rules are always followed because there is no 8-bit sample represented by 8 zeros (see the section "Quantization" earlier in this chapter). Trouble occurs, however, when users utilize unused voice DS0s to carry data. Data has a propensity to transmit varying strings of zeros because the source of the data cannot be controlled. This may result in strings of zeros across multiple DS0s that do not pulse the clocking circuit, thereby causing the clock to drift from reference. The initial attempt at ensuring that the Ones Density rules were followed when carrying data traffic was to force one of the bits in every 8-bit DS0 to a 1. This was done only in DS0s that carry data and not in DS0s that carry voice. Effectively, then, each DS0 bearing data used 7 bits 8000 times per second for a total data rate of 56000 bits per second. This is where the 56000 data rate came from and is still in use today for circuits that use AMI line coding.

If the entire T1 is used for carrying data, the effective user traffic is reduced to 1344000 bits per second due to the ones used to ensure Ones Density rules are followed ($56000 \times 24 = 1344000$). This made early users of T1 quite upset, since they paid for 1544000 and lost 200000 bits each second. AT&T then went back to the engineering labs to develop a modulation method to ensure Ones Density, regardless of user source (voice or data).

A new modulation scheme emerged, called Bipolar with Eight-Zero Substitution Return-to-zero (or B8ZS). In this modulation scheme, a substitution pattern containing intentional BPVs is inserted when eight consecutive zeros need to be transmitted. The substitution pattern contains BPVs in the 4th and 7th bit positions, and the receiver then reads the 8-bit sequence as 8 zeros. Figure 3-11 shows the implementation of B8ZS compared to pulses in AMI. Since the substitution pattern does contain pulses, clock reference is maintained.

Figure 3-11 *AMI and B8ZS Modulation Compared*

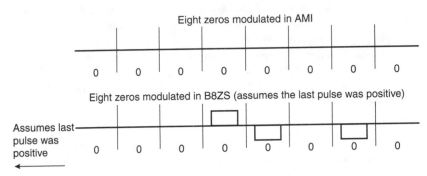

E1 Channel Signaling

Like T1, the European digital transmission method for PCM voice, E1, has physical characteristics and framing characteristics. E1 can terminate at impedances of 75 or 120 ohms, and can use AMI or HDB3 (high density bit 3) line coding. E1 can also run balanced or unbalanced by standard, meaning that the two wires used for transmission can vary independently in voltage to represent ones and zeros, or one wire represents ground and the other wire varies in voltage.

NOTE   HDB3 is the E1 equivalent of B8ZS for ensuring Ones Density. HDB3 inserts a BPV in the fourth bit position of a 4-bit sequence of consecutive zeros. If the receiver sees three consecutive zeros, it checks the next bit. If it's a BPV, then the transmitter inserted it, and it is interpreted as a zero. This ensures enough pulses on the circuit for timing to be maintained.

Like T1, E1 also has structured framing characteristics and signaling methods and line coding. These methods are not always exactly like those used in T1, and vary due to the attempt to recommend a standard that can be deployed worldwide, regardless of infrastructure. These ideas are explained in the following sections.

E1 Framing and Signaling

The E1 framing method defines 32 DS0s as one frame, with 16 contiguous frames considered a Multiframe. Figure 3-12 shows the format of an E1 Multiframe. The user capacity for E1 voice channels is 30 DS0s, with the first full DS0 carrying synchronization patterns (there is no separate framing bit in E1), and the center DS0 used to carry ABCD signaling. The first frame in a Multiframe has its first DS0 carrying a unique 8-bit pattern establishing the beginning of a new frame. The 17th DS0 carries a different unique pattern, signifying the first frame of a Multiframe. The second frame in a Multiframe has its first DS0 carrying the same pattern as in the first frame, while the 17th DS0 is used to carry ABCD for the second DS0 and the 18th DS0. This continues until the 16th frame is completed, thus finishing one Multiframe.

Figure 3-12 *E1 Multiframe*

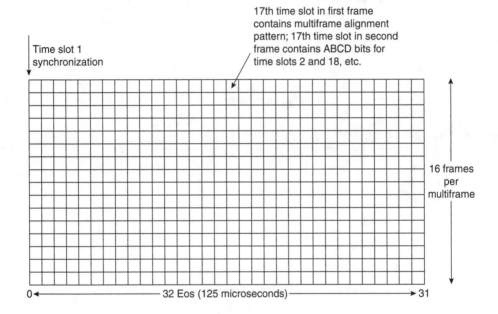

E1 Line Coding

Unlike T1, E1 can be deployed with Return-to-zero line coding or Non-return-to-zero line coding. It also has two modulation formats as noted previously: AMI and HDB3. These variations are necessary to ensure that E1 can be deployed in poor infrastructures (such as those found in third-world countries) without needing expensive twisted pair and coaxial cabling combinations.

Digital Channel Signaling Types: CAS and CCS

The way that T1 carries signaling information within each of the DS0s every 6th frame is called *channel-associated signaling (CAS)*, which is synonymous with robbed bit signaling. A T1 circuit bearing encoded PCM voice channels with robbed bit signaling is said to be running in CAS mode. You will need to configure the Cisco Voice router for CAS mode when connecting the router to a CAS service, since CAS is not default for T1 interfaces.

On the other hand, E1 carries its signaling in DS0 17 all the time, so it is said to run in common channel signaling (CCS) mode. However, since the format of the ABCD bit patterns used for signaling matches that of T1 exactly, some E1 providers will still refer to their signaling as CAS. As an example, a more definitive description of CCS is ISDN BRI and PRI. Each of these services has a definite common channel used for signaling known as the D channel, which is used to set up and tear down B channel connections.

T1 and E1 Digital Telephony Compared

As is shown in Table 3-6, T1 and E1 are similar in their ability to carry PCM voice with signaling. While E1 has a larger carrying capacity than T1, both are arranged in multiple framing formats to allow signaling information to be carried. A variation of E1, called J1, is used in parts of Japan.

Table 3-6   *T1 and E1 Compared*

	T1 (ITU-T G.733)	E1/J1 (ITU-T G.732)
Sampling Frequency	8 kHz	8 kHz
Channel Bit Rate	DS0 – 64K	DS0 – 64K
Timeslots per Frame	24	32
Channels per Frame	24	30
Bits per Frame	$24 \times 8 + 1 = 193$	$32 \times 8 = 256$
Framing	Superframe (12) Extended Superframe (24)	Multiframe (16)
Framing Indicator	193rd bit of frame	First DS0
System Bit Rate	$8000 \times 193 = 1.544M$	$8000 \times 256 = 2.048M$
Signaling	Robbed bit CAS	CAS in timeslot 17

Synchronization of Digital Telephony

Because it is imperative that synchronization be maintained across a digital interface, many facets of maintaining synchronization are at work. Bit synchronization is maintained by using a primary clock reference source drawn from the circuit pulses themselves. Ones Density must be maintained in order to allow enough pulses from which to develop a reliable clock source. Time-slot synchronization is maintained by simply counting off 8 bits for each DS0. Since the entire circuit is time-division multiplexed, this task is relatively easy. Frame synchronization is maintained by having a unique framing pattern occur over the course of numerous frames grouped in a set. Superframe, Extended Superframe, and Multiframe are examples that have been discussed in this chapter. As an example, in Superframing, the 193rd bit from each frame follows a specific pattern over the course of transmitting 12 frames, then repeats itself. Extended Superframe does this over 24 frames, while E1 uses the first DS0 in its entirety.

Maintaining synchronization is even more critical when you consider that one T1 bit time lasts only 648 nanoseconds. Across miles and miles of transmission facilities, that fact becomes amazing to imagine. One 8-bit DS0 is transmitted in 5.18 microseconds, one frame takes 125 microseconds, and one Superframe lasts for only 3 milliseconds. It is amazing, when you think of it this way, that the telco service providers manage to keep it all going!

ISDN

As services to deploy voice and data over the same network become more common, ISDN will arise again, this time as a preferred connection method between PBXs and voice/data transmission equipment (such as Cisco Voice routers). Since ISDN allows access to multiple services through a common interface (2B+D, or 2 Bearer or user channels and one Delta or control channel for Basic Rate Interfaces, or BRI, and 23B+D, or 23 Bearer channels and one Delta channel for Primary Rate Interfaces, or PRI), is standards-based, and can switch channels on and off as needed, it is well suited to mixing voice and data. Data channels can be connected on the B channels when voice is not using them, and voice switches on and off on demand. In the next two sections, we will examine the architecture of ISDN networks as they may apply to voice traffic and other integrated services. We will also take a look at Signaling System 7, or SS7, an out-of-band signaling and control network originally developed by AT&T, and how ISDN lends itself to extending the services of SS7.

ISDN Network Architecture

As Figure 3-13 shows, ISDN PBXs may be installed to invoke different types of services in the telco network. Services such as circuit-switched (on-demand connections, voice or data), dedicated (such as nailed-up data connections), or packet services may be provisioned as needed. A very attractive aspect here is in considering the small office/home office (SOHO) user. Such a user has a BRI connection in the home and may need to connect to the central site for telephony services and/or data services at various times throughout the day. In consideration of these types of connections to the PBX, Cisco Voice routers have the capability to connect to PRI- and BRI-compatible PBXs. These methods are already used extensively in European countries.

Figure 3-13 *ISDN Network Architecture*

ISDN network architecture

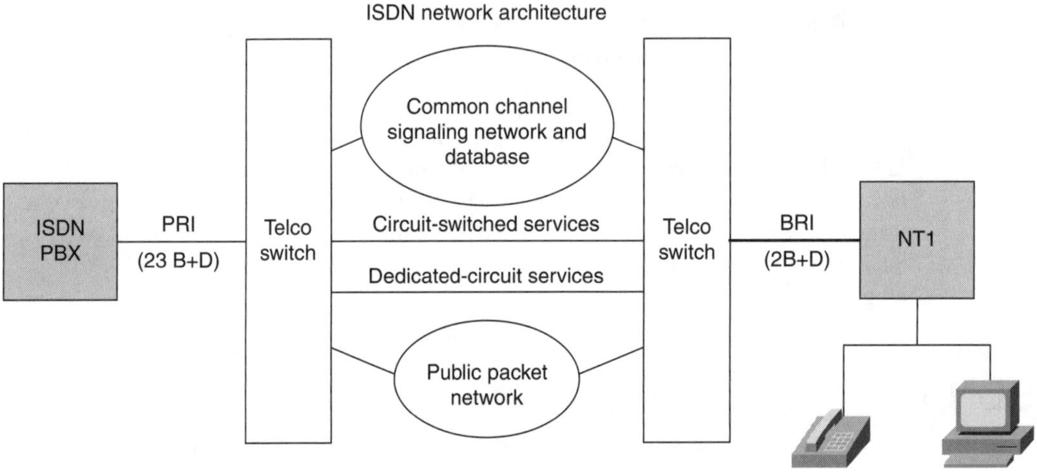

ISDN Network Protocols

The D channel in ISDN is used to set up and tear down switched and permanent calls. At the D channel's layer 3, the ITU-T protocol Q.931 is used for this purpose. Its function is to accomplish the following:

- Establish new calls, both incoming and outgoing.

- Collect and act upon call information during the progress of the call (such as hold and transfer).

- Clear the call when requested.

- Perform miscellaneous functions, such as keeping track of the call duration, and manage and negotiate for such things as reverse charging.

As ISDN has these responsibilities between the user interface and the network interface, some of these signaling aspects need to be propagated to the telco network to accomplish requested connections and disconnections. Unfortunately, the telco network uses a different internal signaling method that is not directly compatible with Q.931.

Signaling System 7

Developed in 1984, Signaling System 7 (SS7) is an internal out-of-band signaling system within the telco network. Originally installed by the Bell System, SS7 is now funded and maintained by the major telco service providers throughout the U.S. It not only performs call setup and teardown tasks, but it also has the capability to detect a caller's source address (thus enabling caller ID) and even the caller's physical location (used by emergency response teams to detect a 911 call source).

SS7 maintains multiple databases throughout its network to keep caller source information redundant in the event of failure, and also serves to reroute call requests around failed network areas. A commercial use of SS7 might be to make a single 7-digit telephone number useful throughout all 50 of the United States, for example, to order pizza. The user need only dial only those seven digits, and SS7 would locate the caller's physical location and direct the call to the nearest of a franchise's pizza shops within the local delivery area. The physical location (that is, the caller's street address) could then be forwarded to the pizza shop on a screen to allow the shop to prepare the pizza for delivery and print directions for the delivery driver.

Q.SIG Protocol

As signaling and digital voice technologies progress, a standard means by which two PBXs can interconnect across private corporate networks is needed. The Q.SIG protocol is a standards-based protocol based on the ITU-T Q.9*XX* family of protocols, which defines a means to set up, tear down, and add additional signaling services between private PBXs. Independent of standard ISDN interconnections, Q.SIG happens to be compatible with public ISDN standards but need not be deployed with ISDN. A major advantage of Q.SIG over standard ISDN interconnections is that it adds the capability to convey nonstandard features between PBXs. This capability, known as Q.SIG GF, or Q.SIG Generic Features, provides for the conveyance of signaling and control information not provided for in standard ISDN connections. Q.SIG GF allows proprietary and nonstandard PBX signaling to be carried across the network transparently. Later versions of Cisco IOS for voice-capable routers allow the use of Q.SIQ on PBX interconnections. Q.SIG supports the following categorized services:

- **Basic service**—This category includes the signaling methods to set up, maintain, and tear down calls, similar to ISDN basic call signaling.

- **Q.SIG GF**—This category allows the conveyance of nonstandard signaling between PBXs. The transportation of this set of signaling is transparent end to end.

- **Supplementary services**—This set of signaling includes such services as call waiting, transfer, reverse charging, and call transfers.

It is important to note that Q.SIG is an alternative to standard ISDN signaling, and is used in place of Q.931 when PBXs need the additional signaling services provided for with Q.SIG. As more corporate voice networking is deployed, Q.SIG will become more prevalent for private PBX installations.

Summary

In this chapter we have discussed digital signaling. There are many different ways to encode an analog signal into digital, and these coding/compression methods yield various levels of voice quality. T1 and E1 are digital transmission methods used to convey coded analog voice channels. As they are time-division multiplexed, they carry a fixed number of channels. They also use different methods to convey signaling end to end. The formats for framing these transmission methods vary, with synchronization being the most important reason for the framing and line-coding methods. ISDN is an emerging voice PBX connection type in the U.S., with deployments already seen in Europe. Merging the capabilities of SS7 and ISDN's Q.931 to form Q.SIG will further enhance the connection options available.

Review Questions

The following questions should help you gauge your understanding of this chapter. You can find the answers in Appendix A, "Answers to Review Questions."

1 What is the rate at which analog voice samples are taken in G.711 PCM coding?

_____

_____

_____

2 What are two examples of hybrid coding?

_____

_____

_____

3 How does the ITU-T express voice quality measurement?

_____

_____

_____

4 How many contiguous frames are contained in a Superframe?

_____

_____

_____

5 What T1 line-coding method guarantees that Ones Density requirements will be met?

_____

_____

_____

6 What is the D4 framing pattern used in T1 Superframes?

_____

_____

_____

7 What is the framing format of an E1 circuit called?

_____

_____

_____

8 What are the two line-coding methods used for E1?

_____

_____

_____

9 How long does one bit time last in T1?

_____

_____

_____

10 What is the name of the external signaling network used in the telephony network?

_____

_____

_____

Cisco Products and Solutions

After reading this chapter, you should be able to perform the following tasks:

- List Cisco voice-capable router products and describe their multiservice voice/data integration features.

- Describe the features (chassis, memory, and interface cards) of Cisco voice-capable routers.

- Compare the features of Cisco voice-capable routers.

CHAPTER 4

Cisco Voice Hardware

Cisco offers a variety of voice products. It is important to understand what these products are and what functions they can offer a network. This chapter describes the various Cisco voice-enabled routers. These routers can perform Voice over Frame Relay (VoFR), Voice over Asynchronous Transfer Mode (VoATM), or Voice over Internet Protocol (VoIP).

The various voice/data products are described in three sections, according to their functionality: data/voice integration access routers, data/voice integration multiservice routers, and large-scale data/voice integration multiservice routers. Each section describes the basic functions and hardware capabilities of these products.

This chapter lays a foundation for the following chapters, where you will learn to place and configure the products described here. While it is not imperative to memorize each function and feature, it is important that you have a basic comprehension of what features and hardware are available to provide these services. Also, it is important to realize that the hardware capabilities for the routers, especially the modular routers, are updated continuously.

Cisco Voice Product Portfolio

The Cisco voice product portfolio consists of many products that serve as the framework for a voice network. These products have many options and features, which are constantly being enhanced. This chapter describes the three major types of voice router platforms as as shown in Table 4-1.

Table 4-1 *Cisco Voice Product Classification*

Data/Voice Integration Access Multiservice Routers	Data/Voice Integration Multiservice Routers	Large-Scale Data/Voice Integration Multiservice Routers
Cisco 1750 Series	Cisco MC3810	Cisco 7200
Cisco 2600 Series	Cisco AS5300	Cisco 7500
Cisco 3600 Series		Cisco AS5800

This chapter covers the data/voice integration routers such as the 1750, 2600, and 3600, as well as the multiservice devices such as the MC3810 and the AS5300 access router, and, finally, large-scale routers such as the 7200, 7500, and AS5800 access router.

The Cisco 1750, 2600, and 3600 series are optimized for LAN-to-LAN, WAN edge devices, and multiservice voice/data integration, as well as for dial applications. The Cisco 1750, 2600, 3600, and 7200 support VoFR and VoIP. The 3600 also now supports VoATM. The Cisco 3600 series, 2600 series, 1700 series, and other modular routes share some common modular components that can be used on various platforms such as WAN interface cards (WICs). This helps end-to-end network management.

The MC3810 routers provide an end-to-end solution for multiservice networks over Frame Relay or ATM, as well as VoIP, VoFR, and VoATM.

Table 4-2 lists Cisco Systems voice-capable routers and some of their features.

Table 4-2   *Standards-Based Voice-Capable Routers*

Router	Supported Voice Interfaces
1750 series	2-port: E&M, FXO, FXS
2600/3600 series	2-port: E&M, FXO, FXS, BRI 1- to 2-port digital T1/E1
MC3810	2-port: E&M, FXO, FXS, BRI 6-port analog and 1-port DVM digital T1/E1
AS5300	2- to 8-port digital T1/E1 VoIP gateways
AS5800	Up to 1344 voice ports for VoIP calls
7200/7500 series	1- to 2-port digital T1/E1 port adapter

The next few sections outline the basic functionality of these platforms and many of the features and options supported by each.

NOTE   The Cisco voice product line is constantly changing and new features are being added regularly. For the most current product lines and features, you should check www.cisco.com or contact your local Cisco sales representative.

Data/Voice Integration Routers

The data/voice integration routers, such as the 1750, 2600, and 3600 series devices, provide a low-cost, entry-level solution to provide voice access in remote and branch offices. These routers are well positioned to allow users to take advantage of multiservice applications such as the integration of voice and data over a single connection. The next few sections outline the features and functions of each of these families of routers.

Cisco 1750 Series Routers

Cisco 1750 routers are multiservice access routers that use Cisco Internetwork Operating System (IOS) technologies with quality of service (QoS) to provide voice/fax/data integration, VPN (Virtual Private Network) access, and secure Internet/intranet access. A Cisco 1750 router is pictured in Figure 4-1.

Figure 4-1   *A Cisco 1750 Series Router*

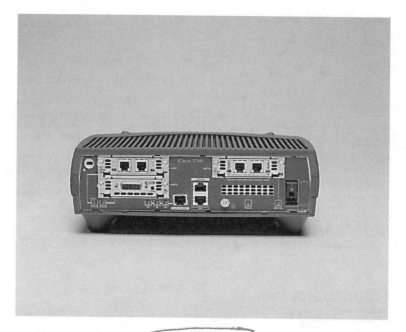

The Cisco 1750 router offers three modular slots for voice and data interface cards: one slot for voice interface cards (VICs) only and two slots that support either a VIC or a WIC. The 1750 also has an autosensing 10/100 Ethernet port and standard console and auxiliary (AUX) ports on the chassis.

The 1750 can support diverse WAN and voice interface types. The Cisco 1750 typically supports the same WAN interface cards as the Cisco 1600, 1720, 2600, and 3600 routers, and most of the same analog VICs and VoIP technology as the Cisco 2600 and 3600 routers.

The WAN interface cards support a wide range of services, including synchronous and asynchronous serial, ISDN BRI, and serial with CSU/DSU options for primary and backup WAN connectivity.

The VICs include support for foreign exchange office (FXO), foreign exchange station (FXS), and Ear & Mouth (E&M).

to Phone.   to PBX   to Co

There is an internal slot for an optional hardware-assisted encryption card. The encryption card provides IPSec DES and 3DES encryption at T1/E1 speeds.

The WAN and voice interfaces support the following applications: multiservice voice/fax/data integration, Frame Relay, ISDN BRI, SMDS, X.25, broadband services, and VPNs.

The 1750 is an integrated router, voice gateway, firewall, encryption device, VPN tunnel server, DSU/CSU, and NT1 in a single device.

The Cisco 1750 series has the features listed in Table 4-3.

Table 4-3   *Cisco 1750 Series Features*

Chassis	• 48-megahertz (MHz) (RISC) processor
	• One autosensing 10/100-megabits-per-second (Mbps) Ethernet interface
	• Console and AUX ports with up to 115 kilobits per second (kbps)
Memory	• 16-megabyte (MB) default internal DRAM, expandable up to 48 MB; IOS runs from DRAM
	• 4-MB default internal Flash memory, expandable up to 16 MB
Three interface card slots	One voice-only and two voice or WAN slots for multiple combinations using supported cards: serial sync/async, CSU/DSU cards, ISDN BRI cards, and VICs
Internal/expansion slot	Used for hardware-assisted data encryption up to T1/E1 speeds

Cisco 2600 Series Routers

The Cisco 2600 series features a high-performance Motorola MCP860 PowerQUICC RISC CPU capable of forwarding traffic between Ethernet interfaces at wire speed (25,000 packets per second [pps]). Figure 4-2 shows the Cisco 2600 series router.

Like the 3600, the 2600 has a run-from-DRAM architecture. Up to 64 MB of DRAM and 16 MB of Flash memory are supported.

Console and AUX ports support up to 115.2 kbps for high-speed dial backup, remote monitoring, and software downloading.

Figure 4-2 *A Cisco 2600 Series Router*

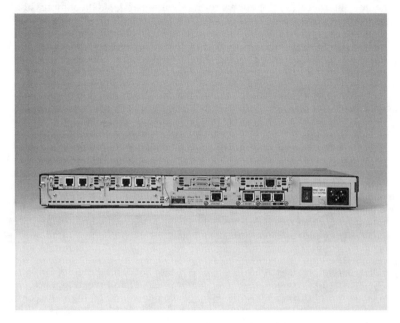

The network module slot is shared with the 3600 series and supports many of the 3600's current modules as well as some of its own. The modules supported provide the following features:

- LAN-to-LAN routing, which includes the ability to pass data traffic between different LANs

- VoFR with FRF.11 and FRF.12

- VoIP with H.323

- Cisco IOS Firewall feature sets with intrusion detection

- Data compression Advanced Integration Module (AIM) and encryption AIM—up to 8 Mbps

- 3DES IPSec encryption for VPNs

- High-density packet voice modules for T1/E1 (up to 60 calls)

- ATM access: T1/E1 IMA and ATM25

- T1/E1 multiflex VWICs supporting data/voice integration

- G.703 interface (unframed E1)

The WAN interface card slots are the same ones used with the 1600 series and the 1720. The new internal AIM slot allows for hardware-assisted data compression and hardware-assisted data encryption.

The Cisco 2600 series router consists of four basic components, as shown in Figure 4-3, and has the features listed in Table 4-4.

Figure 4-3   *The Four Components of the Cisco 2600 Series*

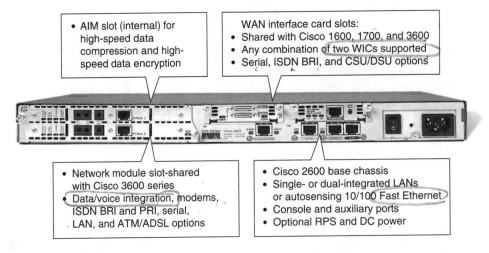

- AIM slot (internal) for high-speed data compression and high-speed data encryption

WAN interface card slots:
- Shared with Cisco 1600, 1700, and 3600
- Any combination of two WICs supported
- Serial, ISDN BRI, and CSU/DSU options

- Network module slot-shared with Cisco 3600 series
- Data/voice integration, modems, ISDN BRI and PRI, serial, LAN, and ATM/ADSL options

- Cisco 2600 base chassis
- Single- or dual-integrated LANs or autosensing 10/100 Fast Ethernet
- Console and auxiliary ports
- Optional RPS and DC power

Table 4-4   *Cisco 2600 Series Features*

Chassis	• 40- or 50-MHz RISC processor
	• Single or dual integrated LAN interfaces (Ethernet, Token Ring, mixed, or autosensing 10/100-Mbps Ethernet)
	• Voice-capable modules available
	• Console and AUX ports with support for up to 115.2 kbps
	• Optional external redundant power supply and DC power options
Memory	• 24-MB DRAM (default), expandable to 64 MB
	• 8-MB internal Flash memory (default), expandable to 16 MB
Network module slot	Multiservice network module support and data/voice integration slots. This slot also supports modems, ISDN BRI and PRI, serial, and LAN and ATM or ADSL options.
WAN interface card slots	Two slots to support WIC cards supported by the 1600, 1700, 2600, and 3600 platforms. Options include serial, ISDN BRI, and 56K and T1 with integrated CSU/DSU.
AIM slot	Slot for high-speed data compression or high-speed data encryption.

Cisco 3600 Series Routers

The 3600 series modular routers provide multifunction access for medium and large offices. These routers provide greater functionality than the 1750 and 2600 series routers. The 3600s use many of the same cards as the 1750 and 2600s. The Cisco 3600 addresses three distinct market segments:

- LAN-to-LAN routing, which includes the ability to pass data traffic between different LANs

- Dial access, which is a service that allows for remote users to access the network

- Multiservice (voice, video, and data) support, which allows the Cisco 3600 series routers to support VoIP, VoFR, and VoATM—which means they allow voice traffic to be transported over existing WAN infrastructures including ISDN, leased-line, ATM, and Frame Relay

Bandwidth efficiencies include:

- Voice compression using standard compression techniques such as G.729

- Silence suppression

- Weighted fair queuing (WFQ), multilink fragmentation, and interleave

The 3600 series of routers can be broken into three distinct product lines: the 3620, 3640, and 3660. Each of these is described in the following sections.

The Cisco 3620 Modular Access Router

The Cisco 3620 router is shown in Figure 4-4 and has the features described in Table 4-5.

The Cisco 3640 Modular Access Router

Not all networks have the same requirements. The 3640 offers more processor power and faster switching than the 2600 and 1750 as well as more slot capabilities. The Cisco 3640 modular access router, shown in Figure 4-5, has the features listed in Table 4-6.

The Cisco 3660 Modular Access Router

The 3660 is the flagship of the 3600 series routers. It has the fastest processor speed and switching capabilities as well as more slots than the other 3600 products. The Cisco 3660, shown in Figure 4-6, has the features listed in Table 4-7.

Figure 4-4   *The Cisco 3620 Router*

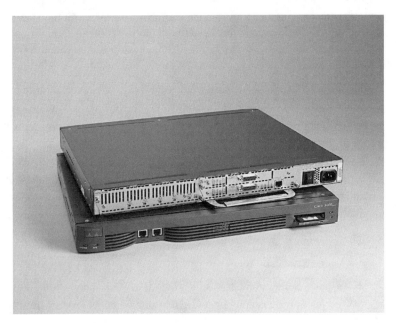

Table 4-5   *Cisco 3620 Router Features*

Chassis	• 80-MHz IDT R7400 RISC processor
	• Two network module slots
	• Two PCMCIA slots for Flash memory
	• Two internal Flash SIMM slots
	• Fast switching performance of 20,000 to 40,000 packets per second (kpps) and process switching of 1.5 to 2 kpps
	• 60-watt (W) AC or DC power with optional external redundant power supply
	• Console and AUX ports with support for up to 115.2 kbps
	• The 3660 supports online insertion and removal (OIR) of modules
Memory	• 32-MB DRAM default, expandable up to 64 MB
	• 8-MB Flash memory default, expandable up to 32 MB
Network module slots	Supports data/voice integration slots, as well as modems, ISDN BRI and PRI, serial, and LAN and ATM options.

Table 4-5   *Cisco 3620 Router Features (Continued)*

WAN module slots	Slots in the LAN/Combo slots support the WIC cards supported by the 1600, 1700, 2600, and 3600 platforms. Options include: serial, ISDN BRI, and 56 kbps and T1 with integrated CSU/DSU.
	Individual WAN modules options include 4- and 8-port BRI, 4-port sync serial, 4- and 8-port sync/async serial, 16- and 32-port async, 1- and 2-port channelized T1, and 1-port HSSI.

Figure 4-5   *The Cisco 3640 Router*

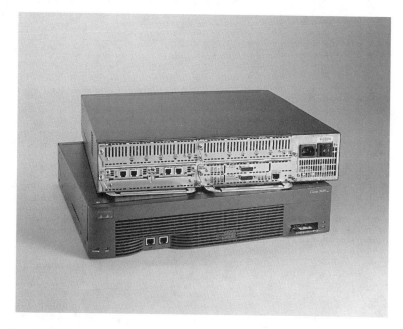

Table 4-6   *Cisco 3640 Router Features*

Chassis	• 100-MHz IDT R4700 RISC processor
	• Four network module slots
	• Two PCMCIA slots for Flash memory
	• Two internal Flash SIMM slots
	• Four internal DRAM SIMM slots
	• Fast switching performance of 50 to 70 kpps and process switching of 3 to 4 kpps
	• 140W AC or DC power supply with optional external redundant power supply
	• Console and AUX ports with support for up to 115.2 kbps

continues

Table 4-6   *Cisco 3640 Router Features (Continued)*

Memory	• 16-MB DRAM default, expandable up to 128 MB
	• 8-MB Flash memory default, expandable up to 32 MB
Network module slots	Supports data/voice integration slots, as well as modems, ISDN BRI and PRI, serial, and LAN and ATM options.
WAN module slots	Slots in the LAN/Combo slots support the WIC cards supported by the 1600, 1700, 2600, and 3600 platforms. Options include serial, ISDN BRI, and 56K and T1 with integrated CSU/DSU. Individual WAN modules options include 4- and 8-port BRI, 4-port sync serial, 4- and 8-port sync/async serial, 16- and 32-port async, 1- and 2-port channelized T1, and 1-port HSSI.

Figure 4-6   *The Cisco 3660 Router*

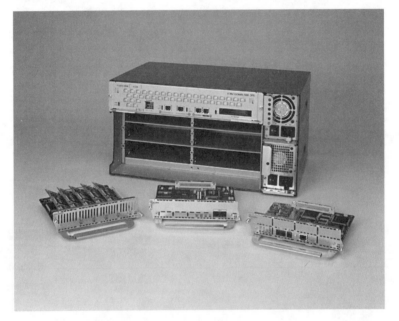

Table 4-7   *Cisco 3660 Router Features*

Chassis	• 255-MHz QED RM5271 RISC processor
	• One or two autosensing 10/100-Mbps Ethernet interfaces
	• Six network module slots
	• Two PCMCIA slots for Flash memory
	• Two internal Flash SIMM slots
	• Four internal DRAM SIMM slots
	• Fast switching performance of 100 to 120 kpps and process switching of 12 kpps
	• Removable motherboard with 2-MB level 2 cache
	• One or two 250W AC or DC power supplies.
	• Console and AUX ports with support for up to 115.2 kbps
	• One TDM controller slot
	• Enterprise chassis and NEBS-compliant telco chassis options
	• NEBS level 3/ETSI compliant
	• Software for telco Data Communications Network (DCN) applications
Memory	• 32-MB DRAM default, expandable up to 256 MB
	• 8-MB Flash memory default, expandable up to 64 MB
Network module slots	Supports data/voice integration slots, as well as modems, ISDN BRI and PRI, serial, and LAN and ATM options
WAN module slots	Slots in the LAN/Combo slots support the WIC cards supported by the 1600, 1700, 2600, and 3600 platforms. Options include serial, ISDN BRI, and 56K and T1 with integrated CSU/DSU.
	Individual WAN modules options include 4- and 8-port BRI, 4-port sync serial, 4- and 8-port sync/async serial, 16- and 32-port async, 1- and 2-port channelized T1, and 1-port HSSI.
AIM slot	Two AIM slots for high-speed data compression or high-speed data encryption.

Modules and Cards of the 2600 and 3600

The 2600 and 3600 series products offer flexible telecommunications options as well as many functions for LAN connectivity. The VIC/WIC cards are compatible across all the 2600 and 3600 platforms. Table 4-8 shows these WICs, along with the WAN module, ATM modules, LAN/Combo modules, and the AIM modules specified in each product's feature table.

Table 4-8   *Modules and Cards of the 2600 and 3600*

VICs	WICs	3600 Series Network Modules	WAN Modules	Voice Modules	3600 Series ATM Modules
2-port voice FXS	1-port ISDN BRI S/T	1- and 4-port 10BaseT module	4- and 8-port BRI (ST or U with NT1)	1- or 2-slot voice/fax	1-port 25-Mbps ATM
2-port voice E&M	1-port ISDN BRI U	1- and 2-port 10BaseT module with two WIC slots	4-port serial	1-port T1 24-channel voice/fax	1-port OC3-155-Mbps ATM
2-port voice FXO	1-port sync serial		4- and 8-port sync/async serial	2-port T1 48-channel voice/fax	4- and 8-port T1 ATM
2-port voice BRI (S/T-TE)	1-port 4-wire 56-kbps CSU/DSU	1-port 10/100BaseTX module	16- and 32-port async	1-port E1 30-channel voice/fax	
	1-port T1/FT1 CSU/DSU	1-port 100BaseFX module	1- and 2-port channelized T1/ISDN PRI	2-port E1 60-channel voice/fax	
	2-port multiflex T1/E1 drop and insert **	1- or 2-port 10/100BaseTX, with two WIC slots	1-port HSSI**		
		1-port 10BaseT, 1 port TokenRing and two WIC modules			
		1-port 10/100BaseTX with 1- or 2-port channelized T1 or E1/ISDN PRI (Balanced or unbalanced E1)			
		Two WIC Slots			

** The 2-port multiflex T1/E1 drop-and-insert WIC module, the ATM 0C3, and the HSSI WAN module are not supported by the 2600 series.

The 2600/3600 Voice/Fax Network Modules

The voice options available for the 3600/2600 routers offer a variety of connections, signaling, and services.

The 2600/3600 voice/fax network module, shown in Figure 4-7, has the following features and benefits:

Figure 4-7   *The Voice/Fax Network Module*

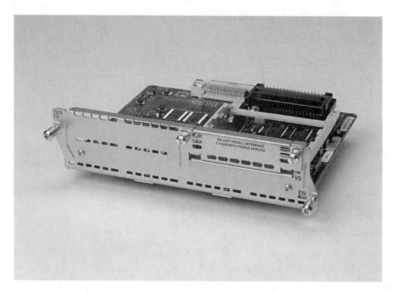

- VICs
  - FXS: 2-wire, 4-wire
  - E&M types I, II, III, IV, and V
  - FXO
  - BRI
- Diverse signaling
  - Delay, immediate, and wink start
  - Loop and ground start
- Voice/fax over IP including compatibility with H.323 standard for audio- and videoconferencing
- Voice over Frame—VoFR, FRF.11, and FRF.12

- High-performance digital signal processor (DSP) architecture TI542:
  - g711alaw—G.711 a-law 64,000 bits per second (bps)
  - g711ulaw—G.711 μ-Law 64,000 bps
  - g723ar53—G.723.1 Annex A 5300 bps
  - g723ar63—G.723.1 Annex A 6300 bps
  - g723r53—G.723.1 5300 bps
  - g723r63—G.723.1 6300 bps
  - g726r16—G.726 16000 bps
  - g726r24—G.726 24000 bps
  - g726r32—G.726 32000 bps
  - g728—G.728 16000 bps
  - g729br8—G.729 Annex B 8000 bps
  - g729r8—G.729 CS-ACELP 8000 bps (uses G.729a for medium-complexity compression and G.729 for high-complexity compression)
  - Group III fax relay

Interface Cards for Analog Voice

The voice network modules (VNM) for the Cisco 2600 and 3600 series support one or two VICs, depending on the module. These VICs come in different types, and each has two ports. Each type provides a different interface for connecting to different types of equipment:

- The foreign exchange office (FXO) interface is an interface that allows an analog connection to be directed at the central office (CO) of the Public Switched Telephone Network (PSTN) or a PBX or key system. The FXO card is the only analog card that is approved to be connected to a line that goes off premises—for example, to a CO. This interface may be used to provide backup over the PSTN or for Centrex-type operations. Because this card will need to be approved by local Post, Telephone, and Telegraph (PTT) agencies, it will not be available in every country at release.

- The foreign exchange station (FXS) interface is an interface that allows connection for analog 2500-set phones (residential phones) and for fax machines, and provides ring voltage, dial tone, and so forth. It will be used where phones are connecting directly to the router.

- The E&M interface is an interface that allows connection for PBX trunk lines (tie-lines). It is a signaling technique for 2- and 4-wire trunk interfaces. This has been a very popular interface for PBX applications.

The Digital T1/E1 Packet Voice Trunk Module

The digital T1/E1 packet voice trunk, shown in Figure 4-8, is placed into the NM-HDV card and provides connectivity for voice and drop-and-insert functionality.

Figure 4-8   *The T1/E1 Packet Voice Trunk Module*

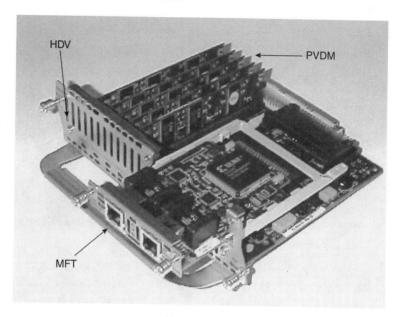

This module complements the existing voice offerings on the 2600 and 3600 by providing higher voice-port densities as well as digital T1/E1 integration with legacy telephony equipment.

There are three components to the Digital T1/E1 Packet Voice Trunk Module:

- **High Density Voice NM (HDV)**—This is the network module carrier card. It has five slots for PVDMs and one slot for the VWIC module. The NM-HDV also contains the onboard PLL for clocking. It can be placed in any 2600 or 3600 network module slot.

- **MultiFlex Trunk VWIC (MFT)**—This is the first module that can be either a VIC or a WIC, depending on whether it is in the NM-HDV. If inserted into the NM-HDV, it provides physical T1/E1 voice connectivity. If used standalone in a WIC slot, it is a T1/E1 WIC module for WAN connectivity. Supports one or two T1/E1 ports. It has a built-in CSU/DSU.

- **Packet Voice DSP Module (PVDM)**—These are the DSP SIMMs. Each PVDM has three TI-549 DSPs per SIMM. Up to five PVDMs can populate the NM-HDV. A minimum of one PVDM is required for Voice.

The following are some of the features supported by the Digital T1/E1 Packet Voice Trunk Module:

- Provides one or two channelized T1/E1 interfaces (via RJ-48 connectors)
- Provides T1/E1 channel-associated signaling (CAS) for connectivity to PBXs and PSTN
- Provides T1/E1 QSIG signaling for connectivity to PBXs
- Supports individual DS0s for connection trunk and connection private line, automatic ringdown (PLAR)
- Provides 60 channels of normal pulse code modulation (PCM) voice (G.711)—four calls per DSP
- Provides 60 channels of medium-complexity compressed voice (G.729a/b, G.726)—four calls per DSP
- Provides 30 channels of high-complexity compressed voice (G.729, G.728, G.723.1)—two calls per DSP
- Provides 60 channels of medium-complexity fax-relay (Group 3 up to 14400 baud)
- Provides 30 channels of high-complexity fax-relay (Group 3 up to 14400 baud)
- Provides signaling: delay, immediate, and wink start
- Provides ground start and loop start

Support for the ISDN BRI signaling type allows a Cisco 2600 or Cisco 3600 to provide voice-access connectivity to either an ISDN telephone network or a digital interface on a PBX/key communications system, as shown in Figure 4-9.

The voice or data also crosses an IP network to which the router connects. This allows branch offices and enterprises to route incoming PSTN or PBX ISDN BRI calls over an IP network.

ISDN BRI VoIP offers direct ISDN network connectivity as well as connectivity to the digital interfaces of PBX and key communications systems. ISDN BRI VoIP provides the following toll-saving benefits for enterprises and branch offices:

- ISDN BRI network connectivity (connects to PBX)
- Two BRI interfaces per VIC (one-slot chassis has two DSPs)
- Up to four voice channels
- Q.931 user side support
- Support for direct inward dial (DID) and caller ID
- Terminal endpoint only; no U interface

Figure 4-9   *VoIP Using the BRI Interface*

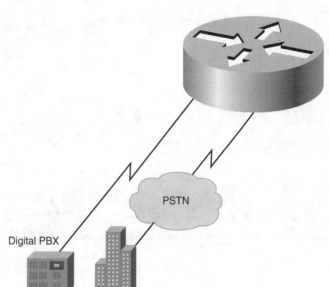

Data/Voice Integration Multiservice Routers

Multiservice routers are specialized devices that provide a specific set of functions. These devices are typically limited in hardware configuration and flexibility, but provide a low-cost solution for data/voice integration. The following sections outline the features and functions of the MC3810 and AS5300 multiservice routers.

The Cisco MC3810 Router

The Cisco MC3810, shown in Figure 4-10, is a fixed configuration multiservice concentrator. This streamlined device allows customers to integrate all traffic, voice, fax, video, legacy (SNA), and LAN over a single unit. This device can utilize leased-line *and* ATM, *and offers* Frame Relay services for branch connectivity.

The MC3810 optimizes network bandwidth by multiplexing voice and data on the same circuit or physical interface. MC3810 has a single Ethernet port and two serial ports that support speeds up to 2 Mbps.

The analog model of the MC3810 has six voice ports (six compressed voice channels), and the digital version houses a single digital voice access port (T1/E1) with up to 24/30 compressed voice channels.

Figure 4-10 *A Cisco MC3810 Router*

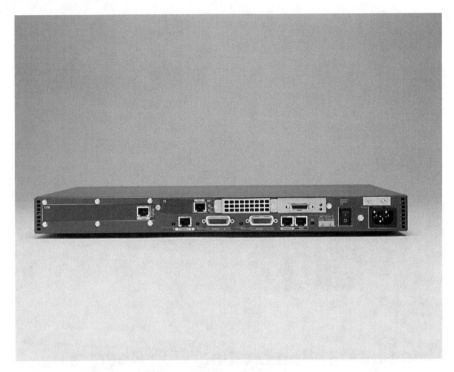

The features of the MC3810 are listed in Table 4-9.

Table 4-9 *MC3810 Router Features*

Chassis	• 40-MHz Motorola 860 PowerPC processor
	• One fixed 10BaseT Ethernet port
	• Two fixed sync serial ports
	• Three modular slots support WAN or voice options
	• AC and DC power supplies with optional external redundant power supply
Memory	• 32-MB default DRAM (not expandable)
	• 8-MB default Flash memory, expandable up to 16 MB
WAN Modules	Support for the following: 1-port channelized T1/E1 multiflex trunk with optional BRI backup
Voice/Video Modules	Support for 6-port analog voice, 1-port digital voice, 4-port BRI voice, and 1-port RS-366 video module

High-performance voice-compression modules (HCMs) support more voice channels than the earlier voice-compression modules (VCMs). Table 4-10 shows the channel capabilities of the MC3810 with the HCM option.

Table 4-10   *HCM Channel Support*

Configurations	Slot 2	Channels	Slot 5	Channels	Total Channels
Analog, 4 ports G.729	HCM2	4	Empty	0	4
Analog, 6 ports G.729a	HCM2	6	Empty	0	6
Digital T1, G.729	HCM6	12	HCM6	12	24
Digital T1, G.729a	HCM6	24	Empty	0	24
Digital E1, G.729	HCM6	12	HCM6	12	24
Digital E1, G.729a	HCM2	8	HCM6	24	32

The Cisco AS5300 Router

The Cisco AS5300, shown in Figure 4-11, is a dialup remote access server and VoIP gateway. When equipped with voice feature cards (VFCs) and voice-enabled Cisco IOS software, the AS5300 supports large service provider or carrier-class VoIP and fax over IP services.

Cisco IOS software offers an array of QoS mechanisms, variable frame sizing, and standards-based H.323 controls, which provide voice quality and call control routing.

In addition to being H.323 v1 & v2 compliant, the Cisco AS5300/Voice Gateway supports a family of industry-standard voice codecs and provides echo cancellation and voice activity detection (VAD)/silence suppression. It offers an integrated interactive voice response (IVR) application that provides voice prompts and digit collection in order to authenticate the user and identify the call destination. Users can readily interface with PSTN digital switches or PBXs, and existing Remote Access Dial-In User Service (RADIUS) authentication and billing servers.

The Cisco AS5300 voice/fax feature cards are coprocessor cards, each with a RISC engine and dedicated, high-performance DSPs to ensure predictable, real-time voice processing. The design couples this coprocessor with direct access to the Cisco AS5300/Voice Gateway routing engine for streamlined packet forwarding.

The Cisco AS5300/Voice Gateway can accept two voice/fax feature cards, so the Cisco AS5300/Voice Gateway can scale up to 96/120 voice connections within a single chassis.

Figure 4-11   *The Cisco AS5300 Router*

The features of the AS5300 are listed in Table 4-11.

Table 4-11   *AS5300 Router Features*

Chassis	• 150-MHz R4700 RISC processor
	• One fixed 10BaseT Ethernet port
	• One fixed 100BaseT Ethernet port
	• 4-port sync serial
	• Four T1 (AS5300-96 VoIP)
	• Four E1 (AS5300-120VoIP)
	• Three modular slots—one for Quad T1/E1, two for VoIP feature cards
	• Single AC or DC power supply with an optional internal redundant supply
Memory	• 64-MB DRAM default
	• 16-MB system Flash memory default
Module slots	The AS5300 Gateway comes with VoIP modules installed.

Access and Multiservice Product Line Overview for Data/Voice Integration

You have seen the numerous choices for data/voice integration. The product you choose depends on the environment. Table 4-12 gives an overview of the basic differences in the products discussed in the previous sections.

Table 4-12 *Access and Multiservice Product Overview*

Feature	Cisco AS5300	Cisco MC3810	Cisco 3600	Cisco 2600	Cisco 1750
Cisco IOS LAN-to-LAN routing	Yes	Yes	Yes	Yes	Yes
VoIP	Yes	Yes	Yes	Yes	Yes
VoFR	No	Yes	Yes	Yes	Yes
VoATM	No	Yes	Yes	No	No
ATM WAN access	No	Yes	Yes	Yes	No
Integrated dial	Yes	No	Yes	Yes	Yes
LAN media	E/FE	E	E, TR, FE	E, TR	E/FE
WAN media	PRI	Serial, BRI, ATM	PRI, MBRI, serial, ATM	PRI, MBRI, serial, ATM	BRI, serial
Maximum analog/digital voice density	0/120	6/24	24/360	4/60	4/0

The MC3810 is Cisco's lowest-cost, highest-density fixed configuration voice/data integration solution for branch offices.

The Cisco 2600 and 3600 have field-upgradable voice modules for customers who see voice/data integration in the future, but aren't ready to invest in voice technologies yet.

The Cisco 2600 supports up to four analog voice ports, two LAN ports, two serial ports, and ISDN backup for voice and data in one cost-effective package.

The 5300 is a dialup remote access server and supports carrier-class VoIP and fax over IP services.

Cisco also offers products designed for larger customers requiring integration with carrier-class systems. The next few sections describe the voice products offered in the high-end Cisco equipment.

Large-Scale Data/Voice Multiservice Integration Routers

The final set of devices discussed in this chapter is the large-scale data/voice routers. These devices provide connectivity for large-scale networks on the order of service provider networks. The devices featured here are the AS5800, the 7200, and the 7500 series routers.

The Cisco AS5800 Access Device

The Cisco AS5800, shown in Figure 4-12, is a carrier-class, dial access concentrator that provides a high-density dial access system specifically designed for large service providers, such as regional Bell operating companies (RBOCs), inter-exchange carriers (IXCs), and large ISPs.

Figure 4-12 *The Cisco AS5800 Router*

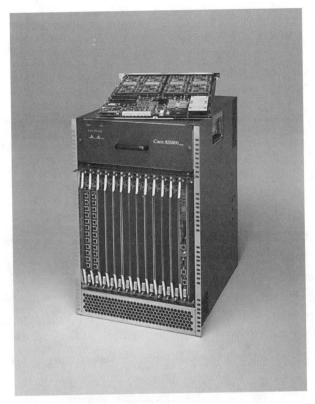

The new AS5800/Voice Gateway feature enables highly scalable deployment of toll-quality voice and fax services over data networks. The Voice Gateway supports features such as prepaid and postpaid calling card, toll-free call redirect, voice-activated dialing, and voice and fax mail.

The AS5800 supports up to 1344 voice ports in a single system, providing the highest concentration of VoIP DSPs available in a single voice gateway.

Major AS5800 applications include toll bypass, universally accessible voice-mail and fax-mail services, PSTN voice- and fax-traffic offload, aggregation for intracompany phone calling and faxing, phone to phone through PBXs and key systems, real-time fax to fax, computer-phone to PSTN, and more.

Cisco AS5800 Voice Feature card supports up to 192 DSP-based voice ports. Voice-processing capabilities include VAD, comfort noise generation, adaptive jitter buffering, programmable 16- and 32-millisecond (ms) echo cancellation, programmable frame size, and DTMF (dual tone multifrequency) detection and generation. The AS5800 Voice Feature card offers industry-leading DSP density and a wide range of codecs, including G.711, G.729, G.729a, G.723.1, and Group III real-time fax support, on any port at any time. The AS5800/Voice Gateway is fully interoperable with standard H.323 gatekeepers, including Cisco gatekeeper platforms, as well as third-party gatekeepers, including VocalTec, NetSpeak, and RADVISION.

The features of the AS5800 voice gateway are listed in Table 4-13.

Table 4-13   *AS5800 Router Features*

Chassis	• 300 MHz (R4700 RISC processor)
	• Fourteen expansion slots
	• Redundant load sharing AC or DC power supplies:
	— Router shelf: 300,000 packet-per-second fast switching; one processor card, four customer configurable slots for egress port adapters.
	— Dial shelf: Maximum of eight DSP-based voice feature cards; maximum of ten DSP-based modem cards; maximum of four T1/E1 trunk cards; Maximum of two CT3 trunk cards. T3 provides a maximum of 672 trunks via a single T3 connection with up to two cards (1344 trunks).
	• Hot-swapability:
	— Dial shelf: All cards (modem cards, trunk cards, chassis interconnect [CICL] cards)
	— Router shelf: All cards except processor (all port and service adapters)

continues

Table 4-13   *AS5800 Router Features (Continued)*

Memory	• 256 MB DRAM (RS726VXR) • 128 MB DRAM (RS7206) • 4 MB boot Flash memory • 16 MB PCMCIA
Data shelf/ingress interfaces	Voice Feature card: 192 DSP-based codecs per card giving 192 to 1344 VoIP ports; 12-port T1/E1/PRI trunk card; T3 ingress card; 144-port DSP-based MICA technologies card, 144 modems per card; maximum of 1440 modems per shelf
Router shelf/egress interfaces	Fast Ethernet, FDDI, ATM, HSSI, V.35 serial, 100VG-AnyLAN, Token Ring, 10BaseT, and 10BaseFL (multiple supported for redundant configurations)
Modem/ISDN sessions	Any call can be either async or ISDN, because AS5800 has full analog and digital coverage—1440 analog calls, 1440 digital calls (B channels), or any combination

Cisco 7200 Series Routers

The Cisco 7200 series, an example of which is shown in Figure 4-13, provides both enterprise and service provider customers with a choice of either a four- or six-slot chassis for interface density.

The 7200 provides service providers with the foundation for network services, with multiservice support for data, voice, and video integration.

The 7200 offers redundancy and high availability with support for dual power supplies and OIR.

Support for Cisco IOS features such as QoS, Cisco Express Forwarding (CEF), and tag switching are also available with the 7200 series routers.

Figure 4-13 *A Cisco 7200 Series Router*

The 7200 series routers come in various configurations, offering the features listed in Table 4-14.

Table 4-14 *Cisco 7200 Features*

Chassis	• 200/250- or 263-MHz processor
	• 4- or 6-slot chassis
	• Various network processing engines
Memory	• 64 MB default DRAM, expandable to 256 MB
	• 20 MB default PCMCIA Flash memory, expandable up to 110 MB
Network module slots	Multiple LAN modules including Ethernet, Fast Ethernet, Token Ring, and Gigabit Ethernet
WAN module slots	Multiple adapters, including ISDN, serial, ATM, HSSI, and SONET
Digital voice trunk port adapter	2-port T1/E1 high- or medium-capacity digital voice

The 7200 multifunction platform provides significant value in five distinct networking applications:

- The distributed routed network, supporting high-speed LAN backbones such as FDDI or Fast Ethernet
- The collapsed backbone, where the router provides Layer 3 capabilities to a network of Cisco Catalyst switches
- IBM data center integration, with support for the full suite of Cisco IOS IBM features and the new 7200 channel interface port adapter for Bus and Tag or ESCON data center connectivity
- Support for service provider networks
- Private intranets and Internets

Benefits of Cisco 7200 Routers

The Cisco 7200 supports multiservice aggregation of lower-end—or branch/CPE—applications for a cost-effective, end-to-end solution.

The 7200 provides:

- Multichannel technology, eliminating stacked routers
- Scalable transport for:
    - VoIP: IP, QoS mechanisms, cRTP, LFI
    - VoFR: FRF.11 and FRF.12
- Tandem PBX bypass

Cisco 7500 Series Routers

The Cisco 7500 advanced router system (ARS) has many chassis, as demonstrated in Figure 4-14. This is the high-end router for both collapsed-backbone LAN and enterprise WAN applications.

The Cisco 7500 series router is Cisco's high-end platform of multiprotocol routers. This router family is highly flexible and offers a variety of interfaces and services including voice functionality. Table 4-15 briefly describes a few of the options available for this family of routers.

Figure 4-14 *A Cisco 7500 Router*

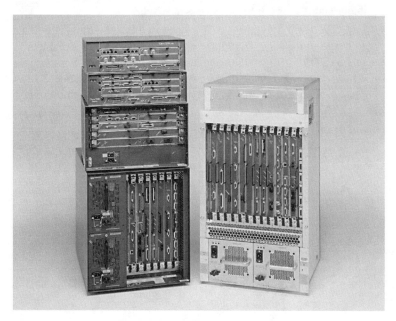

Table 4-15 *Cisco 7500 Features*

Chassis	• 100-MHz, 200-MHz, or 250-MHz processor • 5-, 7-, or 13-slot chassis • Various route switch processors with optional redundancy
Memory	32-MB default DRAM with options of up to 256 MB (processor dependent)
Network module slots	Multiple LAN modules including Ethernet, Fast Ethernet, Token Ring, and Gigabit Ethernet
WAN module slots	Multiple adapters, including ISDN, serial, ATM, HSSI, and SONET
Digital voice trunk port adapter	2-port T1/E1 high-capacity digital voice

Cisco 7500 ARS multifunction technology includes the following:

• Data integration with VoIP support.

• Versatile Interface Processors (VIPs) providing multigigabit networking and extensive application of WAN services.

- Gigabit Ethernet and OC-12 (622 Mbps) connections for high-speed LAN and WAN interconnects.

- Multichannel WAN interfaces and increased WAN port density. A single T3/E3 (or T1/E1) port from a Cisco 7500 can be channelized into individual DS0s, nxDS0s, PRI, and DS1s. T1s can be combined via Multilink PPP. For example, a multichannel T3 interface could be configured to provide ten clear-channel T1 circuits, forty DS0 circuits, five nxDS0 circuits, and two PRI circuits through a single T3 interconnect.

- Distributed services for internetworking, scaling services such as QoS, and security to 155 Mbps.

The digital T1/E1 high-capacity voice-port adapter shown in Figure 4-15 provides large-scale, high-capacity voice termination to PBXs and the PSTN.

Figure 4-15 *The Digital T1/E1 Voice-Port Adapter*

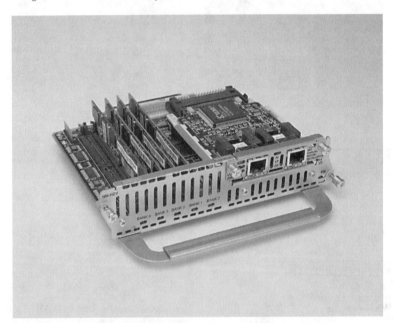

The following are features of the digital T1/E1 voice-port adapter:

- Two software-selectable, RJ-48 T1/E1 universal ports
  - PBX or PSTN connectivity
- Large-scale voice termination
  - High capacity (30 DSPs)
  - 60 channels
  - Voice compression from 64 kbps down to 5.3 kbps

- MIX
  - Drop-and-insert: Port-to-port
  - DS0 cross-connect: Port-to-backplane
- Integrated CSU/DSUs, G.703/G.704 PA-VXC-2TE1
- Codec features:
  - G.711 (a-law/μ-law) G.729, G.729.a (with b variant)
  - G.723, G.728, G.726
  - DTMF tone detection/generation
- Cisco Voice Manager 2.0 Support
- T1/E1 BERT and MIBs

Summary

Many options are available for integrating voice into an existing data network or for designing a new voice/data network using Cisco equipment. The key factors to remember are what services you need and what you will require in the near future. Remember to always configure and choose for the long haul. The products featured in this chapter represent the majority of the router-based voice products currently offered, but new products and services are always being developed, Therefore, it is imperative that you check with Cisco to get a more current listing of these service and products.

Review Questions

The following questions should help you gauge your understanding of this chapter. You can find the answers in Appendix A, "Answers to Review Questions."

1 Which Cisco product provides an end-to-end integrated multiservice solution in a single fixed configuration chassis?

   _____

   _____

   _____

2 True or False: The 3660 router supports analog and digital voice ports.

   _____

   _____

   _____

3 True or False: The 3640 can perform VoIP/VoFR/VoATM functions.

_____

_____

_____

4 List the three components of the digital T1/E1 packet trunk module.

_____

_____

_____

5 Which analog interface card allows connection to a central office of the PSTN?

A) FXS

B) FXO

C) E&M

_____

_____

_____

After reading this chapter, you should be able to perform the following tasks:

- Describe the different connection types provided by Cisco voice-capable routers.
- Identify where Cisco voice-capable routers fit in a multiservice network.
- Evaluate the business applications enabled by Cisco voice-capable routers.
- Compare voice-capable routers in actual network applications of VoIP, VoFR, and VoATM.

Applications for Cisco Voice-over Routers

Voice-over routers, which give you the ability to control expensive communications costs, have many applications. The key to using voice-over technology is identifying where and how it is useful in a network. This chapter gives a brief overview of where Cisco Voice over Frame Relay (VoFR), ATM (VoATM), and IP (VoIP) solutions fit into the overall network picture. The chapter also gives examples of Cisco voice-capable routers in actual network applications.

Placing Cisco Voice-Capable Routers in a Network

The first task in placing voice-capable routers in a network is to understand how they fit into the overall voice architecture. Voice signaling, or communicating with voice devices, is accomplished by the router interpreting and responding to the different types of voice signaling. Cisco voice-capable routers are capable of interpreting and terminating different types of signaling, including ABCD bits, E&M, ground start, and loop start.

Instead of passing the signaling information end-to-end, in its native form, the Cisco voice-capable router locally acknowledges the signaling and terminates the signaling, as shown in Figure 5-1. The Cisco router can then use a standards-based signaling to contact the remote Cisco router. The remote Cisco voice-capable router then generates the proper local signaling to complete the call. Different types of hardware and connections interpret the different protocols in voice applications.

Figure 5-1 *Routers Intercepting Voice Signals*

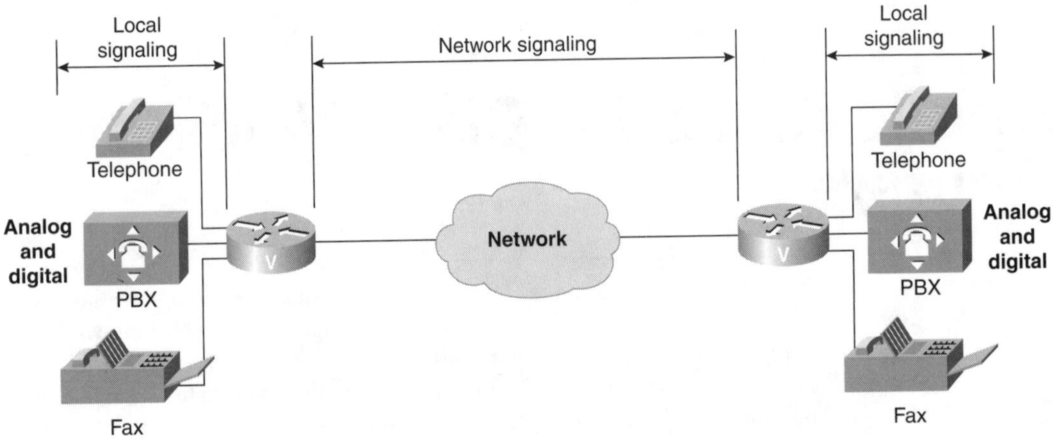

Cisco Voice Connection Types

Table 5-1 lists the connection types and shows the voice functions of Cisco routers and how Cisco voice-capable routers fit in a multiservice network. These connections allow for the interpretation of standard voice protocols to provide end-to-end voice-over services. The following sections discuss these connection types and outline applications of each.

Table 5-1 *Cisco Voice Connection Types*

Connection Type	Description
Local	Calls within the same router
On-net	Calls within or between offices using enterprise facilities
Off-net	Calls between offices using public facilities (Public Switched Telephone Networks, or PSTNs)
Private line, automatic ringdown (PLAR)	Automatically dials an extension
PBX-to-PBX	Provides a tie-line type of connection between PBXs
On-net to off-net	Reroutes calls off-net when on-net is unavailable

Local Connections

Cisco voice-capable routers permit phone or fax calls between two foreign exchange station (FXS) ports attached to the same voice-capable router. These calls are referred to as *local* because they do not traverse any network facility, and they remain local to the fabric of the router. Figure 5-2 shows an example of a local call; a standard analog phone, shown on the left, is calling another phone, 338-8801, that is physically attached to the same router.

Figure 5-2   *A Local, or Phone-to-Phone, Call on the Same Router*

On-Net Connections

Although local calls can be useful, it is often more desirable to call devices that are not physically connected to the same voice-capable router. Calls can be made from a phone attached to a Cisco voice-capable router, across the data network, to another phone also attached to a voice-capable router. When this type of call is placed between two voice-capable routers within the same enterprise network it is referred to as an *on-net call*. Figure 5-3 shows such a call being placed across the data network. The call is routed based on its *destination pattern*. The destination pattern is the phone number of the end device. When sending destination patterns between devices, extension digits can also be passed to the PBX.

Figure 5-3   *An On-Net Call Between Two Local Routers*

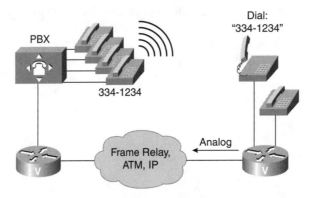

Off-Net Connections

It is also desirable to call devices that are not part of the same company, such as a customer or vendor in the same city as a remote office. In off-net dialing, the user dials an *access code* from an analog phone that is directly connected to a Cisco voice-capable router to gain access to the Public Switched Telephone Network (PSTN), or other service provider voice network. For example, in the United States the access code to get to the PSTN is usually 9. When the call reaches the PSTN, a second dial tone can be heard. At this point, the user dials the destination pattern (that is, phone number) of the party it wishes to reach. Figure 5-4 shows a typical off-net call.

PLAR Connections

PLAR automatically connects a telephone to a second phone as soon as the first phone is lifted from the cradle (that is, goes off-hook)—just like the Batphone. There is no need to dial any digits, because the first phone is automatically connected to the second phone. This application is useful for applications such as connecting customers to sales representatives, taxi dispatchers, or your friendly neighborhood superhero. Figure 5-5 gives an example of PLAR.

Figure 5-4   *An Off-Net Call Using Voice-Capable Routers*

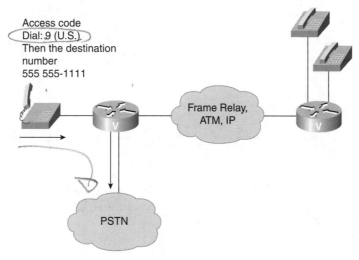

Figure 5-5   *A PLAR Connection*

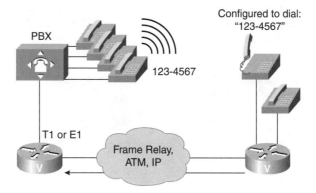

Tie-Line Trunks

Tie-line trunks allow the router to connect PBX to PBX across the data network. This is accomplished over a leased-line trunk that connects the PBXs together. Signaling used for tie-line trunks includes PRI, QSIG, E&M, and T1/EQ, although others may also be supported. Figure 5-6 demonstrates the functionality of the tie-line trunk.

Figure 5-6   *A Tie-Line Trunk*

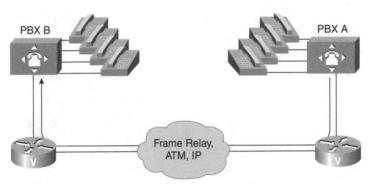

When PBX A seizes a channel, the Cisco voice-capable router inserts the correct
destination pattern into the signal before dialing PBX B. The call is PBX to PBX, instead
of handset (phone) to handset (phone).

On- to Off-Net Connections

Another advantage of the Cisco voice-capable router is that it offers the ability to handle
many different situations. Because on-net resources are limited, the Cisco voice-capable
router must handle various types of call demands. One way to do so is to utilize off-net
resources when on-net services are not available. Figure 5-7 shows how the router is
capable of using off-net services to complete the call:

1 The phone on the left attempts to make an on-net call. The Cisco voice-capable router
   can respond in different ways.

2 Resources are not available to complete the call.

   — The Cisco far-end router is out of trunks to the PBX to which it is connected.
     This router on the far end can send "fast-busy" to a user to inform the user
     that resources are not available.

   — The router on the near end, using its QoS features, realizes that there isn't
     enough bandwidth to complete the call.

3 The Cisco router on the near end can then choose to seize an off-net trunk and place
   the call over the PSTN.

Figure 5-7   *Completing a Call Using Off-Net Service*

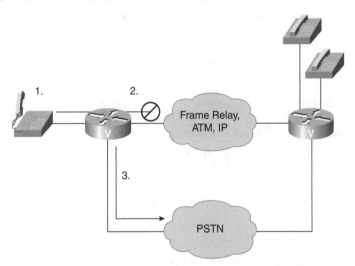

Using Cisco Voice-Capable Router Applications

Different types of hardware or connection types interpret the different signaling and protocols used in voice applications. Chapter 4, "Cisco Voice Hardware," reviews the cards supported by the voice-over routers. This section explains how the voice interface cards (VICs) and connection types are used, and it describes the applications and signaling they support. It also discusses the different types of applications and connections for using these services.

The E&M (for Earth & Magneto or Ear & Mouth) VIC is used for analog trunking connections of PBX-to-PBX or switch-to-switch connections. As shown in Figure 5-8, the E&M VIC connects to PBX services.

Figure 5-8   *An E&M VIC Connection to a PBX*

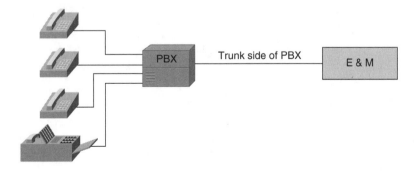

E&M is a trunking arrangement generally used for:

- Tie-lines between telephone switches (PBXs)
- Centrex connections between a PBX and a central office (CO)

Cisco's E&M interfaces use RJ-48 (8-pin modular) connectors that allow connections to PBX trunk lines (that is, tie-lines).

The foreign exchange office (FXO) emulates a phone and connects to a station side of a PBX or directly to a central office switch at the phone company. This type of interface allows for an analog connection to the PSTN, which allows signals to be sent into that network. This effectively lets users access the PSTN from within the enterprise network. Figure 5-9 shows the different connection types used for an FXO.

Figure 5-9   *An FXO Connection*

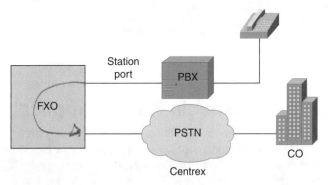

FXO interfaces use RJ-11 connectors, the same interfaces you will see on telephony devices, which connect directly to the PSTN central office or a PBX.

The FXS provides connections for standard phones, fax machines, or PBXs. Figure 5-10 shows a typical application with an FXS interface.

Figure 5-10   *An FXS Connection*

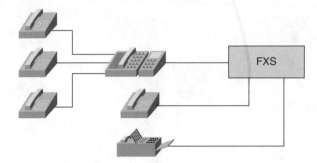

FXS connects directly to a standard telephone, fax machine, or similar device and supplies ring 48 voltage, line power, and dial tone. Cisco's FXS interfaces use RJ-11 connectors that allow connections to basic telephone service equipment, key sets, and PBXs.

Connection Applications

A typical FXS application is to provide connectivity to the voice network for end stations (telephones, faxes, etc). Figure 5-11 shows the FXS ports on two voice-capable routers used for on-net calls.

Figure 5-11 *An FXS-to-FXS Voice Call*

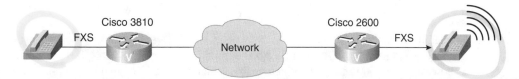

A Cisco voice-capable router uses an E&M VIC to connect tie-line types of connections (that is, PBX-to-PBX, or switch-to-switch). Figure 5-12 shows how Cisco voice-capable routers handle tie-line type connections.

Figure 5-12 *A Tie-Line Connection Using an E&M VIC*

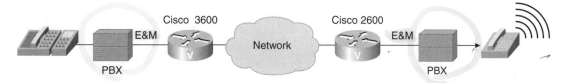

Most applications use a combination of cards in the routers to provide solutions. In the application shown in Figure 5-13, a Cisco router in a branch office uses an FXS card to connect via the key system to the network. On the other end of the network, a Cisco router in the main headquarters uses an E&M card to connect to the corporate PBX system.

Figure 5-13 *A Multicard Application*

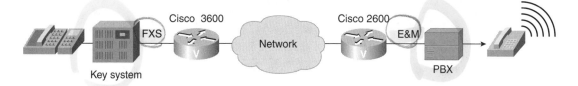

Another multicard application is fax relay, as shown in Figure 5-14. In this application example, the fax machine on the left, connected to an FXS port on a Cisco voice-capable router, transmits a fax across the corporate intranet to a fax machine in a small branch office that is connected to a key system. The key switch is connected by an E&M port to a Cisco voice-capable router.

Figure 5-14 *Fax-to-Fax Connection*

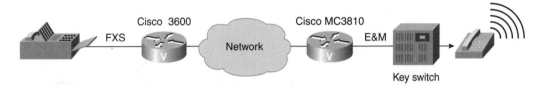

An application that is becoming extremely popular is the H.323 application known as an IP Phone or Ethernet Phone, which connects across the network to the PSTN. Figure 5-15 shows a Cisco 2600 and 3600 voice-capable router running VoIP, handling a phone-to-phone (or phone-to-CO) call.

Figure 5-15 *An Ethernet Phone-to-PSTN Connection*

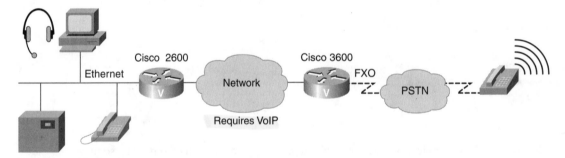

You use an FXO VIC when connecting to the station side of a PBX or directly to the telco CO.

By using the services described, a company can build integrated multiservice networks utilizing VoIP, VoFR, or VoATM for a small- to medium-sized business. Figure 5-16 shows a typical configuration that may be implemented to provide these services.

Figure 5-16 *Voice/Data Integration (Multiservice) Example*

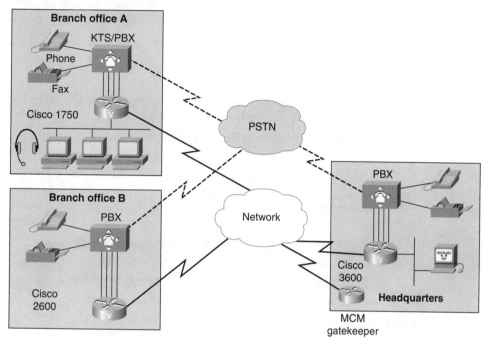

The business in Figure 5-16 has a main office, two branch offices connected by the PSTN for off-net local area calls, and a leased-line network via a secure tunnel for on-net voice and data. This provides the customer with cost savings by utilizing the data network for PSTN bypass. There would be even more savings for the customer if the company were a multinational organization with a large telephone network using PBXs. The PSTN could be used as a backup connection for the PBX-to-PBX connections if the leased lines failed.

Separate networks are inefficient and difficult to administer. In addition, they are not positioned well for the multimedia applications, such as unified messaging and collaborative data sharing, that are becoming increasingly prominent. By combining voice and data traffic, integrated voice/data networks enable cost savings through leveraging cheaper facilities, more efficient bandwidth utilization and the capability to reuse bandwidth between voice and data applications, and a migration path toward multimedia applications.

The future is on the Web. Internal and external communications will increasingly require multimedia capabilities that only a converged network can deliver.

When preparing to implement multiservice networks, an organization must make some decisions—particularly which of the available services will be used to transport the application traffic.

It is more realistic to take an evolutionary approach to multiservice migration, enabling customers to take advantage of emerging multimedia applications while leveraging their investment in old-world equipment.

Which technology you choose depends on services available to the customer, familiarity of the users with the technology, and the existing network environment. Several voice-over options are available:

- VoIP, regardless of the transport technology, offers advanced H.323-based, Web-applications interworking to the desktop. Methods of VoIP include the following:

  — VoIP over serial (leased lines) using HDLC encapsulation is a practical, easily implemented, and configured choice for T1 link speeds and higher.

  — VoIP over serial (leased lines) using MLPPP (Multilink Point-to-Point Protocol) encapsulation offers fragmentation and interleaving (LFI) to ensure acceptable voice quality on slower link speeds, down to 56 kbps. Analysts predict that VoIP will have the best long-term growth opportunity.

  — VoIP over Frame Relay (leased lines or Frame Relay circuits) using Frame Relay encapsulation is an easy-to-implement choice for slower links.

  — VoIP over ATM (ATM circuits) using ATM circuits to provide connectivity between sites can be used to carry high traffic volumes.

- VoFR (leased lines or Frame Relay circuits) provides leveraging of an existing Frame Relay network. It can be used as a technique over point-to-point leased lines or a Frame Relay circuit—a full-fledged Frame Relay network or service is not required to make use of this technology.

- VoATM offers all the links of T1 or higher, often leveraging an existing ATM network.

The following sections describe examples of existing WANs that implement the voice-over applications.

VoIP Applications and Examples

VoIP gateway routers are deployed in the remote locations—1750 routers for small office/ home office (SOHO), 2600 for a branch office, and 3600 for a larger branch office. Back at headquarters, a 7200, an AS5300, Cisco Works Voice Manager, and Multimedia Conference Manager (MCM) may be working together to handle call control. Figure 5-17 shows an example of how VoIP can be implemented.

Figure 5-17 *VoIP Example*

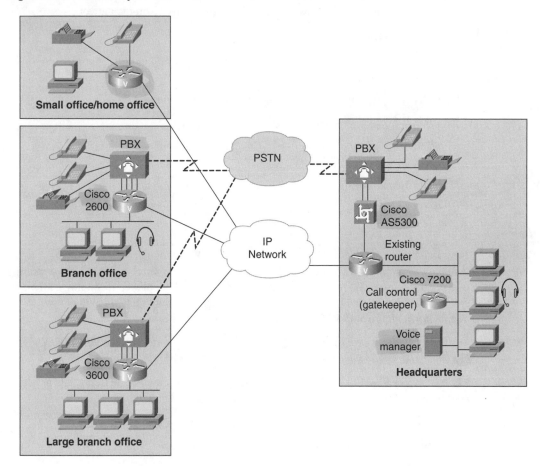

The Cisco MCM provides gatekeeper capabilities to this network example. Gatekeepers allow for the centralization of the dial plan.

In the VoIP example in Figure 5-17, the gatekeeper works with existing telephony architecture by translating phone numbers to IP addresses for routing of calls through the data network. This allows the office-to-office telephone and fax calls to be sent across the data network and reduces toll charges.

To further manage the VoIP network, Cisco Works Voice Manager provides detailed information, including the following, regarding voice activity on the network:

- Bandwidth and traffic management
- Call accounting
- Misuse reporting
- Activity reports by company, division, department, and user
- Dial plan cost allocation
- Cost
- Tariff allocation
- Security and alarms

VoFR

For customers who have implemented a Frame Relay backbone and face growing bandwidth requirements, VoFR provides a good migration option. In general, this topology consumes the least amount of bandwidth. VoFR is a widely deployed technology, all major service providers offer Frame Relay services, and it works with existing telephony architecture. Figure 5-18 shows a typical VoFR enterprise example.

VoFR networks are easy to manage. Cisco devices that support VoFR (or VoIP) also provide routing, CSU/DSU services, and security functions all within one device.

VoFR and VoIP over Frame Relay typically leverage an existing Frame Relay data network, and the cost savings derived from a joint voice and data backbone include:

- Toll-bypass savings from voice and fax calls between offices
- Reduced management and operation costs
- Equipment savings

Figure 5-18 *VoFR Enterprise Example*

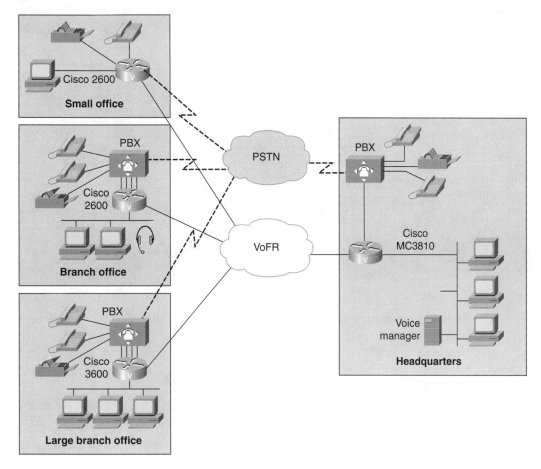

VoATM

VoATM in a Cisco network is supported by ATM adaptation layer 5 and either CBR (constant bit rate) or VBR (variable bit rate) class of service. The QoS capabilities of ATM allow the specific requests of CBR traffic with bandwidth and delay variation guarantees. The use of virtual circuit (VC) queues allows each traffic stream to be treated uniquely. In the case of voice traffic, priority can be given for its transmission. Figure 5-19 shows a VoATM enterprise example.

Figure 5-19 *VoATM Enterprise Example*

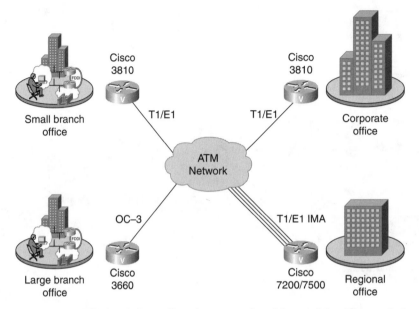

The use of small, fixed-size cells reduces queuing delay and the delay variation associated with variable-sized packets.

Voice-over Solutions in Enterprise Networks

Many solutions are offered by the various voice-over applications. The most common solutions are:

- Toll bypass—Because of its inherent cost savings, toll bypass may be the best reason for deploying an integrated multiservice network.

- PBX and OPX—OPXs and PBXs offer a powerful way of providing remote users with the services of a centrally located PBX.

- Hoot-and-holler—Hoot-and-holler provides the capability to support one-to-many or many-to-many voice conferencing, typically found as an integral part of financial service brokerage firms.

- H.323 interoperability (VoIP)—The H.323 standard provides a foundation for audio, video, and data communications across IP-based networks, allowing users to interoperate without concern for compatibility.

- Managed voice/data services—Managed voice and data services provide complete command and control of your multiservice network.

Toll Bypass

An integral part of providing multiservice connectivity to remote sites is the ability to switch voice between sites. Some vendors' VoFR/VoIP products rely on the PBX to perform this tandem switching of voice calls from one location to another. However, tandem switching is generally perceived as undesirable by telecom managers for two reasons. First, it ties up two trunks on the PBX by taking a call in on one line and switching it to another— PBX real estate is expensive. Second, tandem switching of VoIP/VoFR through a PBX can cause degraded voice quality because voice coming into the PBX must be uncompressed and then recompressed going out of the PBX.

Cisco's integrated multiservice architecture allows VoFR/VoIP/VoATM calls to bypass the traditional tandem PBX switch. Cisco performs this tandem PBX bypass in the router, avoiding the use of the valuable PBX trunk lines and multiple voice encodings. Figure 5-20 shows a tandem PBX bypass using VoFR.

Figure 5-20 *Tandem PBX Bypass*

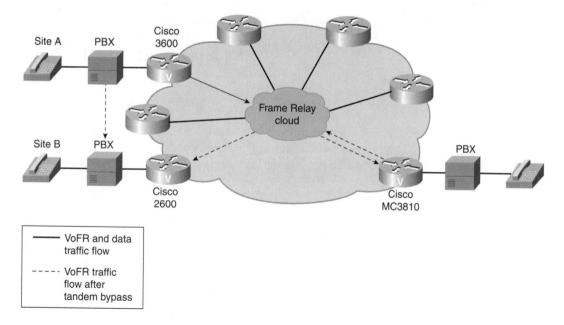

The benefits of tandem bypass are:

- It allows large-scale VoFR/VoIP/VoATM networks.
- It switches voice without decompression/recompression.

OPX Connections

An OPX passes telephone and fax calls across a data network to remote devices, reducing tolls. Figure 5-21 shows an OPX connection.

Figure 5-21 *OPX Connections*

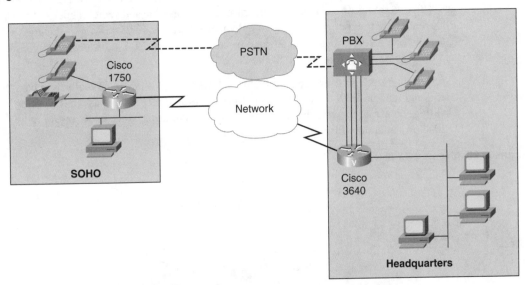

OPX uses a connection trunk feature to deliver analog PBX services to remote phones and faxes. It creates a permanent voice-over call between two endpoints that allows it to pass some supplemental signaling between the two telephony devices, like hookflash or point-to-point hoot-and-holler.

Connection trunk allows OPX functionality between FXO and FXS ports. In this situation, remote stations (connected to FXS ports) appear to the PBX as a physically connected station. If this remote station does not answer a call, it can be rolled over to centralized voice mail (if configured on the PBX).

Hoot-and-Holler

The IP- or VoFR-based hoot-and-holler solution allows customers to eliminate expensive leased lines by enabling them to run the hoot-and-holler applications on their data networks.

Point-to-point connections are depicted in Figure 5-22. The implementation of hoot-and-holler is ideal for 5 to 10 sites.

There are several limitations of a point-to-point unicast implementation, such as:

Figure 5-22 *Hoot-and-Holler*

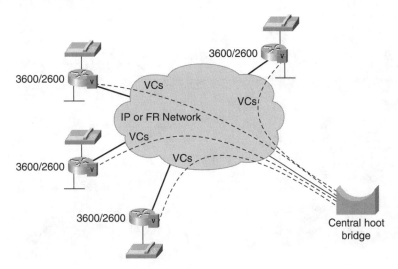

- Point-to-point is a bandwidth hog.
- You have a tandem encoding problem because multiple compression/decompression cycles degrades voice quality, particularly with G.729, and adds load to the router. However, G.711 holds up fine.
- Delay is introduced because of the centralized bridge; each call must traverse the bridge for mixing.

Future implementations with multicast will be enabled with a simple software upgrade and will scale to large network sizes.

H.323

The H.323 standard provides a foundation for audio, video, and data communications across IP-based networks, including the Internet. By complying to H.323, multimedia products and applications from multiple vendors can interoperate. Figure 5-23 shows a typical H.323 application.

Figure 5-23 *H.323 Interoperability*

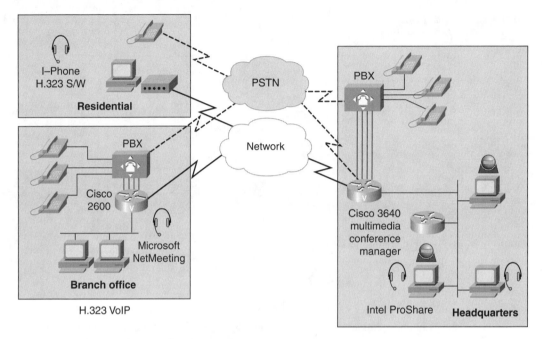

The benefits of H.323 include the following:

- It is a flexible application configuration for VoIP, desktop video conferencing (such as NetMeeting), collaborative computing, and electronic whiteboard applications.

- There are multimedia standards for existing multivendor IP-based infrastructures.

- Network loading enables administrators to prioritize available bandwidth for conferencing, eliminating the congestion and controlling other types of traffic using the network at the same time.

- H.323 works with the existing telephony architecture.

- Office-to-office telephone and fax calls are passed across a data network, reducing tolls.

Cisco MCM has two principal functions: It acts as a gatekeeper and as a proxy.

The gatekeeper provides admission control, bandwidth management, address resolution, user authentication and authorization, call accounting, and call routing functions for H.323 connections.

A primary use for a proxy is hiding addresses in one zone from view by devices in another zone (this is typically the case between two companies). Figure 5-24 shows the location of the H.323 gateway.

Figure 5-24 *H.323 Gateway*

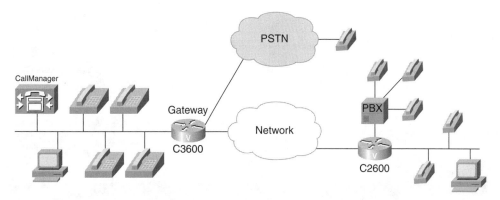

Internetworking with Cisco H.323 gateway networks means that Cisco 2600 and 3600 H.323 gateways can be the PSTN and PBX gateway for Cisco IP Telephony (CIPT) call manager.

The H.323 standard allows for the IP phone or computer phone. Phone internetworking does the following:

- It uses H.323 VoIP.
- It interoperates with existing data networking.
- It interoperates with existing telephony architecture.
- It passes office-to-office telephone and fax calls across a data network, reducing tolls.

One of the major advantages of integrating data and voice is unified messaging, which brings together all the message systems—including voice mail, e-mail, fax, and video messages—into one access point for end users. Figure 5-25 shows how all these services can be integrated by unified messaging.

If you can access everything through some application such as an exchange inbox on your PC, it's easier to organize everything according to your preferences. Today's voice-mail systems cannot display or preview which messages are from whom—you have to listen to them all and then decide which ones to call back first. Unified messaging puts everything in one place and lets you sort things by priority before reviewing any of them.

Figure 5-25 *Unified Messaging*

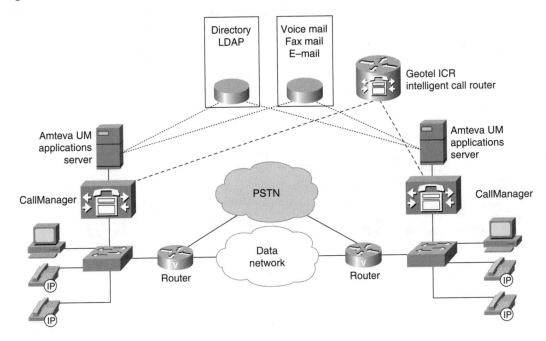

Summary

In this chapter you have learned about the different voice connection types, including on-net, off-net, PLAR, and tie-line connections. This chapter also discusses the VICs used for establishing these different connections. You have also had an opportunity to see some examples of using these connections to integrate data and voice in enterprise network environments. The chapter discusses the advantages of toll bypass, off-premise extension calling, hoot-and-holler, H.323 interoperability, and managed voice/data services. The chapter also shows how an organization can deploy these services across an existing WAN using VoIP, VoFR, and VoATM.

Review Questions

The following questions should help you gauge your understanding of this chapter. You can find the answers in Appendix A, "Answers to Review Questions."

1 What is the connection type that allows for calls between two devices connected to the same router?

_____

_____

_____

2 What is a connection that automatically dials when a phone is lifted off-hook?

_____

_____

_____

3 Which interface allows for what connections to the PSTN?

_____

_____

_____

4 Which interface connects directly to a phone handset and provides dial tone, line power, and ring voltage?

_____

_____

_____

5 Which of the voice-over technologies do analysts predict will have long-term growth potential?

_____

_____

_____

6 The H.323 standard provides a foundation for what?

_____

_____

_____

7 Hoot-and-holler allows customer to save money by eliminating what, using VoIP or VoFR networks?

_____

_____

_____

8 The use of small, fixed-size frames in VoATM reduces what in voice networks?

_____

_____

_____

9 Which standard described in this chapter is designed to operate on various vendor equipment?

_____

_____

_____

10 What application allows users to get all of their messaging from a single point?

_____

_____

_____

After reading this chapter, you should be able to perform the following tasks:

- Describe chassis I/O ports, hardware components, and interface cards for Cisco 2600, 3600, MC3810, and AS5300 routers.

- Install and verify modules, interface cards, and components for Cisco 2600, 3600, MC3810, and AS5300 routers.

- Install, connect, and verify the hardware voice ports and cable connections necessary to transport a voice-over call over the network.

- Using Cisco IOS commands, access each voice port on a Cisco voice-capable router.

CHAPTER 6

Setting Up Cisco Routers

After you receive Cisco voice-over equipment, one of the first things you to have to do is install the equipment into the network before you can begin configuring it for voice operations.

Chapter 4, "Cisco Voice Hardware," describes a variety of Cisco voice-over routers, and this chapter describes how to set up those routers for operation in a typical network environment. The focus of this chapter is on the differences in the chassis types and the installation of the voice and voice-related interface cards. This chapter also discusses the use of Cisco IOS commands and how to interact with the different voice ports on the router.

Setting Up a 2600 Router

One of the Cisco voice-over products on which you need to be able to identify the components and install interface cards is the Cisco 2600 series router. Figure 6-1 shows the features of a basic 2600 series router.

Figure 6-1   *The Back View of a 2600 Router*

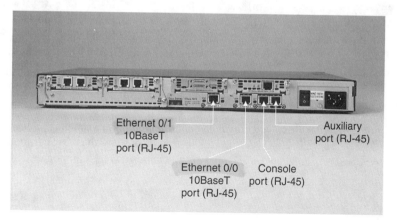

There are several routers in the 2600 series, and the difference between them is the number and speed of the Ethernet ports:

- 2610—One Ethernet 10BaseT port, 0/0
- 2611—Two Ethernet 10BaseT ports, 0/0 and 0/1
- 2620—One Ethernet 10/100BaseT port, 0/0
- 2621—Two Ethernet 10/100BaseT port, 0/0 and 0/1

You can use a 2600 router to transport voice calls across a network. However, to do so you must install additional hardware components into a router to transport voice messages. To make analog voice calls on a 2600 router, you need to install two hardware devices into the router:

- Voice network module (VNM)—Converts and compresses digital voice signals into voice packets that can be transmitted in an IP environment. The VNM is installed into the 2600 chassis in the NM slot.
- Voice interface card (VIC)—Provides the connection from the telephony device to the router. An analog VIC also converts analog signals into a digital format. This card is inserted into the VNM.

A VNM must be installed in the network module slot before the router can be used for voice calls. The VNM is required to handle VICs. A 2600 router also has two WAN interface card (WIC) slots available. Although the slots are the same in size, WICs cannot be installed in VIC slots.

A 2600 supports up to four analog voice ports, and up to 60 digital voice channels in its single network module slot, slot 1. Figure 6-2 shows how the FXS VIC is placed into the VNM. The VNM is then placed into the network module slot in the router.

To make digital voice calls on a 2600 router, you can install the new high-density packet voice/fax network module. This single network module with dual port supports up to 60 simultaneous telephony connections with two E1 cards and up to 48 using two T1 cards. The packet VNMs support T1/E1 CAS, QSIG, PRI, BRI, E1 R2, and SGCP signaling.

Setting Up a 3600 Router

Another popular Cisco voice-over router is the 3600 series router. The 3600 series has three products, the 3620, 3640, and 3660. This section outlines the basic components of the 3600 series products and explains how to install the basic voice components.

Figure 6-2   *VNM Placement*

Voice network module

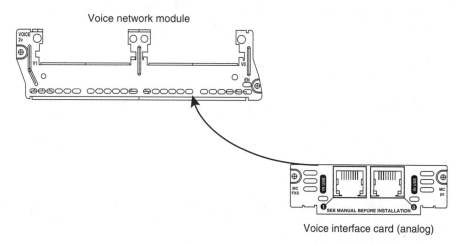

Voice interface card (analog)

Identifying 3600 Router Components

The Cisco 3620 router is a two-slot modular access router whose LAN and WAN connections can be configured by means of interchangeable network modules and WICs. The modular design of the router provides flexibility, allowing you to configure the router to your needs and to reconfigure it if your needs change.

The 3620 and 3640 series routers include the following visible features on the front of each router (see Figure 6-3):

Figure 6-3   *Front View of a 3620*

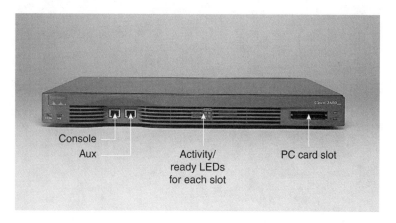

Console
Aux
Activity/
ready LEDs
for each slot
PC card slot

- Console port
- Auxiliary port
- Activity/ready LED lights for each module slot
- PC Flash card slot

The back of the 3620 and 3640 includes the following:

- Connector for the power cord
- Power on/off switch
- Two chassis slots used to mount network modules (on a 3620 router)
- Four chassis slots used to mount network modules (on a 3640 router)
- Six chassis slots used to mount network modules (on a 3660 router)

On the 3620, module slot 0 is on the right, near the power supply. Slot 1 is on the left. Slot numbers correspond to the two sets of LEDs on the front of the chassis. The slot number is used as part of the identification of the network interfaces installed in the router. One slot should have a WAN network module installed to provide the connection to the IP WAN. The other is used for the VNM.

As shown in Figure 6-4, a 3640 router has four module slots. Facing the router, module slot 0 is on the bottom right, slot 1 is on the bottom left, slot 2 is on the upper right, and slot 3 is on the upper left.

Figure 6-4   *The Rear View of a 3640*

Unlike the 3620 and 3640, the 3660 does not have any ports or LEDs on the front; these items are on the back of a 3660. Figure 6-5 shows the rear view of the 3660 router.

Figure 6-5   *Rear View of a 3660*

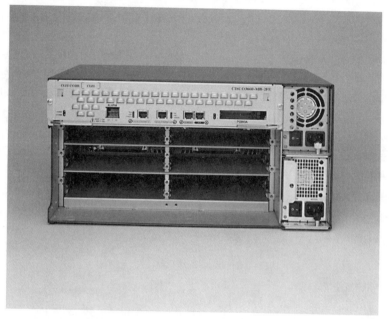

The chassis of the 3660 includes all of the following:

- One or two integrated 10/100 autosensing Ethernet ports
- Six expansion slots for network module support
- Two advanced integration module (AIM) slots for hardware acceleration and increased processing power
- Chassis support for redundant AC or DC power supplies
- One AUX port
- One console port
- Two PCMCIA card slots for software and configuration backup

NOTE   For more information on available module interfaces and WICs, refer to the *Cisco 3600 Router Installation and Configuration Guide* that came with your specific router model.

Selecting a VNM

Cisco manufactures three VNMs designed to convert telephone voice signals into a form that can be transmitted over an IP network (see Figure 6-6). You must install at least one of these on a 2600 or 3600 series router to transport voice traffic:

- One-slot NM-1V—You can install one VIC into a one-slot VNM. The single VIC slot is slot 0.
- Two-slot NV-2V—You can install two VICs into a two-slot VNM. Facing the rear of the router, the VIC slots are slots 0 and 1 from right to left, respectively.
- One-port NM-HDV—You can install one VWIC for multiflex trunk support.
- Two-port NM-HDV—You can install two VWICs for multiflex trunk support.

Because each VIC interfaces with up to two telephony devices, one-slot and two-slot VNMs can support a maximum of two and four telephony devices, respectively. Device selection depends upon the number of telephony devices you wish to connect to the router.

Each VNM has an enable LED that indicates that the module has passed its self-tests and is available to the router.

Figure 6-6   *VNMs*

One-slot VNM (NM-1V)

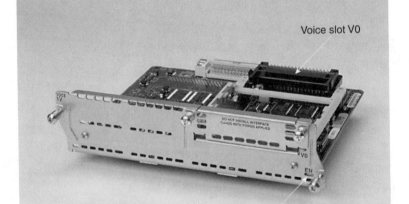

Voice slot V0

Enable LED

Figure 6-6   *VNMs (Continued)*

Two-slot VNM (NM-2V)

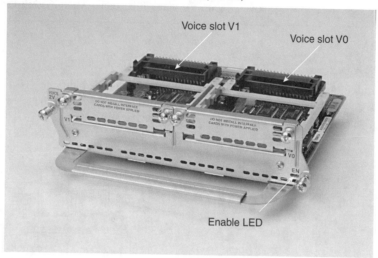

Voice slot V1

Voice slot V0

Enable LED

High-density VNM (NM-HDV)

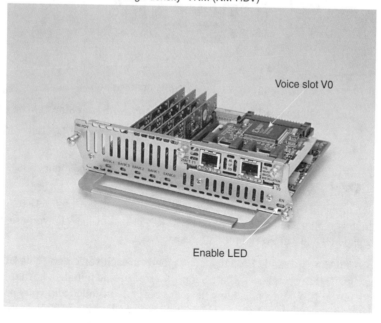

Voice slot V0

Enable LED

NOTE   Each router comes with panels covering the module slots, called *module blank filler panels*. For safety and proper ventilation, remove the filler panels only if you install a module into the slot. Keep this slot covered with the blank filler panel when modules are not installed.

Selecting a VIC

In addition to the VNMs, at least one VIC must be installed into the VNM to process voice for transport over an IP network. VICs provide the connection to the telephone equipment or telephone network. Figure 6-7 shows the available VICs.

Figure 6-7   *VICs*

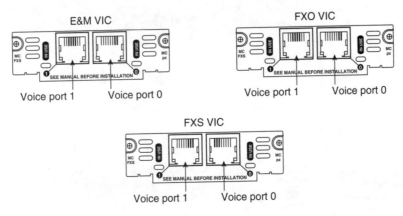

Three VICs, each providing two voice ports, are available:

- Ear & Mouth (E&M) VIC—Used for PBX-to-PBX tie-line connections
- Foreign exchange office (FXO) VIC—Used for connections to central office equipment
- Foreign exchange station (FXS) VIC—Used for direct connections to telephony devices

Each of the listed VICs includes two voice ports. Each port is an RJ-style connection into which you plug a telephone, fax, or PBX. Regardless of the VIC you use, the voice ports are called 0 and 1 from right to left, respectively.

Each VIC has an enable LED that corresponds to each voice port. This LED has an amber flash while the router and VIC are being powered up and initialized. After initialization, the light turns off. When a telephony device is properly installed and you operate it (for example, pick up the telephone receiver to make a call), the LED light turns on and turns green to show that the voice port is in operation. An amber signal indicates that the voice port is not in operation.

NOTE   Although VICs physically resemble WICs, which install in a Cisco 3600 series two-slot network module to provide WAN interfaces and in the 2600 WAN interface slot, VICs and WICs are not interchangeable. VICs cannot be installed in these slots, and WICs cannot be installed in a VNM. For information on available WICs, refer to the installation and configuration guide that accompanied the router.

NOTE   Similar to the router, the VNM has panels covering the VIC slots, also called *blank filler panels*. For safety and proper ventilation, remove the filler panels only if you install a VIC into the slot. Keep this slot covered with the blank filler panel when there are no VICs installed.

Voice Port Addressing on a VIC

When you have installed the router and you begin configuring the voice services, you must be able to correctly identify each voice port on the VIC. The locations of the VICs and VNMs are important for understanding how to identify the ports with Cisco IOS software commands such as **voice-port** and **port.** These commands identify each port with a three-part specification, each of which is delimited with a forward slash (/):

- The module slot on the router
- The VIC slot, also known as the subslot
- The actual voice port for the telephony device

For example, if a telephone were plugged into voice port 1/0/1 on a 2600 or 3600 router, it would be in module slot 1, VIC slot 0, and voice port 1.

Your choice of VIC will depend on the way the network is configured. Each card is used for specific applications.

Router, VNM, and VIC Assembly

Once you have chosen the VNM and VICs for a Cisco voice-capable router, you need to install them. The 2600, 3620, or 3640 router, VNM, VIC, and telephony device fit together using these steps:

1 Install the VNM into the router.

2 Install the VIC into the VNM.

3 Plug the telephony device into a VIC port like 1/0/1.

Verifying Hardware Installation

After the cards have been installed in the router, you need to ensure that the modules are installed properly. When the router's power switch is in the ON position, check that the LED light corresponding to the VNM slot you installed is on. If the light is on, you most likely installed the VNM and VIC properly. If the light is off, turn off the machine and check that the VNMs and VICs are securely positioned. You may need to remove and reinstall them.

NOTE   FXS VICs provide a dial tone. If you are using an FXS VIC, check for dial tone on the telephony device by lifting the receiver and listening for a dial tone.

Setting Up an MC3810 Router

As you recall from Chapter 4, the MC3810 is a multiservice device designed to transport voice and data over the network. This section describes how to set up an MC3810 for operation in an integrated network. You will read about the components and their operation and learn how this device functions in an integrated environment.

MC3810 Interface Ports

Beginning in May 1999, Cisco began shipping a revised version of the standard MC3810 chassis. This new version is designated as the "V-chassis" because of design changes added to accommodate the new video dial module (VDM).

You can easily identify the new chassis by the new cutouts on the rear panel. These new cutouts are to the right of the existing cutout for the T1/E1 multiflex trunk module (MFT). In addition, there are two ground lugs located next to the power receptacle. Figure 6-8 shows the back of a MC3810 V-chassis.

The rear panel of the V-chassis MC3810 has permanent installation of the following:

- One RJ-45 10BaseT Ethernet/IEEE 802.3 interface
- Two 60-pin Cisco universal input/output (UIO) serial interfaces
- The RJ-45 console and auxiliary interfaces
- The electrical power on/off switch
- Two grounding lugs
- Either the grounded 3-pin AC or a DC power receptacle

Figure 6-8   *The Rear View of an MC3810 Chassis*

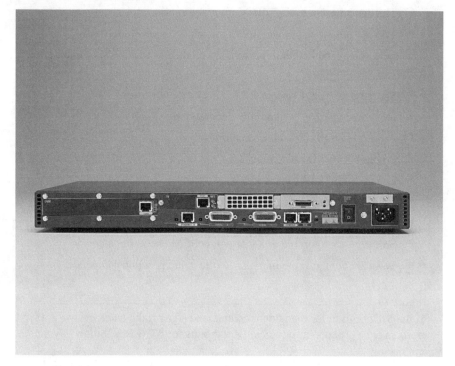

The cutout in the upper-left side of the rear panel is for access to interfaces on one of three daughter cards that provide the following:

- Six analog voice port interfaces with RJ-45 jacks without inserts for E&M connections, or with plastic RJ-11 inserts for adapting the 8-wire RJ-45 interfaces to FXS or FXO 6-wire connectors
- Four Basic Rate Interface (BRI) CB-1 jacks providing eight generic QSIG voice ports (two bearer [B] channels) per BRI interface
- One T1/E1 digital voice interface with either one RJ-48 8-wire connector or a dual BNC connector

The second rear panel cutout, located above the Ethernet/IEEE 802.3 10BaseT interface, is for access to the T1/E1 MFT with one of the following connectors:

- A single RJ-48 8-wire connector
- Dual BNC connectors
- A single RJ-48 or dual BNC T1/E1 connectors plus a CB-1 connector for the BRI data backup interface co-mounted on the MFT

Two cutouts are on the V-chassis rear panel:

- The one adjacent to the MFT cutout is a cooling slot.
- The one next to the power on/off switch is for access to the VDM.

The audio voice module/audio personality module (AVM/APM) combination provides the following features:

- Six ports of FXS, FXO, or E&M (in any combination)
- Integrated talk battery and ring generator
- Adjustable transmit and receive levels
- 2-wire FXS/FXO voice interface
- 2- and 4-wire E&M interface
- Wink start, immediate start, and delayed start
- Software-configurable ground start, loop start, or battery-reverse signaling
- Software-configurable a-law or μ-law PCM encoding
- Software configurable impedance (the value is country dependent)

NOTE   If the MFT is used for the network trunk, and ATM is the trunk protocol, then the serial ports may be grouped into logical DS0 groups for ATM Adaptation Layer 1 (AAL1) connections.

A diagram of the AVM and APM is shown in Figure 6-9. A voice channel is associated with each APM. Voice signals are digitized into PCM-encoded voice samples by the codec on the APM. The codec may be configured for either μ-law (T1) or a-law (E1) PCM coding. The PCM samples are then passed to the voice-compression services in the Cisco MC3810.

Figure 6-9   *An AVM with APM*

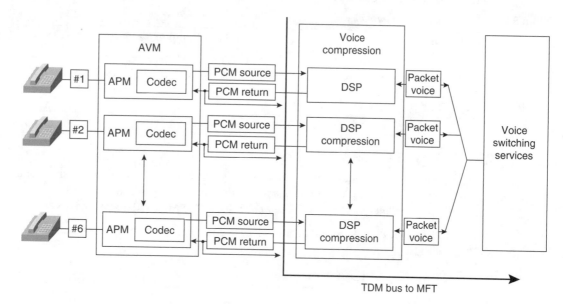

Cisco MC3810 series concentrators now contain high-performance voice-compression modules (HCMs). HCMs provide voice compression according to the voice-compression coding algorithm (codec) specified when the Cisco MC3810 is configured. Table 6-1 shows the number of voice channels each type of compression module can support.

Table 6-1   *High-Performance Compression Modules*

Configuration	Slot 2	Number of Channels	Slot 5	Number of Channels	Total Number of Channels
Analog 4 ports, G.729	HCM 2	4	Empty	0	4
Analog 6 ports, G.729a	HCM 2	6	Empty	0	6
Digital T1, G.729	HCM 6	12	HCM 6	12	24
Digital T1, G.729a	HCM 6	24	Empty	0	24
Digital E1, G.729	HCM 6	12	HCM 6	12	24
Digital E1, G.729a	HCM 2	8	HCM 6	24	32

Table 6-2 shows the codecs supported in high and medium complexity.

Table 6-2   *Codec Packaging Information*

High Complexity	Medium Complexity
G.711ulaw, G.711alaw, G.723.1(r5.3), G.723.1 Annex A(r5.3), G.723.1(r6.3), G.723.1 Annex A(r6.3), G.726(r16), G.726(r24), G.726(r32), G.728, G.729, G.729 Annex B, fax relay	G.711ulaw, G.711alaw, G.726(r16), G.726(r24), G.726(r32), G.729 Annex A, G.729 Annex B with Annex A, fax relay

The MC3810 Digital Chassis

The digital chassis provides digital voice interface capabilities. The chassis has slots for a digital voice module (DVM), which enables T1/E1 connections to a digital PBX, compression modules, and an MFT. Figure 6-8 shows the MC3810 digital chassis.

The MFT provides for voice and data services over a single WAN interface running at T1/E1 speeds.

Digital Voice Architecture

The DVM provides connectivity to digital PBXs or channel banks on a T1/E1 interface for up to 24 digital phone lines. Signaling types supported on a per-channel basis include FXS, FXO, E&M, and CAS (Mercury). Figure 6-10 outlines this architecture.

Figure 6-10   *The DVM Architecture*

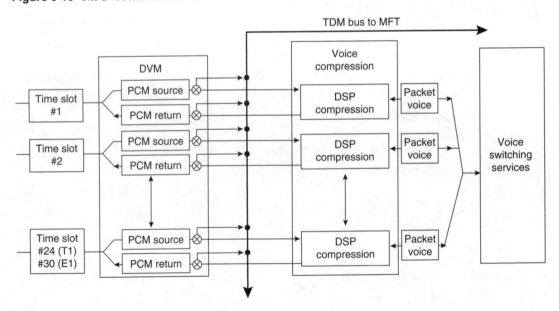

NOTE	The digital signaling mentioned in this section is also available on the 2600/3600 series routers using the T1/E1 module.

The MC3810 will work with PBXs conforming to the standard specification, EIA/TIA-464-A (February 1989), for PBX switching equipment. Channel-associated signaling (CAS) is supported on the DVM. Both North American T1 CAS (ABCD bits robbed from each channel) and European CAS are supported. The chassis supports FXS, FXO, or E&M VICs.

Depending on the configuration used, channels from a DVM may be directed to either the voice-compression services or the time-division multiplexing (TDM) bus. Figure 6-10 shows how the DVM operates with the TDM bus if the MFT is installed.

NOTE	It is not possible to configure an MC3810 with both analog and digital voice modules.

MC3810 WAN Trunk Options

There are two WAN trunk options:

- Two universal I/O serial interfaces, which support the following:
  - Frame Relay
  - HDLC
  - Fractional T1 of speeds to nx64 up to 2 Mbps
- The MFT option, which supports the following:
  - Frame Relay
  - ATM over T1/E1
  - TDM (voice/video/data)
  - nx64 kilobits per second (kbps) up to E1 speeds

The two universal I/O serial interface (also known as a 5-in-1 serial port with DB-60) connections can handle voice and data for Frame Relay or just data for HDLC, or can run in "transparent" mode.

The MFT has all the features listed, and is discussed in more detail in the next section.

MC3810 MFT Options

The MFT provides a multiservice, T1/E1 trunk with built-in, long-haul CSU/DSU. The MFT is software configurable to support either ANSI T1.403 (T1) or ITU G.703 (E1). It supports connectivity to ATM, Frame Relay, or leased-line carrier services. Figure 6-11 shows how the MFT option can be used to carry voice traffic over the carrier network. To connect to a T1 ATM network, you need to use a T1/E1 trunk interface. The MFT provides a RJ-48 connector jack for the network interface and a T1.403-compliant, onboard CSU/DSU to support a T1 trunk.

Figure 6-11 *MC3810 MFT Options*

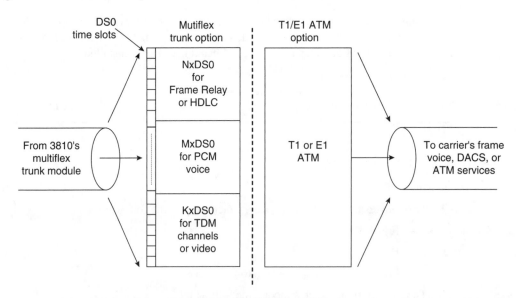

The MFT can be ordered with either RJ-48 or BNC connections, though not all versions will be available at first release. The MFT derives clocking and distributes it to the UIO serial and DVM.

The MFT has two WAN trunk options, multiflex mode and T1/E1 ATM mode.

The features of the MFT in multiflex mode are as follows:

- Frame Relay, HDLC, PPP, IP
- Compressed voice
- MxDS0 for PCM voice (TDM)
- DACS channels:
    - Bit serial data from UIO
    - Video or data

The MFT T1/E1 ATM WAN services are as follows:

- Data and video in structured AAL1 format
- Compressed voice or data in AAL5 format

The BRI Voice Module

The BRI voice module (BVM) is a standoff, mounted daughter card that installs in slot 4 and covers slot 5 on the SCB. The BVM provides four CB-1 8-wire interfaces with two voice-only B channels and one signaling channel per interface. Figure 6-12 shows the BVM module.

Figure 6-12 *A BVM*

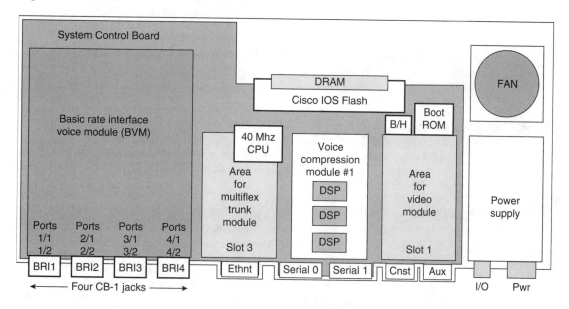

The BVM is designed to provide single-line BRI S/T connections to private integrated services network exchange (PINX) devices providing ECMA standard QSIG call setup signaling to the QSIG call handler in the Cisco MC3810.

The BVM and the associated Cisco MC3810 QSIG call handling software supports standard ECMA QSIG voice call setup messages only.

Currently the BVM does not support ISDN Q.921 and Q.931 BRI call setup messaging.

The BVM interface supports eight voice calls due to the following:

- Each of the four BRI S/T CB-1 interfaces identified as interface BRI 1 through BRI 4 supports two voice ports per interface with two B and one D channel per voice port. The voice ports are identified as port x/y, where x is the BRI S/T interface (1 through 4) and is either the first (1) or second (2) voice port on the interface (that is, voice port 3/2 is the second voice port on BRI S/T interface 3).

- Each BRI S/T interface defaults to a terminal endpoint (TE) (also identified in the Cisco IOS CLI as an end user) configuration. Each interface can be reconfigured to a network terminal (NT) (also identified in the Cisco IOS CLI as a network) to emulate a PINX.

The production BVM is not intended to directly attach to an ISDN telephone. In addition, circuitry for generating line power *is not* built into the BVM, which means that an NT-configured BRI S/T interface will not supply line power to a TE device.

The BVM can be specified as a clock source using the **network-clock-select** *priority* [**serial 0** | **system** | *controller*] command. This command is covered in detail in later chapters.

NOTE   Cisco C2600 and C3600 platforms have BRI Voice over IP capabilities provided by a 2-port VIC part number VIC-2BRI-S/T-TE. This VIC supports ISDN Q.921 and Q.931 call setup signaling and currently does not interwork the Cisco MC3810's BVM QSIG BRI S/T.

The VDM shown in Figure 6-13 provides an RS-366 dialing interface to an H.320 video codec. The MC3810 automatically accepts dial-out requests from the video system and then reconciles the dialed address (E-164) with a standard 20-octet ATM network service access point (NSAP) address.

As illustrated in Figure 6-8, the RJ-366 connection on the top right is for the establishment of an ATM switched virtual circuit (SVC) for AAL1 circuit-emulation service (CES) traffic originating on the H.320 video V.35 port that is connected to the MC3810's serial 0 interface.

Figure 6-13 *A VDM*

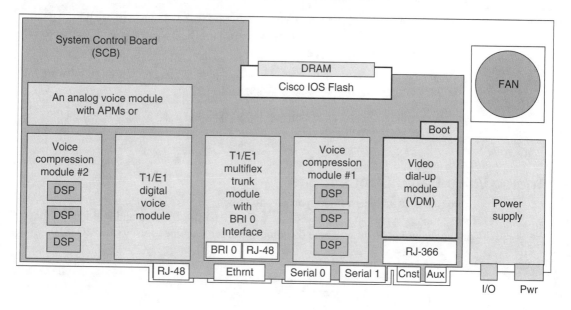

Setting Up an MC3810 Router for VoFR, VoATM, or VoIP

To set up, install, and configure the MC3810 for VoFR, VoATM, or VoIP communications, complete the following steps:

1 Connect the input power (AC or DC and including the redundant power system).

2 Connect the MC3810 console port to either an ASCII terminal or a PC running terminal emulation software (or a modem to the auxiliary port if you plan to configure the MC3810 remotely).

3 On the ASCII terminal, connect the light-blue cable from the console port (light blue) to the female DTE adapter.

4 Plug the female DTE adapter into the DB-25 I/O port on the terminal.

5 Set the terminal for 9600 baud, 8 data bits, 1 stop bit, no parity, and no flow control.

6 For the PC, plug the adapter into a DB-9 serial port on the PC.

7 At the AUX port, connect the light-blue cable from the auxiliary port (black) to the male DCE adapter.

8 Plug the male DCE adapter into the DB-25 port on the modem.

9 Configure the modem to match the transmission speed of the auxiliary port (default is 9600 baud), and set the hardware flow control for DCD (data carrier detect) and DTR (data terminal ready) operation.

10 The baud rate for the auxiliary (and console) port can be configured in software for 4800, 1200, 2400, 19200, 38400, 57600, and 115200.

11 Connect the cable from the Ethernet 0 port (yellow) to an available port on the Ethernet hub (if required).

Analog Voice Cable Connections

To provide analog voice services you need to provide physical connectivity to the MC3810. This section provides an overview of the MC3810 hardware setup and interface cabling options. Figure 6-14 shows some common voice cable connections to the MC3810.

Figure 6-14 *Voice Cable Connections*

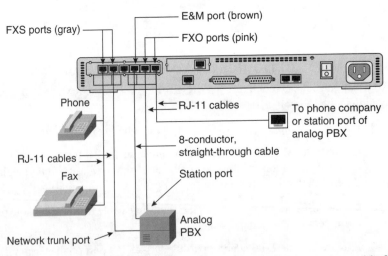

If you are making analog voice connections, connect the appropriate cable from an analog voice port to the equipment or line:

- FXS port (gray)—To telephone or fax equipment or network trunk port of analog PBX or Key System

- FXO port (pink)—To central office line or to station port of analog PBX

- E&M port (brown)—To analog PBX or key system

NOTE	Locations shown in Figure 6-14 for FXS, FXO, and E&M ports are for illustration only. Actual port types and locations will vary.

Use Table 6-3 as a guide for the cables you will be installing by cable color, connector type, and which port on the MC3810 connects to which interface.

Table 6-3   *MC3810 Cable Connections*

MC3810 Port	Port Color	Connector	Interface
Console cable	Light blue	RJ-45	Console port
AUX port	Black	RJ-45	Modem
Analog voice—FXS	Gray	RJ-11	Analog phone or fax
Analog voice—FXO	Pink	RJ-11	Telephone central office
Analog voice—E&M	Brown	RJ-1CX	Analog PBX
T1/E1 digital voice	Tan	RJ-48	Digital PBX
T1/E1 trunk—MFT	Light green	RJ-48	T1/E1 trunk (telco demarc)
Universal IO serial 2	Dark blue	DB-60	Serial Trunk
Ethernet	Yellow	10BaseT RJ-45	Ethernet hub

If you are making digital voice connections, connect the RJ-48 straight-through cable from the MC3810's T1/E1 digital voice port to a PBX or channel bank.

To make an MFT cable connection, connect the RJ-48 cable from the T1/E1 trunk port (marked T1/E1 on a light-green label) to the RJ-48 jack in the network demarcation device (telco demarc).

To make a synchronous serial connection, connect the appropriate serial interface cable (DB-60) from the serial port you are using (0 or 1) to the CSU/DSU or modem. Both serial ports are color-coded dark blue. The serial interface cable must match the signaling protocol being used. Connections may be made for EIA/TIA-232, -449, 530, V.35, or X.21.

Verifying the MC3810 Hardware Installation

To verify MC3810 hardware installation, perform the following steps:

1   Flip the I/O (on/off) switch to the O (on) position so that the router's power LED indicator emits a green color, indicating successful initiation of the MC3810. Make sure all interface cabling LEDs emit a green color, indicating successful connectivity to the network.

2   Using a PC laptop configured with Telnet or Hyper Term software attached to the MC3810 using the standard rollover console cable, perform basic initial router configuration using Cisco IOS software.

3   Verify and troubleshoot setup, configuration, and connectivity as appropriate to achieve a stable VoFR, VoATM, or VoIP connection using the MC3810. If you have not been able to successfully install the MC3810 hardware and cables, consult Table 6-4 for troubleshooting and correcting any problems that may have occurred.

Table 6-4   *Troubleshooting MC3810 Startup Procedures*

Symptom	Possible Cause	Corrective Action
Power LED and fan are off	• Power source switched off • Faulty power cable • Faulty power source • Faulty internal power supply	• Switch power source on • Check/replace power cable • Check/correct input power • Contact Cisco Technical Assistance Center (TAC) or a Cisco reseller
Power LED on; fan off	• Faulty Cisco MC3810	• Contact Cisco TAC or a Cisco reseller
Power LED off; fan on	• Faulty Cisco MC3810	• Contact Cisco TAC or a Cisco reseller
No initialization response from Cisco MC3810	• Faulty modem console terminal • Faulty cabling to terminal • Faulty Cisco MC3810	• Check/replace modem/terminal • Check/replace cable • Contact Cisco TAC or a Cisco reseller
Unit shuts off after operating for some time	• Overheating • Faulty Cisco MC3810	• Check ventilation • Contact Cisco TAC or a Cisco reseller
Console screen display freezes	• Console fault • Software error • Faulty Cisco MC3810	• Reset/replace console • Repeat power-up procedure • Contact Cisco TAC or a Cisco reseller

Setting Up an AS5300 Router

The AS5300 is a high-performance universal access product. This device, like the others discussed in this chapter, provides access for analog and ISDN as well as voice services. Figure 6-15 shows the AS5300. This section describes how to set up an AS5300 router.

Figure 6-15 *An AS5300 Router*

The AS5300 has three feature card slots, 0 (bottom), 1 (middle), and 2 (top). Up to two of the slots are available to install VoIP feature cards. One feature card slot must have either a quad T1/PRI or a quad E1/PRI installed for telephony device connections.

The full features of the AS5300 are as follows:

- One 10-megabyte (MB) and one 10/100-MB Ethernet interface
- Four high-speed sync serial interfaces
- Fast 150-megahertz (MHz) R4700 RISC CPU
- Either four or eight CT1/PRI or CE1/PRI cards
- Selectable for balanced/unbalanced (75/120 ohms)
- Supports E&M, link start, wink start, loop start, immediate start, and ground start
- Port selecting bantam jack interfaces to provide monitoring capability
- An integral AC or DC power supply

Modem ISDN Channel Aggregation

The AS5300 has up to 240 modem ISDN channel aggregation (MICA technologies) digital signal processor (DSP) universal port engines, or 60 VOX DSP engines, onboard and also has the following:

- Six independent DSPs capable of 12 total modem sessions each
- High-performance Intel i960 RISC processor per module
- High-speed, parallel bus architecture
- Extensive modem management capabilities

MICA is a modem module and card used in the Cisco AS5300 universal access servers. A MICA modem provides an interface between an incoming or outgoing digital call and an ISDN telephone line; the call does not have to be converted to analog, as it does with a conventional modem and an analog telephone line. Each line can accommodate, or aggregate, up to 24 (T1) or 30 (E1) calls.

The A55300 has the following features and functions:

- Four high-speed serial ports at 2 MB to 8 MB: (4×2 MB, 2×4 MB, 1×8 MB)
- Supports up to 12 different protocols, including V.35, RS232, X.21, RS449, RS530, IPCP, IPXCP, G.711, G.729, G.729a, G.723.1, H.323, ARA 1.0 and 2.0, ATCP, X Remote, PPP, and SLIP
- Uses the same cables as 3600 and 2600 series routers
- Allows for aggregate backhaul in distributed environments and eliminates the need for external switches and routers, thereby enhancing scalability and investment protection
- Allows lease-line termination (branch-offices, backup router links, and so forth)

Router, Feature Card, and DSP Module Assembly

The AS5300 router, VoIP feature card, and DSP modules fit as shown in Figure 6-16. First you place up to five DSPs into the VoIP feature card, and then you install the feature card into one of the three slots in the AS5300.

Figure 6-16 *DSP Module Assembly*

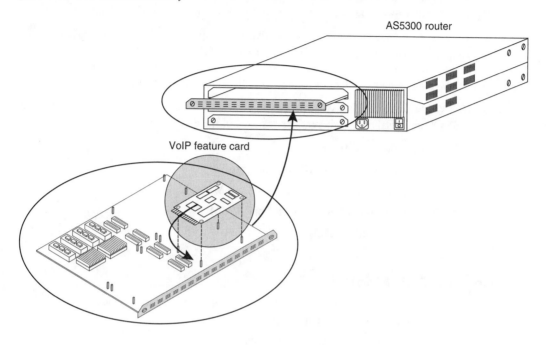

NOTE   This section describes only the additional components you must install to transport VoIP. For information on installing other feature cards, including the quad TI/Primary Rate Interface (PRI) or quad E1/PRI feature cards, refer to the *Installation and Configuration Guide* that accompanied the router.

Selecting the VoIP Feature Card

The Cisco AS5300/voice gateway can accept two voice/fax feature cards, so it can scale up to 96/120 voice connections within a single chassis.

Like the other voice cards discussed in this chapter, the Cisco AS5300 voice/fax feature cards offer toll-quality voice. Cisco AS5300 voice/fax feature cards are coprocessor cards with a RISC engine and dedicated DSPs for each voice channel. The card's design couples this coprocessor with direct access to the routing engine for packet forwarding.

Incoming calls are terminated on the voice/fax feature card, where the voice is encoded, using standard algorithms including G.711 and G.729, compressed, and encapsulated in Real-Time Transport Protocol (RTP) packets.

A call is placed to the remote voice gateway using the standard H.323 protocol and the remote gateway decodes the voice and delivers it to the receiver.

High-performance DSP architecture delivering 50 millions of instructions per second (mips) per DS0 to support high-compression/low-delay codecs; adaptive jitter buffer, voice activity detection, comfort noise generation, echo cancellation, and concealment fax carriage make packet telephony gateway calls sound just like PSTN calls.

The VoIP feature card is a voice processing card that resides in one of the slots in the Cisco AS5300 universal access server. The features of the card are as follows:

- CPU: 4700 mips, 100 MHz
- Support chipset: GT-64010 system controller
- DRAM: Standard 72-pin SIMM (4, 8, 16 MB)
- Flash: Cisco proprietary Flash 80-pin SIMM
- Five DSP module sockets

Selecting the DSP Module

The DSP module provides voice-compression and packetization services to the VoIP feature card in a configurable and expandable fashion.

The features of the DSP module are as follows:

- Six DSPs per DSP module.
- Uses T1 TMS320C542 50 MHz.
- DSP SRAM supports 120 kilo (16 bit). A kiloword is 16 Kb, so 120 kilowords is 1920 Kb.

NOTE   If you are running T1 applications, use 4-DSP modules for a total of 24 DSPs. If you are running E1 applications, use 5-DSP modules for a total of 30 DSPs.

Summary

In this chapter we have reviewed the installation and selection of the Cisco voice-capable routers and the feature cards that support these functions. You have seen how the VNM allows for VIC cards to be installed into 2600 and 3600. This chapter discusses how the port addressing for the cards is associated with the slots in which they are installed. The chapter discusses the MC3810 multiservice device and how to install this router into the network and provide voice-capable connections. Finally, the chapter discusses the voice services of the AS5300 and how cards are installed to provide voice connectivity in conjunction with data services.

Review Questions

The following questions should help you gauge your understanding of this chapter. You can find the answers in Appendix A, "Answers to Review Questions."

1 Which module is required in the 3600 and 2600 series routers to provide a slot for VICs?

_____

_____

_____

2 On the 2600, how many voice ports can be installed in a single-network module slot?

_____

_____

_____

3 What are the three types of WICs that can be installed in the VNM?

_____

_____

_____

4 True or False: All 3600 series routers have a console connection on the front panel.

_____

_____

_____

5 What do you check on the 2600/3600 to verify proper installation of a VIC module?

_____

_____

_____

6 MC3810 design chassis were modified to accommodate which module?

_____

_____

_____

7 If the MC3810 shuts down after operating for some period of time, what is a possible cause?

_____

_____

_____

8 When installing voice modules in the AS5300 at least one feature card must be what?

_____

_____

_____

9 What does MICA stand for?

_____

_____

_____

10 True or False: The AS5300 uses the same cables as the 2600/3600 routers.

_____

_____

_____

PART III

Configuring Cisco Voice Solutions

After reading this chapter, you should be able to perform the following tasks:

- Given an IP, Frame Relay, or ATM network, configure and verify the voice ports so that you properly interface to the attached telephony devices.

- Given a Cisco router capable of transporting voice, configure and verify POTS dial peers so you can place calls to others connected to the router.

- Given an IP, Frame-Relay, or ATM network, configure and verify VoIP, VoFR, or VoATM dial peers so you can place calls to others connected to your router.

- Employ Cisco IOS commands to tune voice-port voice quality on Cisco voice-capable routers.

- Configure POTS, VoFR, VoATM, and VoIP dial peers and verify their correct operation.

CHAPTER 7

Configuring Voice Ports and Dial Peers for Voice

So far in this book we have discussed the technologies behind voice services and the hardware required to perform these services on Cisco equipment. This chapter discusses the configuration of voice services on Cisco voice-capable routers. In order to place an end-to-end call, you must configure voice ports accurately so that proper signaling occurs between the telephony device and the router. You must also configure routers to establish logical voice connections with dial peers. The *voice port* is the physical connection to the telephony equipment. *Dial peers* provide a logical mapping between these voice ports, they provide the capability to implement a dialing plan and determine the routing of calls. The voice ports and dial peers are fundamental portions of this configuration and establish entry points and transport services for voice traffic.

This chapter describes procedures applicable to voice ports and dial peers for Voice over IP (VoIP), Voice over Frame Relay (VoFR), and Voice over ATM (VoATM) on Cisco voice-capable routers. The commands presented in this chapter apply to both analog and digital voice ports unless otherwise indicated. This chapter also lists many of the voice-port configuration commands available for voice-capable routers. Some commands and command features may vary slightly from application to application and between router types. For information specific to your router and application, reference the *Command Reference Guide* applicable to your product.

NOTE

To learn more about configuring voice ports and dial peers, refer to the following related Cisco CCO documents: *Voice over IP over ATM Networks on the Cisco MC3810*, *Feature Module SGCP Feature Module for Sprint Ion*, and *Voice over ATM on the Cisco 2600 and 3600 Routers*. Most configuration commands for the Cisco 2600, 3600, and MC3810 series platforms have been made usable on all three platforms. Differences in usage are noted for individual commands in the command reference section of the Cisco documentation or online at CCO.

Configuring Voice Ports

There are basically two functions that need to be configured to support voice services: connectivity to the voice port and a transport between the voice ports. As discussed in earlier chapters, voice ports provide connectivity to telephony devices.

You need to configure Cisco voice-over routers to support voice ports. In general, voice-port commands define the characteristics associated with a particular voice-port signaling type. Voice ports on Cisco routers support three basic voice signaling types:

- Foreign exchange office (FXO) interface—The FXO interface is an RJ-11 connector that allows a connection to be directed at the Public Switched Telephone Network's (PSTN's) central office or to a standard private branch exchange (PBX) interface. This interface is of value for off-premise extension applications.

- The foreign exchange station (FXS) interface—This interface is an RJ-11 connector that allows connection for basic telephone equipment, keysets, and PBXs, and supplies ring, voltage, and dial tone.

- The "Ear and Mouth" (E&M) interface (or "receive and transmit" interface)—This interface is an RJ-48 connector that allows connection for PBX trunk lines (tie lines). It is a signaling technique for 2-wire and 4-wire telephone and trunk interfaces.

Under most circumstances, the default voice-port command values are adequate to configure FXO and FXS ports to transport voice traffic over your existing network. Because of the inherent complexities involved with PBX networks, E&M ports might need specific voice-port values configured, depending on the specifications of the devices in your telephony network.

The next few sections discuss how to configure the FXO, FXS, and E&M voice ports. These sections discuss the details of how to identify which ports are needed for specific applications and what parameters can be modified. There are also some examples of configurations for use of these ports.

Identifying POTS and Voice-over Dial Peers

Dial peers are an addressable call endpoint. Dial peers establish logical connections called *call legs* to complete an end-to-end call. There are two kinds of dial peers: POTS and voice-over (that is, VoFR, VoATM, VoIP) dial peers. Figure 7-1 shows the relationship between POTS dial peers and voice-over dial peers.

Figure 7-1   *POTS and Voice-over Dial Peers*

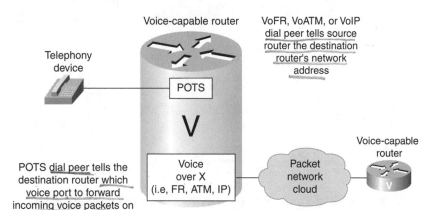

- *POTS* (plain old telephone service) dial peers are those connected to a traditional telephony network that includes, but is not limited to, telephone handsets. POTS peers point to a particular voice port on a voice network device. POTS dial peers tell the router to which interface port each telephony device connects. The router will then know where to forward incoming calls because the POTS dial peer specified the voice port. Multiple POTS dial peers can point to the same voice port. The POTS dial peers also implement the dialing plan.

- Voice-over dial peers are those connected via a packet network backbone. In the case of VoATM, for example, the X is an ATM network. Voice-over peers point to specific network devices, like the router. These voice-over dial peers at the source router are associated with the destination address of the destination router.

To place a voice-over call, the source router must be configured with the appropriate voice-over dial peer specifying the recipient's phone number or E.164 address. The recipient's router must be configured with a POTS dial peer specifying to which voice port and telephony device to forward the voice call.

Establishing Dial-Peer Call Legs

A *call leg* is a logical connection between the router and either a telephony endpoint or another endpoint using a session protocol like Frame Relay, ATM, or IP. Figure 7-2 illustrates dial-peer call legs.

Figure 7-2 *Dial-Peer Call Legs*

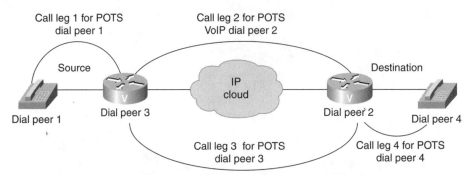

Call legs are router-centric, which means that when an inbound call arrives on the router, it is processed as a separate call until the destination is determined. A second outbound call leg is then established. The two call legs compose an end-to-end call through the router.

The connections are made when dial peers are configured on each interface. An end-to-end call comprises four call legs, two from the perspective of the source router, as shown in Figure 7-2, and two from the perspective of the destination router. To complete an end-to-end call and send voice packets back and forth, all four dial peers must be configured.

Figure 7-2 illustrates an end-to-end call with the four call legs. The source router must have dial peers 1 and 2 configured to specify call legs 1 and 2. To send voice packets back to the source router, the destination router must have dial peers 3 and 4 configured to specify call legs 3 and 4.

In order to configure these call legs we need to configure the voice ports and then the dial peers. The following sections describe the configuration commands for accessing and configuring an analog voice-port connection for Cisco voice-capable routers. These sections include how to configure options on analog voice ports, digital voice ports, FXS, FXO, and E&M subcommands as well as examples of configurations.

Analog Voice Port Commands

The following syntax describes how to access analog voice ports for configuration on the 2600/3600 analog routers:

```
Router(config)#voice-port slot-number/subint-number/port
```

The command used for the MC3810 is as follows:

```
Router(config)#voice-port slot/port
```

You use the **voice-port** global configuration command to enter the voice port configuration mode and configure a specific voice port. The **voice-port** configuration commands are nested so that all subsequent commands affect only the specified voice port unless otherwise noted.

Tables 7-1 and 7-2 describe the **voice-port** command options.

Table 7-1   **voice-port** *Command Options for Voice-capable Analog Routers*

Command Option	Description
slot	Specifies a router slot number in which the voice network module is installed. Valid entries are router slot numbers for the particular platform.
subunit	Specifies a voice interface card (VIC) where the voice port is located. Valid entries are 0 and 1. (The VIC fits into the voice network module.)
port	Specifies an analog voice port number. Valid entries are 0 and 1.

Table 7-2   **voice-port** *Command Options for MC3810 Analog Routers*

Command Option	Description
slot	Specifies the physical slot in which the analog voice module (AVM) is installed. The slot number is always 1 for analog voice ports in the Cisco MC3810.
port	Specifies an analog voice port number. Valid entries are 1–6.

Digital Voice Port Commands

Analog voice ports are not the only module types available in the voice-capable routers. It is also possible to have digital ports like the multiflex trunk and ISDN voice WAN interface cards (VWICs). To access these voice ports, a different set of specifications is used with the **voice-port** commands. The syntax for the digital ports on modular routers is as follows:

```
Router(config)#voice-port slot/port:[ds0-group | pri-group]
```

The command structure on the MC3810, fixed configuration, digital ports would be formatted like the following:

```
Router(config)#voice-port slot:ds0-group
```

Similar to the analog voice ports, you would use the **voice-port** global configuration command to enter the voice port configuration mode and configure a specific voice port. The **voice-port** configuration commands are nested so that all subsequent commands affect only the specified voice port unless otherwise noted. Table 7-3 shows the command options for the **voice-port** command and Tables 7-4 and 7-5 show the options for the **voice-port** command on a fixed configuration MC3810.

Table 7-3   **voice-port** *Command Options for Modular Routers*

Command Option	Description
slot	Specifies a router slot number in which the packet voice trunk network module (NM) is installed. Valid entries are router slot numbers for the particular platform.
port	Specifies an E1 or T1 physical port in the VWIC. Valid entries are 0 and 1. (One VWIC fits in a NM.)
ds0-group	Specifies an E1/T1 logical port number. Valid entries are 0–30 for E1, and 0–23 for T1.
pri-group	Specifies an ISDN PRI logical port number.

Table 7-4   **voice-port** *Command Options for MC3810 Digital Routers*

Command Option	Description
slot	Specifies the module (and controller). Valid entries are 0 for the MFT (controller 0) and 1 for the DVM (controller 1).
ds0-group	Specifies an E1/T1 logical port number. Valid entries are 0–30 for E1, and 0–23 for T1.

Table 7-5   *Device Slot Numbers for MC3810 Routers by Port Type*

Device	Slot
MFT	0
AVM	1
DVM	1

Voice Port Subcommands

In the voice port mode, the prompt will look like Example 7-1. In this mode, commands are entered that change the values for the parameters that affect the operation of the physical voice port.

Example 7-1   *Voice Port Mode*

```
Router(config-voiceport)#
```

Some of the basic **voice port** commands you can enter are described in the following sections.

cptone *country*

The **cptone** *country* command sets the call progress tone. Using this command, you configure the voice port for the local territory's call progress tone setting. The call progress tone setting determines the settings for dial tone, busy tone, fast-busy (reorder tone), and ring-back tone. These are the local dial tone, the remote side ring-back, and busy tones. So if you call a device that has the cptone set for the United Kingdom, you will hear the ring-back tone for that country on your local handset. The call progress tone is different from the DTMF tone generated for signaling and dialing. These call progress tones only communicate tones for the user.

NOTE   Cisco routers comply with the ISO 3166 country name standards, which use a two-letter code to represent a country. The default for the **cptone** *country* command is **us** (United States). For the complete list of supported countries, refer to the *Command Reference Guide* specific to your router.

description *string*

The **description** *string* command allows you to include a string description for the voice port. This description can tell what type of device or which device is connected to the port. The option *string* is a character string describing port connections. This string will show up when you view the voice port from the command prompt.

shutdown

The **shutdown** command is typically used to disable a port. However, when making changes to a voice port, you use the **shutdown** command to allow these changes to take effect. In order to activate these changes and begin using the port, you activate the voice port with the **no** form of the **shutdown** command (that is, **no shutdown**).

TIP   If you have a voice port that will not be used, you can disable the port by using the **shutdown** command.

signal {loop-start | ground-start}

The **signal {loop-start | ground-start}** command is used to specify the type of signaling for the specific voice port. The **no** form of this command is used to restore the default value for this command.

NOTE   You may wish to confirm the type of signaling you should use by consulting your telephone vendor. Once you know the type of signaling to use, you can specify the proper signaling with the **signal** command.

signal command options for FXO and FXS interfaces are shown in Table 7-6.

Table 7-6   signal *Command Options*

Command Option	Description
loop-start	Specifies loop-start signaling. Used for FXO and FXS interfaces. With loop-start signaling, only one side of a connection can hang up. This is the default setting for FXO and FXS voice ports. You should find out from your telecommunications vendor if you are using **loop-start** or **ground-start**. The default is **loop-start**.
ground-start	Specifies ground-start signaling. Used for FXO and FXS interfaces. **ground-start** allows both sides of a connection to place a call and to hang up.

NOTE   Configuring the **signal** command for a voice port will change the signal value for both voice ports on a Cisco 2600 or 3600 VIC. This is also true for several other commands on voice cards, including **shutdown**.

loop-start and **ground-start** are the choices for FXS and FXO. **wink-start**, **immediate**, and **delay-dial** are the choices for E&M; they are covered later in the chapter, in Table 7-9.

ring number *number*

The **ring number** *number* command allows you to specify the maximum number of rings to be detected before answering a call over a FXO voice port. Valid entries are from 1 to 10; the default is 1. This command applies to FXO ports only.

Normally, this command should be set to the default so that incoming calls are answered quickly. If you have other equipment available on the line to answer incoming calls, you might want to set the value higher to give the equipment sufficient time to respond. In that case, the FXO interface would answer if the equipment on the line did not answer the incoming call in the configured number of rings.

NOTE	The **ring number** *number* command is not applicable to FXS or E&M interfaces because they do not receive ringing to receive a call.

dial-type {pulse | dtmf}

You use the **dial-type** {**pulse** | **dtmf**} command to set the dial type for out-dialing to pulse or tone on the FXO ports only.

FXS Subcommands

The different types of voice ports support commands that are relevant to that particular voice port and its function. For example, the commands described in the following sections are specific to FXS ports.

ring frequency *number*

The **ring frequency** *number* command allows you to specify a ring frequency for a FXS voice port. The ring frequency you select must match the connected telephony equipment. If the frequency is set incorrectly, the attached telephony device may not ring or it may buzz. In addition, the ring frequency is usually country dependent, and you should take into account the appropriate ring frequency for your area before configuring this command. Table 7-7 shows the **ring frequency** command option.

Table 7-7   **ring frequency** *Command Option*

Command Option	Description
number	Ring frequency (hertz) used in the FXS interface. Valid entries on the Cisco 3600 series are 25 and 50. Valid entries on the Cisco MC3810 are 20 and 30. Default values are: • 25 Hz on the Cisco 3600 series • 20 Hz on the Cisco MC3810

This command does not affect ring-back, which is the ringing a user hears when placing a remote call.

An Example of Configuring FXO Voice Ports

You would want to configure voice-port parameters to meet the needs of the devices connected. Figure 7-3 shows a central office PSTN loop connected to a voice port on the router named Katie. The configurations in Example 7-2 enable loop-start signaling on modular routers voice port 1/0/0. **cptone** is set to **us** (the default), and the description tells what is connected to this voice port.

Figure 7-3 *A Central Office Connected to an FXO Voice Port*

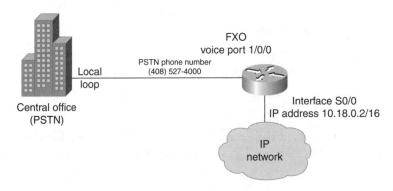

Example 7-2 *Configuration for an FXO Voice Port*

```
Katie#configure terminal
Katie(config)#voice-port 1/0/0 !This command enters voice port configuration mode
Katie(config-voiceport)#signal loop-start  !This enables loop start signaling
Katie(config-voiceport)#cptone us !This command sets the cptone to United States
Katie(config-voiceport)description To PSTN !This command describes the usage
  of the port
```

An Example of Configuring FXS Voice Ports

Using the FXS commands previously discussed, Figure 7-4 shows a station connection to the FXS port of the router named Logan. The configuration in Example 7-3 shows how to enable loop-start signaling on a 2600 or 3600 FXS voice port 1/0/0. The ring frequency is set to 25 Hz.

Figure 7-4 *A Telephone Connected to a FXS Voice Port*

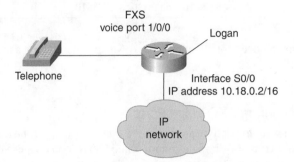

Example 7-3   *Configuration for an FXS Voice Port*

```
Logan#configure terminal
Logan(config)#voice-port 1/0/0 !Enters voice port configuration mode
Logan(config-voiceport)#signal loop-start  !Enables loop start signaling
Logan(config-voiceport)#ring frequency 25  !Sets the ring frequency to 25 Hz
```

Configuring E&M Voice Ports

When configuring a voice port, you should be prepared to spend some time determining the wiring scheme for the PBX to which you plan to connect. Understanding the timing parameters of the PBX is also imperative. You can get this information from your PBX vendor or reference the manuals that accompany your PBX. After you understand the PBX's wiring scheme, you can select a specific cabling scheme and operational parameters for E&M ports with the **operation**, **signal**, **type**, **auto-cut-through**, **codec**, and **compand-type** voice-port configuration commands.

operation {2-wire | 4-wire}

The **operation** {**2-wire** | **4-wire**} command only affects voice traffic and not the signaling. Signaling is independent of 2-wire versus 4-wire settings. If the wrong cable scheme is specified, the user might get voice traffic in only one direction. This command applies to analog connections only.

Configuring the **operation** command on a voice port changes the operation of both voice ports on a VIC. The voice port must be **shut/no shutdown** for the new value to take effect.

This command is not applicable to FXS or FXO interfaces because they are, by definition, 2-wire interfaces. On the Cisco MC3810, this command only applies to the AVM. Table 7-8 describes the command options.

Table 7-8   **operation** *Command Options*

Command Option	Description
2-wire	Specifies a 2-wire voice path (default).
4-wire	Specifies a 4-wire voice path.

signal {wink-start | immediate | delay-dial}

With the **signal** {**wink-start** | **immediate** | **delay-dial**} command, signaling is independent of 2-wire versus 4-wire settings. To specify the type of signaling for an E&M voice port, use the **signal** voice-port configuration command shown below. Use the **no** form of this command to restore the default value for this command. Table 7-9 describes the command options.

Table 7-9   **signal** *Command Options*

Command Option	Description
wink-start	Indicates that the calling side seizes the line by going off-hook on its E-lead, then waits for a short off-hook "wink" indication on its M-lead from the called side before sending address information as DTMF digits or dialing pulses. Used for E&M tie trunk interfaces. This is the default setting for E&M voice ports.
immediate	Indicates that the calling side seizes the line by going off-hook on its E-lead and sends address information as DTMF digits (or dialed pulses on the 2600 and 3600 routers).
delay-dial	Indicates that the calling side seizes the line by going off-hook on its E-lead. After a timing interval, the calling side looks at the supervision from the called side. If the supervision is on-hook, the calling side starts sending information as DTMF digits; otherwise, the calling side waits until the called side goes on-hook and then starts sending address information.

Sometimes a PBX will miss initial digits if the E&M voice port is configured for immediate signaling. If digits are missed, you can use delay-dial signaling instead. Some non-Cisco devices have a limited number of DTMF receivers. This type of equipment must delay the calling side until a DTMF receiver is available.

type {1 | 2 | 3 | 5}

For an analog E&M interface you may need to also specify the interface type for a particular voice port with the **type** {**1** | **2** | **3** | **5**} command. Table 7-10 describes the options for this command.

Table 7-10   **type** *Command Options*

Command Option	Description
1	Indicates the following default lead configuration:
	• E—output, relay to ground
	• M—input, referenced to ground
	This is the default.
2	Indicates the following lead configuration:
	• E—output, relay to signal ground (SG)
	• M—input, referenced to ground
	• Signal battery (SB)—feed for M, connected to –48V
	• SG—return for E, galvanically isolated from ground

Table 7-10   **type** *Command Options (Continued)*

Command Option	Description
3	Indicates the following lead configuration: • E—output, relay to ground • M—input, referenced to ground • SB—connected to –48V • SG—connected to ground
5	Indicates the following lead configuration: • E—output, relay to ground • M—input, referenced to –48V

You would select the type based on the following descriptions:

• Type 1—The tie-line equipment generates the E-signal to the PBX type grounding the E-lead. The tie-line equipment detects the M-signal by detecting current flow to ground. If you select (type) 1, a common ground must exist between the line equipment and the PBX. This type is common in the United States.

• Type 2—The interface requires no common ground between the equipment, thereby avoiding ground-loop noise problems. The E-signal is generated toward the PBX by connecting it to SG. M-signal is indicated by the PBX connecting it to SB. Although (type) 2 interfaces do not require a common ground, they do have the tendency to inject noise into the audio paths because they are asymmetrical with respect to the current flow between devices.

• Type 3—The interface operates the same as (type) 1 interfaces with respect to the E-signal. The M-signal, however, is indicated by the PBX connecting it to SB on assertion and alternately connecting it to SG during inactivity. If you select 3, a common ground must be shared between equipment.

• Type 5—Type 5 line equipment indicates an E-signal to the PBX by grounding the E-lead. The PBX indicates an M-signal by grounding the M-lead. A (type) 5 interface is quasi-symmetrical in that while the line is up, current flow is more or less equal between the PBX and the line equipment but noise injection is a problem. This type is common in Europe and Africa.

auto-cut-through

The **auto-cut-through** command enables call completion when a PBX does not provide an M-lead response. This command has no arguments or keywords.

NOTE	The **auto-cut-through** command is new for Release 12.0(7)XK and applies to E&M voice ports only.

Example 7-4 shows how to enable call completion on a Cisco MC3810 or a 2600/3600 when a PBX does not provide an M-lead response.

Example 7-4 *Using* **auto-cut-through**

```
router3810(config)# voice-port 1/1
router3810(config-voiceport)# auto-cut-through
```

codec {g711alaw | g711ulaw | g723ar53 | g723ar63 | g723r53 | g723r63 | g726r16 | g726r24 | g726r32 | g728 | g729br8| g729r8 } [bytes *payload_size*]

The **codec** command specifies the voice coder rate of speech and payload size for a dial peer. For toll quality, use **g711alaw** or **g711ulaw**. These values provide high-quality voice transmission but use a significant amount of bandwidth. For almost toll quality (and a significant savings in bandwidth), use the **g729r8** value.

Table 7-11 describes the **codec** command options for the 2600 and 3600 router.

Table 7-11 codec *Command Options*

Command Option	Description
g711alaw	G.711 a-law at 64000 bits per second (bps).
g711ulaw	G.711 μ-law at 64000 bps.
g723ar53	G.723.1 ANNEX A at 5300 bps.
g723ar63	G.723.1 ANNEX A at 6300 bps.
g723r53	G.723.1 at 5300 bps.
g723r63	G.723.1 at 6300 bps.
g726r16	G.726 at 16000 bps.
g726r24	G.726 at 24000 bps.
g726r32	G.726 at 32000 bps.
g728	G.728 at 16000 bps.
g729br8	G.729 ANNEX B at 8000 bps.
g729r8	G.729 at 8000 bps. This is the default codec. • 30-byte payload for VoFR and VoATM g729r8 • 20-byte payload for VoIP

Table 7-11   **codec** *Command Options (Continued)*

Command Option	Description
bytes	(Optional) Parameter word to specify the number of bytes in the voice payload of each frame.
payload_size	(Optional) The number of bytes in the voice payload of each frame. Acceptable values are from 10–240 in increments of 10 (for example, 10, 20, 30, and so on). Any other value is rounded down (for example, from 236 to 230). Enter **?** after **bytes** to get a list of valid payload values.

For the MC3810 with a VCM DSP card, the g729ar8 compression mode can support a maximum of 24 simultaneously active on-net voice calls, while the g729r8 value can only support a maximum of 12. Both compression modes have a nominal data rate of 8 kbps.

Specify the voice coder rate of speech and payload size for the dial peer. The default dial-peer codec is g729r8. Specifying the payload size by entering the bytes value is optional. Each codec type defaults to a different payload size if you do not specify a value. To obtain a list of the default payload sizes, enter the **codec** command and the bytes option followed by a question mark (?).On the Cisco MC3810, you can also assign codec values to the voice port. If configuring calls to a Cisco MC3810 running software versions prior to 12.0(3)XG, configure the **codec** command on the voice port. For versions later than 12.0(7)XK, you must assign the values on the dial peer.

If configuring Cisco-trunk permanent calls, configure the **codec** command on the dial peer. If you configure the **codec** command on the dial peer for VoFR permanent calls on the Cisco MC3810, the dial-peer codec command setting overrides the codec setting configured on the voice port.

The **codec** command applies to both analog and digital voice ports on the Cisco MC3810.

Each codec has different compression values, and some have built-in voice activity detection (VAD) that aids in conserving bandwidth over time. Table 7-12 gives the type of compression and the compression size for various codecs, including VAD. Table 7-13 shows the support on the 2600/3600 and MC3810 platforms.

Table 7-12   *Payload Compression Codecs*

Codec	Coding	Compression	Built-in VAD
G.711	PCM	64 kpbs	—
G.726	ADPCM	32 kbps	—
G.726	ADPCM	24 kbps	—
G.726	ADPCM	16 kbps	—
G.728	LD-CELP	8 kbps	—

continues

Table 7-12 *Payload Compression Codecs (Continued)*

Codec	Coding	Compression	Built-in VAD
G.729	CS-ACELP	8 kbps	—
G.729A	CS-ACELP	8 kbps	—
G.729B	CS-ACELP	8 kbps	Yes
G.729AB	CS-ACELP	8 kbps	Yes
G.723.1	MP-MLQ	6.3 kbps	—
G.723.1	ACELP	5.3 kbps	—
G.723.1A	MP-MLQ	6.3 kbps	Yes
G.723.1A	ACELP	5.3 kbps	Yes

Table 7-13 *Sample of Codecs and Compression by Platform*

Codec	Compression	Cisco 2600/3600	Cisco MC3810
G.729	8 kbps	Yes	Yes
G.711 (a-law & μ-law)	64 kbps	Yes	Yes
G.726	16, 24, 32 kbps	Yes	Yes
G.728	16 kbps	Yes	Yes, with HCM DSP card
G.723.1	5.3, 6.3 kbps	Yes	Yes, with HCM DSP card

The VAD characteristics are as follows:

- B versions of G.729 contain a built-in IETF VAD algorithm, so there is no need to configure VAD.
- As a rule of thumb, 30 percent to 35 percent reduction in bandwidth over a long period of time occurs—this is a more valid assumption for larger pipes (T1 and above).
- The actual savings depends on applications and traffic (for example, Music-on-Hold makes VAD 0 percent because there is no inactivity).

The variable payload size characteristics are as follows:

- Specifies the number of samples per packet.
- Changes the BW, delay, and pps characteristics of the call.
- Usability depends on the delay budget of the network.
- Values > default: decreases BW and increases delay.
- Values < default: increases BW and decreases delay.

compand-type {u-law | a-law}

The **compand-type** {**u-law** | **a-law**} command allows you to specify the companding (that is, compressing-expanding) standard used to convert between analog and digital signals in PCM systems. The **compand-type** command is only available on the MC3810 digital router. Table 7-14 shows the **compand-type** options.

Table 7-14   **compand-type** *Command Options for the MC3810 Router*

Command Option	Description
u-law	Specifies the North American μ-law ITU-T PCM encoding standard. This is the default for T1 digital.
a-law	Specifies the European a-law ITU-T PCM encoding standard. This is the default for E1 digital.

NOTE   On the Cisco 3600 series, the **u-law** and **a-law** settings are configured using the **codec dial-peer** configuration command.

An Example of Configuring E&M Voice Ports

An organization can see considerable cost savings by using a data network to connect the PBX infrastructure. Figure 7-5 illustrates the connection for the E&M port on the router Cameron. Example 7-5 shows this port has wink-start signaling configured on voice port 1/0/0. The cabling is set to 4-wire and E&M Type I is enabled.

Figure 7-5   *A Connection to an E&M Voice Port*

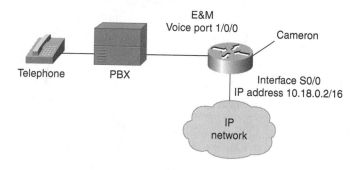

Example 7-5   *Configuration for an E&M Voice Port*

```
Cameron#configure terminal
Cameron(config)#voice-port 1/0/0   !This command enters voice port configuration
  mode
Cameron(config-voiceport)#signal wink-start    !This enables wink start signaling
```

continues

Example 7-5   *Configuration for an E&M Voice Port (Continued)*

```
Cameron(config-voiceport)#operation 4-wire !This command specifies a 4-wire
  wiring scheme
Cameron(config-voiceport)#type 1  !This enables E & M type 1 Wiring(Scheme)
```

For the MC3810 routers some other options are available, such as the **codec** command and the **command-type** command. These commands specify signaling and standards to be used when connecting across dial legs.

Using Other Voice Port Commands

There are a number of other port level commands for the FXS, FXO, and E&M ports. These commands specify how the hardware is suppose to function. These commands are entered at the (config-voiceport)# prompt and are called **voice port** commands. Once these commands are entered, they are loaded to the DSP during a call and establish characteristics of port operation.

You use commands such as **echo cancel enable**, **coverage**, and **non-linear** to set the echo canceling characteristics of the port. The **input** and **output** commands set the decibel bias level applied by the DSP during the call. Table 7-15 lists the **voice port** commands for FXS, FXO, and E&M voice ports.

Table 7-15   *Miscellaneous Voice Port Commands*

Command	Function and Options
busyout*	Configures busyout trigger event and procedure. The following options are available: • *forced*—Force the voice port in busyout • *monitor*—The down event of this interface triggers busyout • *seize*—Option of busyout seize procedure
comfort-noise	Uses the fill-silence option.
default	Sets a command to its defaults.
disconnect-ack	Allows the FXS port to send disconnect acknowledgements.
exit	Exits from voice port configuration mode.
idle-voltage	Sets FXS interface voltage when idle. The following options are available: • *high*—High voltage when idle • *low*—Normal lower voltage when idle
music-threshold	Sets the decibel threshold for music on hold. The option is: • *WORD*—Enter a number between –70 and –30

Table 7-15   *Miscellaneous Voice Port Commands (Continued)*

Command	Function and Options
playout-delay	Configures voice playout delay buffer. The options are: • *maximum*—Maximum playout buffer delay, in milliseconds • *nominal*—Nominal playout buffer delay, in milliseconds
snmp	Modifies SNMP voice port parameters. The option is *trap*, which allows a specific SNMP trap.

* The **busyout** command syntax differs on releases between 12.0.5T and 12.1.3T. Please check the appropriate documentation for your release.

connection Commands and Options

Besides the commands that deal with port parameters, other voice commands, like the **connection** commands, specify how a port is supposed to operate. The following sections describe these port commands.

connection {plar | tie-line | plar-opx} *digits* | {trunk *digits* [answer-mode]}

The **connection** {**plar** | **tie-line** | **plar-opx**} *digits* | {**trunk** *digits* [**answer-mode**]} command allows you to specify a connection mode or type for a specified voice port and the destination telephone number. Table 7-16 lists the purpose of each option.

Table 7-16   **connection** *Command Options*

Command Option	Description
plar	Provides a private-line automatic ring-down connection, which allows an extension to be automatically dialed once the handset is lifted.
tie-line	Establishes a tie-line connection to a PBX.
plar-opx	Allows for PLAR off-premises extension, almost the same as **plar**, but gives the flexibility of the call being answered before connecting the call.
trunk	Emulates a permanent trunk connection to a PBX. A trunk connection remains "nailed up" in the absence of any active calls.
answer-mode	(Optional; used only with the **trunk** keyword.) Specifies that the router should not attempt to initiate a trunk connection, but should wait for an incoming call before establishing the trunk.
digits	Specifies the destination telephone number. Valid entries are any series of digits that represent telephone numbers.

voice confirmation-tone {plar | plar-opx}

The **voice confirmation-tone** command is used when the **plar** or **plar-opx** option is set as the connection type. This command specifies the two-beep confirmation tone that a caller hears when they pick up the phone. Table 7-17 shows the options for which the voice confirmation tone is valid.

Table 7-17 *Confirmation Tones Generated with Connection PLAR Options*

Command Option	Description of Tone
plar	Enables the two-beep confirmation tone that a caller hears when picking up the handset or PLAR and PLAR OPX connections.
plar-opx	Enables the two-beep confirmation tone that a caller hears when picking up the handset.

Examples of Specifying Voice Port Connection Types

Connection type options vary depending on the voice port that is being configured. For example, the tie-line option is used only for E&M ports, whereas PLAR is typically used for FXS ports. It is important to note that the voice port features discussed in this section were unified across the Cisco MC3810, 2600, and 3600 platforms in IOS Release 12.0(7)XK. These commands are used to automatically place a call to another voice port, across a WAN connection.

Private Line Automatic Ring-Down

The **connection plar** command specifies a private line automatic ring-down (PLAR) connection. PLAR is an autodialing mechanism that permanently associates a voice interface with a far-end voice interface, allowing call completion to a specific telephone number or PBX without dialing. These steps show how PLAR works:

1 The calling telephone goes off-hook.

2 The predefined network dial peer is automatically matched, which sets up a call to the destination telephone or PBX.

3 The destination station rings.

The string you configure for this command is used as the called number for all incoming calls over this connection. The destination peer is determined by the called number. This means that anything the user dials on this handset is ignored.

PLAR makes no provision for far-end answer supervision. PLAR connections can operate as either a switched or permanent call, both of which are illustrated in Figure 7-6 and are described by the steps in the following sections.

Figure 7-6   *PLAR Example*

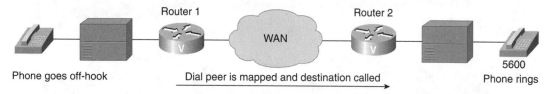

Switched Calls   For switched calls, do the following:

1   The PBX user dials the PBX access code, and the PBX seizes a trunk (voice port 1/1 on the locally attached Router 1).

2   Router 1 recognizes the seized trunk at voice port 1/1. The connection type **plar**, using the specified digits (5600), immediately calls the far-end router, Router 2, which routes the call.

3   The PBX thinks the call is complete and starts billing. The PBX does this because Router 1 has returned an off-hook indication to the PBX. Consequently, at the PBX, **ring timeout** may not operate; no far-end answer supervision will be passed from Router 1 back to the PBX since the call is complete as soon as voice port 1/1 is opened by Router 1.

The voice port configurations for Routers 1 and 2 in Figure 7-6 are shown in Example 7-6.

Example 7-6   *An Example of a PLAR Configuration Between Two Routers*

```
Router1(config)#voice port 1/1
Router1(config-voiceport) Connection plar 5600
Router1(config-voiceport) dial-peer voice 1 vofr
Router1(config-dial-peer)dest-pattern  5...
Router1(config-dial-peer)session target serial 2 123

Router2(config)#dial-peer voice 1 pots
Router2(config-dial-peer)#destination pattern 5
Router2(config-dial-peer)#forward digits 4
```

Permanent Calls   For permanent calls, do the following:

1   On power up, the **ext-sig-master plar** voice port on Router 1 places a call to the far-end router, Router 2 (using the digits 5600).

2   Once the router-to-router link is established, an end user at either PBX dials the PBX access number and the digits are passed over the permanent talk path to the far-end PBX for call completion. (**dtmf-relay** should always be enabled for the low bit rate codecs.)

PLAR Off-Premises Extension

Figure 7-7 shows an off-premises extension for PLAR.

Figure 7-7   *PLAR OPX Example*

The **connection plar opx** command configures a switched, call-by-call connection that provides far-end answer supervision to the local PBX, PBX 1, from the originating router, Router 1—only after a call setup complete signal (that is, the call is connected [ringing]), is received from the far-end router, Router 2 connected to station 5600.

The voice port configurations for Routers 1 and 2 in Figure 7-7 are shown in Example 7-7.

Example 7-7   *OPX PLAR configurations*

```
Router1(config)#voice port 1/1
Router1(config-voiceport) Connection plar opx 5600
Router1(config-voiceport) dial-peer voice 1 vofr
Router1(config-dial-peer)dest-pattern 5….
Router1(config-dial-peer)session target serial 0 456

Router2(config)#dial-peer voice 1 potxs
Router2(config-dial-peer)#destination pattern 56..
Router2(config-dial-peer)#port 1/1
```

NOTE   Station 5600 could also be a PBX connected to the far-end router, Router 2.

To make a plar opx connection, do the following:

1   PBX user dials 5600. The PBX seizes the trunk (voice port 1/1 on Router 1) and waits for return seizure/off-hook signals from Router 1. Digits may not be forwarded by the PBX.

2   Router 1 does not respond with a return seizure/off-hook immediately, but the PLAR voice ports place the call, sending a call setup message across the WAN to Router 2 with voice port 1/1, destination pattern 5600, for that station.

3 Router 1 waits for the far-end router, Router 2, call setup complete message before returning a "trunk seized" indication and opening the audio path to the originating PBX, PBX1. At the same time, the far-end router, Router 2, begins ringing station 5600 and opens the audio path to Router 1, which provides ringing to the originating PBX 1. The PBX 1 ring-out timer begins to wait for a far-end answer.

4 When station 5600 answers, the originating PBX 1 starts call accounting.

NOTE   The **connection plar opx** operation is designed for PBX to off-premises extension (station/FXO device) connections.

Tie Lines

The **connection tie-line** command emulates a temporary tie-line trunk to a PBX. A tie-line connection is automatically set up for each call and torn down when the call ends. Figure 7-8 illustrates a typical tie-line configuration.

Figure 7-8   *Tie-line Usage*

Use the **connection tie-line** command when the dial plan requires that additional digits be added in front of any digits dialed by the PBX, and that the combined set of digits be used to route the call onto the network.

The voice port configurations for Routers 1 and 2 in Figure 7-8 are shown in Example 7-8.

Example 7-8   *Tie-line Configuration*

```
Router1(config)#voice port 1/1
Router1(config-voiceport) connection tie-line 8

Router1(config-voiceport) dial-peer voice 1 vofr
Router1(config-dial-peer)destination-pattern 8...
Router1(config-dial-peer)session target serial 1 789

Router2(config)#dial-peer voice 1 pots
Router2(config-dial-peer)#destination pattern 8
Router2(config-dial-peer)#forward digits 4
Router2(config-dial-peer)#Port1/1
```

The following operations are provided by a tie-line connection:

- Digit store-and-forward rather than auto cut-through operation.
- The source device provides local dial tone and local call setup spoofing.

To make a tie-line connection, do the following:

1 PBX user goes off-hook and dials the PBX tie-line access code and digits. Then PBX 1 opens a trunk to the router and passes the digits.

2 Router 1 responds to the PBX 1 seizure of voice port 1/1 and accepts the digits. The connection type tie-line 8 argument starts the call setup process, and digits equal to the gather digits delimiter (the . . . string after the 8 in the VoFR dial peer) are then gathered by Router 1.

NOTE   The gather digits delimiter (. . .) is discussed in greater detail in the section "Configuring Dial Peers" later in this chapter.

---

3 Router 1 then proceeds to route the call to the far-end router, Router 2. Upon an ACK from the far-end router, Router 2, Router 1 cuts through the end-to-end audio path, and the far-end PBX, PBX 2, begins to ring back to the PBX user.

4 As station 5600 answers, the calling PBX, PBX 1, begins billing. Ring timeout parameters at the source PBX, PBX 1, should operate in a normal manner.

connection trunk [*digits*] [**answer mode**]

The **connection trunk** [*digits*] [**answer mode**] command establishes a two-way, permanent, "nailed up" tie-line connection to a PBX. You can use the connection trunk command for E&M-to-E&M trunks, FXO-to-FXS trunks, and FXS-to-FXS trunks. Signaling will be transported for E&M-to-E&M trunks and FXO-to-FXS trunks; signaling will not be transported for FXS-to-FXS trunks. Figure 7-9 shows a trunk example.

Figure 7-9   *A Trunk Connection*

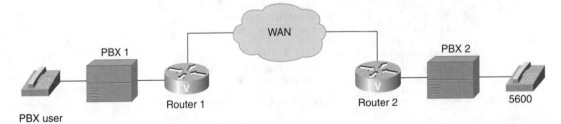

To configure one of the devices in the trunk connection to act as slave and only receive calls, use the **connection trunk** [*digits*] [**answer mode**] option with the **connection trunk** command when configuring that device. The "master" is the default (call originator), and holds the connection open in both directions.

The voice port configurations for Routers 1 and 2 are shown in Example 7-9.

Example 7-9   *PLAR Configurations for Figure 7-9*

```
Router1(config)#voice port 1/1
Router1(config-voiceport) Connection trunk A11
Router1(config-voiceport) dial-peer voice 1 vofr
Router1(config-dial-peer)dest-pattern A….
Router1(config-dial-peer)session target serial 0:4.1 234

Router2(config)#dial-peer voice 1 pots
Router2(config-dial-peer)#destination pattern A11
Router2(config-dial-peer)#connection trunk B4 answer
```

Some general guidelines for trunk connections are:

- A signal type must be specified.
- The DS0 that the master is on must be paired with the answer DS0.
- The **connection type trunk** command may be used for any voice-over packet connection.

NOTE   As with any voice port, extensions must be configured using the **dial-peer voice** [*tag*] **pots** command stream.

Adjusting Voice Quality

An important aspect of voice-over technology is quality. Unlike data transmission, voice transmissions require high-quality transmissions. Voice quality is adjusted by manipulating the transmit and receive power levels on a particular voice port. Figure 7-10 shows where quality tuning is relevant.

Voice quality adjustment commands operate between a voice port and its codec. There are no voice-quality-adjustment commands that modify the encapsulated voice samples routed between DSPs across a WAN.

Figure 7-10 *Voice-Quality Tuning Points*

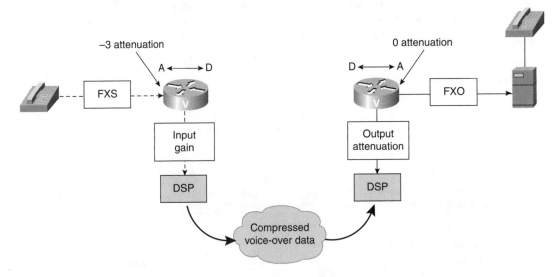

Depending on the specifics of your particular network, you may need to adjust voice parameters involving timing, input gain, and output attenuation for voice ports. Collectively, these commands are referred to as *voice-port tuning commands.* Note that in most cases, the default values for voice-port tuning commands are sufficient.

To configure voice-port tuning commands, perform the following tasks:

- Identify the voice port and enter the voice-port configuration mode.
- For each of the following parameters, select the appropriate value:
    - Input gain
    - Output attenuation
    - Echo cancel coverage
    - Nonlinear processing
    - Initial digit timeouts
    - Timeouts interdigit
    - Timing (other than timeouts)
    - Impedance, ring number (FXO voice ports only)
    - Ring frequency, ring cadence (FXS voice port only)

Calculating Decibel Levels

Calculating network decibel levels is often an exercise in simple number-line arithmetic. Decibel levels are based on what the customer says their transmit and receive levels are, or need to be. See the examples in Table 7-18.

Table 7-18   *Decibel Adjustments*

Source 1 out/in	Router 1 Adjustment	Net Gain at Router 1	WAN N/A	Net Gain at Router 2	Router 2 Adjustment	Destination 1 in/out
0 dB (out)	−3 dB (out)	−3 dB	N/A	−3dB	(+) − 6 dB (in)_	(in) −9 dB
−9 dB (in)	(in)(+) −6 dB	−3 dB	N/A	−3 dB	−3 dB (out)	(out) 0 dB

Tips for Baselining Input and Output Power Levels

The following guidelines can be used for getting a baseline for the input and output power levels:

- Analog voice routers operate best with receive levels from an analog source set in the area of −3 dB.

- In the United States and most of Europe, the receive (transmit) level normally expected by an analog telephone is in the area of −9 dB. Asian and South American countries expect receive levels closer to −14 dB. To accommodate these differences, the output levels to the router can be set over a wide range.

- Overdriving the circuit can cause analog clipping. This happens when the power level is above available PCM codes and a continuous repetition of the last PCM value is passed to the DSP.

- Echo occurs when impedance mismatches cause power to be reflected back to the source.

- If digits are being played out to the far end device before it is ready to accept them a , (comma) may be prefixed to the receiving voice ports **dial-peer voice** [*tag*] **pots** record and the **playout-delay** command (which configures voice playout delay buffer) may be issued on the receiving voice port to modify the how long voice is held in the de-jitter buffer.

Voice Port Voice Quality Tuning Options

The commands described in the following sections deal with quality tuning options for the voice ports. Typically the default settings work well, but when they do not and there is a quality issue, these settings can sometimes help eliminate problems.

input gain *value*

The **input gain** *value* command allows you to set the input gain value (in decibels) to be inserted at the receiver side of the interface.

Integer values range from –6 to 14, and the default value is 0. This default value assumes that a standard transmission loss plan is in effect, meaning, there must be –6 dB attenuation between phones. Other equipment (including PBXs) in the system must be taken into account when creating a loss plan.

You can't increase the gain of a signal going out into the PSTN, but you can decrease it. If the voice level is too high, you can decrease the volume by either decreasing the input gain or by increasing the output attenuation. If the voice level is too low, you can increase the input gain of a signal coming into the router.

output attenuation *value*

The **output attenuation** *value* command allows you to configure the output attenuation value in decibels for the transmit side of the interface. The value represents the amount of loss to be inserted at the transmit side of the interface. Integer values range from 0 to 14, and the default is 0.

NOTE	**input gain** and **output attenuation** commands are used to accommodate network equipment and not used as end-user volume controls for user comfort.

loss-plan {plan1 | plan2 | plan3 | plan4 | plan5 | plan6 | plan7 | plan8 | plan9}

The **loss-plan** command (analog only) sets the analog-to-digital gain offset for an analog FXO or FXS voice port on the analog interface. This command sets the analog signal level difference (offset) between the analog voice port and the digital signal processor (DSP). Each loss plan specifies a level offset in both directions—from the analog voice port to the DSP (A–D) and from the DSP to the analog voice port (D–A).

The defaults for the FXO: A–D gain = 0 dB, D–A gain = 0 dB (loss plan 1). The defaults for the FXS: A–D gain = –3 dB, D–A gain = –3 dB (loss plan 1).

The specified level offsets for each loss plan are listed in Table 7-19.

Table 7-19   *Loss Plans*

Plan Option	A–D Gain	D–A Gain
plan1	FXO: A–D gain = 0 db	D–A gain = 0 dB
	FXS: A–D gain = –3 db	D–A gain = –3 dB

Table 7-19   *Loss Plans (Continued)*

Plan Option	A–D Gain	D–A Gain
plan2	FXO: A–D gain = 3 db	D–A gain = 0 dB
	FXS: A–D gain = 0 db	D–A gain = –3 dB
plan3	FXO: A–D gain = –3 db	D–A gain = 0 dB
	FXS: N/A	
plan4	FXO: A–D gain = –3 db	D–A gain = –3 dB
	FXS: N/A	
plan5	FXO: N/A	
	FXS: A–D gain = –3 db	D-A gain = –10 dB
plan6	FXO: N/A	
	FXS: A–D gain = 0 db	D–A gain = –7 dB
plan7	FXO: A–D gain = 7 db	D–A gain = 0 dB
	FXS: A–D gain = 0 db	D–A gain = –6 dB
plan8	FXO: A–D gain = 5 db	D–A gain = –2 dB
	FXS: N/A	
plan9	FXO: A–D gain = 6 db	D–A gain = 0 dB
	FXS: N/A	

NOTE   The **loss-plan** command is supported only on Cisco MC3810 series concentrators and on FXO and FXS analog voice ports.

Example 7-10 shows how to configure an FXO voice port 1/6 for a –3 dB offset from the voice port to the DSP and a 0 dB offset from the DSP to the voice port.

Example 7-10   *Adjusting Quality on an FXO Port*

```
router(config)# voice-port 1/6
router(config-voiceport)# loss-plan plan3
```

Example 7-11 configures FXS voice port 1/1 for a 0 dB offset from the voice port to the DSP and a –7 dB offset from the DSP to the voice port.

Example 7-11   *Adjusting Quality on an FXS Port*

```
router(config)# voice-port 1/1
router(config-voiceport)# loss-plan plan6
```

impedance {600r I 600c I 900r I 900c I complex1 I complex2}

The **impedance** command allows you to specify the terminating impedance of an FXO voice port interface. If the impedance is set incorrectly and there is an impedance mismatch, a significant amount of echo will be generated that may be masked if you have echo cancellation enabled. The **impedance** command syntax is described in Table 7-20.

Table 7-20   *Impedance Levels*

Command Option	Description
600r	Specifies 600 ohm resistive (real) (default).
600c	Specifies 600 ohm complex (600 ohms in series with 2.15 µf).
900c	Specifies 900 ohm complex (900 ohms in series with 2.15 µf).
900r	Specifies 900 ohm resistive (real); (available only on the MC3810).
complex1	Specifies complex1; (available only on the 2600 and 3600 routers); 220 ohms in series with the parallel combination of 820 ohms and 115 nf.
complex2	Specifies complex2; (available only on the 2600 and 3600 routers); 270 ohms in series with the parallel combination of 750 ohms and 150 nf.

You use the **impedance** command to specify the terminating impedance of an FXO voice-port interface. The impedance value selected needs to match the specifications from the specific telephony system to which it is connected. Different countries often have different standards for impedance. CO switches in the United States are predominantly 600r. PBXs in the United States are normally either 600r or 900c.

If impedance is set incorrectly (that is, an impedance mismatch), there will be a significant amount of echo generated (which could be masked if the **echo-cancel** command has been enabled). In addition, gains might not work correctly if there is an impedance mismatch.

NOTE   The **impedance** command is applicable to FXS, FXO, and E&M voice ports on both the Cisco 3600 series and the Cisco MC3810 (FXO analog only). This section is not applicable on the AS5300 access server.

echo-cancel enable

The **echo-cancel enable** command enables cancellation of voice that is sent out of the interface and is received back on the same interface; sound that is received back in this manner is perceived by the listener as an echo.

Disabling echo cancellation might cause the remote side of a connection to hear an echo. Because echo cancellation is an invasive process that can minimally degrade voice quality, this command should be disabled if it is not needed.

The **echo-cancel enable** command does not affect the echo heard by the user on the analog side of the connection.

There is no echo path for a 4-wire E&M interface. The echo canceller should be disabled for that interface type.

NOTE   The **echo-cancel enable** command is valid only if the **echo-cancel coverage** command has been configured.

---

echo-cancel coverage {8 | 16 | 24 | 32}

The **echo-cancel coverage** {**8** | **16** | **24** | **32**} command allows you to adjust the size of the echo canceller. This command enables cancellation of voice that is sent out of the interface and received back on the same interface within the configured amount of time. If the local loop is longer (that is, the distance from the analog interface to the connected equipment producing the echo), the configured value should be extended.

The *value* is the number of milliseconds the echo canceller will cover on a given signal. Valid values are 16, 24, and 32, and the default is 8.

If you configure a longer value for this command, the echo canceller will take longer to converge; in this case, the user might hear a slight echo when the connection is initially set up. If the configured value for this command is too short, the user might hear some echo for the duration of the call because the echo canceller is not canceling the longer delay echoes. Also, the larger the value, the more memory that is used by the router.

There is no echo or echo cancellation on the network (for example, nonPOTS) side of the connection.

NOTE   The **echo-cancel coverage** command is valid only if the **echo cancel** feature has been enabled.

---

non-linear

The **non-linear** voice port configuration command is associated with the echo canceller operation. The **echo-cancel enable** command must be enabled for the **non-linear** command to take effect. Use the **non-linear** command to shut off any signal if no near-end speech is detected.

The function enabled by the **non-linear** command is also generally known as *residual echo suppression.*

Enabling the **non-linear** command normally improves performance, although some users might perceive truncation of consonants at the end of sentences when this command is enabled.

NOTE   Do not use the quality commands or adjust voice quality unless you are experienced in doing so. Arbitrarily adjusting these parameters could make the voice quality worse.

In an untuned network, such as the one shown in Figure 7-10, a port configuration that delivers perceived "good quality" for a call between A and B may sound terrible when A calls C or D.

Voice quality adjustment needs to be a defined, step-by-step procedure that is implemented after the network is up and running. It would be a waste of time to start changing default voice-port configurations until full cross-network calls can be established. This is because a correctly implemented procedure will result in a "quality" compromise between various sources and sinks which the customer accepts as "good overall quality."

loss-plan

The **loss-plan** command is used to correct input levels, from any source in the network, to the same nominal input level.

NOTE   Correcting input levels results in uniform signal strength before the data is passed to the DSP for compression and transmission over the WAN.

The **loss-plan** command works only on analog voice ports. Example 7-12 gives the default choices for **loss-plan**.

Example 7-12   loss-plan *Command*

```
Router(config)#voice-port 1/1
  Type of VoicePort is FXS
Router(config-voiceport)#loss-plan ?
  plan1  A-D gain = -3db, D-A gain -3dB
  plan2  A-D gain = 0db, D-A gain -3dB
```

Example 7-12   **loss-plan** *Command (Continued)*

```
plan5  A-D gain = -3db, D-A gain -10dB
plan6  A-D gain = 0db, D-A gain -7dB
plan7  A-D gain = 0db, D-A gain -6dB
```

impedance

The **impedance** command allows you to set FXO loop impedance parameters to match that of the source FXS device. Example 7-13 shows the default choices for the **impedance** command:

Example 7-13   *Using the* **impedance** *Command*

```
Router(config)#voice-port 1/4
 Type of VoicePort is FXO
Router(config-voiceport)#impedance ?
 600c    600 Ohms complex
 600r    600 Ohms real
 900c    900 Ohms complex
 900r    900 ohms real
 complex1  complex 1
 complex2  complex 2
```

The **input gain** and **output attenuation** commands work in the DSP and bias PCM codes, coming from the source AVM codec, up or down. The samples are biased to a level which, after passing through the network, are presented at the adjusted power level needed to drive the correct output level to the called device.

The **echo-cancel enable**, **echo-cancel coverage** (8 to 32 ms), and **non-linear** (echo cancel type of operation) commands also work in the DSP and are configured on a port-by-port basis. These commands are used to set the echo cancellation functions to match those of the device to which the port is connected.

NOTE   If digits are being played out to the far-end device before it is ready to accept them a , (comma) may be prefixed to the receiving voice port's **dial-peer voice** [*tag*] **pots** record and the **playout-delay** command (which configures voice playout delay buffer) may be issued on the receiving voice port to modify the how long voice is held in the de-jitter buffer.

timeouts initial *seconds*

The **timeouts initial** *seconds* command specifies the number of seconds the system will wait for the caller to input the first digit of dialed digits. Valid entries are integers from 0 to 120, and the default is 10.

The timeouts initial timer is activated when the call is accepted and is deactivated when the caller inputs the first digit. If the configured timeout value is exceeded, the caller is notified through the appropriate tone and the call is terminated.

timeouts interdigit *seconds*

The **timeouts interdigit** *seconds* command specifies the number of seconds the system will wait (after the caller has input the initial digit) for the caller to input a subsequent digit of the dialed digits. The timeouts interdigit timer is activated when the caller inputs a digit and restarted each time the caller inputs another digit until the destination address is identified.

If the configured timeout value is exceeded before the destination address is identified, the caller is notified through the appropriate tone and the call is terminated. Valid entries are integers from 0 to 120 and the default value is 10.

There are additional **timeouts** *value* commands. Table 7-21 shows how these commands are used.

Table 7-21   **timeouts** *Value Commands*

Command Option	Description
ringing	Configures the timeout value for ringing. The range is 5–60000. The default is 180.
wait-release	Configures the delay timeout before the system starts the process for releasing voice ports. The range is 3–3600. The default is 30.

NOTE   To disable the **timeouts initial** or the **timeouts interdigit** timer, set the seconds value to 0.

timing *value*

The **timing** *value* command allows you to set certain timing parameters on your analog voice port. Table 7-22 is a partial list of all the possible timing variable commands.

Table 7-22   *Timing Variables*

Command Option	Description
clear-wait *milliseconds*	Indicates the minimum amount of time, in milliseconds, between the inactive seizure signal and the call being cleared.
delay duration *milliseconds*	Indicates the delay signal duration for delay dial signaling, in milliseconds.

Table 7-22 *Timing Variables (Continued)*

Command Option	Description
delay-start *milliseconds*	Indicates the minimum delay time, in milliseconds, from outgoing seizure to out-dial address.
delay-pulse min-delay *milliseconds*	Indicates the time, in milliseconds, between the generation of wink-like pulses.
digit *milliseconds*	Indicates the DTMF digit signal duration, in milliseconds.
interdigit *milliseconds*	Indicates the DTMF interdigit duration, in milliseconds.
pulse *pulses per seconds*	Indicates the pulse dialing rate, in pulses per second.
pulse-inter-digit *milliseconds*	Indicates the pulse dialing interdigit timing, in milliseconds.
wink-duration *milliseconds*	Indicates the maximum wink-signal duration, in milliseconds, for a wink-start signal.
wink-wait *milliseconds*	Indicates the maximum wink-wait duration, in milliseconds, for a wink-start signal.

NOTE Some of the **timing** command options are not available on all routers and all interfaces (that is, FXO, FXS, and E&M). The valid entries and defaults may also vary. Refer to the *Command Reference Guide* for features that apply to your router and applications.

voice vad-time *milliseconds*

The **voice vad-time** *milliseconds* command started with release 12.07XK. It is not a **voice-port** command, but affects all voice ports on a router or concentrator.

You can use this command in transparent CCS applications where you want VAD to activate when the voice channel is idle, but not during active calls. That is, it does not affect calls already in progress.

With a longer silence detection delay, VAD reacts to the silence of an idle voice channel, but not to pauses in conversation.

This command does not affect voice codecs that have ITU-standardized built-in VAD features; for example, G.729B, G.729AB, and G.723.1A. The VAD behavior and parameters of these codecs are defined exclusively by the applicable ITU standard. Table 7-23 shows the command options for the **voice vad-time** command.

Table 7-23   **voice vad-time** *Command Option*

Command Option	Description
milliseconds	Indicates the waiting period in milliseconds before silence detection and suppression of voice-packet transmission. The range is 250–65536. The default is 250.

voice local-bypass

The **voice local-bypass** command is a new command avaliable on the MC3810 starting with IOS Release 12.0(7)XK. Figure 7-11 shows how this command allows you to pass uncompressed voice traffic for local POTS calls, bypassing the DSP. Local calls (that is, calls between voice ports on a router or concentrator) normally bypass the DSP to minimize use of system resources.

Figure 7-11   *Local Bypass*

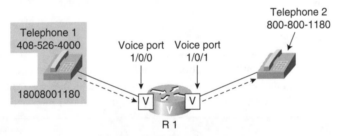

Call connectivity between dial peer
1/0/0 and dial peer 1/0/1 on the same router

The **voice local-bypass** command applies to VoFR, VoATM, and Frame Relay–ATM interworking on the Cisco voice-capable routers. There are no optional parameters with this command. Use the **no** form of this command if you need to direct local calls through the DSP.

Example 7-14 shows how to configure a Cisco voice-capable router to pass local calls through the DSP.

Example 7-14   *Local Bypass Configuration*

```
router(config)# no voice local-bypass
```

Other Quality Commands

The commands listed in this section help to improve quality and performance of the voice-over routers. Table 7-24 lists some other quality-related commands.

Table 7-24   *Quality-Related Commands*

Command	Description
input gain	Configures the receive gain value for a voice port.
output attenuation	Configures the transmit attenuation value for a voice port

NOTE   The **input gain** and **output attenuation** commands can be configured only if calls are directed through the DSP.

Configuring Controller Settings

This section describes the commands you use to configure controllers. Controllers consist of T1 or E1 links that allow for multiple voice connections.

Controller Commands

The commands shown in this section configure the MFT for T1/E1 controller settings. These commands are accessed in controller configuration mode.

At the router> prompt, enter **enable** to enable privileged EXEC command mode. Next, at the router prompt, enter **conf t** to enter global configuration mode. The controller, clock source, loop time, and description commands will then allow you to set parameters for the T1 or E1 controller.

controller {t1 | e1}

You use the **controller {t1 | e1}** *number* command to enter controller configuration mode. Specify whether your controller is E1 or T1, and enter the controller number. If the DVM is installed, the controller number can be either 0 or 1. If the MFT is installed, the number must be 0.

clock source {internal | line | loop-timed}

You use the **clock source {internal | line | loop-timed}** command to configure the controller clock source for a DS1 link. The default for the DVM is *internal*, while the default for the MFT is *line*.

loop-timed

The **loop-timed** parameter specifies that the T1/E1 controller will take the clock from the Rx (line) and use it for Tx. This setting decouples the controller clock from the system-wide clock set with the *network-clock-select* parameter. The loop-timed clock enables the DVM to connect to a PBX and to connect the MFT to a central office (CO) when both the PBX and the CO function as DCE clock sources. This situation assumes that the PBX also takes the clocking from the CO, thereby synchronizing the clocks on the DVM and the MFT.

description *line*

You use the **description** *line* command to enter a description of the controller, such as the destination or its application, for the *line* value. The description for the controller interface can be up to 80 characters long.

Configuring T1 Controller Settings

The commands described in the following sections show controller settings for a T1 ATM. You access these commands in controller configuration mode.

cablelength short {133 | 266 | 399 | 533 | 655}

The **cablelength short** {**133** | **266** | **399** | **533** | **655**} command configures the cable length for the T1 controller if the cable length is 655 feet or shorter.

cablelength long {gain26 | gain36} {-15db | -22.5db | -7.5db | 0db}

The **cablelength long** {**gain26** | **gain36**} {**-15db** | **-22.5db** | **-7.5db** | **0db**} command configures the cable length if the length is longer than 655 feet.

framing {sf | esf}

The **framing** {**sf** | **esf**} command configures the DS1 link framing format.

linecode {ami | b8zs}

The **linecode** {**ami** | **b8zs**} command configures the line encoding format for the DS1 link.

NOTE	Extended Superframe (**esf**) format is required for ATM traffic, and the **b8zs** setting is required for ATM (and Frame Relay) traffic.

After configuring these parameters for the T1 controller, enter the following commands to activate the settings and leave configuration mode and return to the privileged prompt and verify the settings:

- Use the **no shutdown** command to activate the T1 controller.
- Use the **end** command to exit configuration mode.
- Use the **show controller T1 0** command if you want to verify your T1 controller configuration settings. (Remember that controller 0 is always serial port 2.)

Configuring E1 Controller Settings

To configure E1 controller settings, use the commands described in the following sections in controller configuration mode.

framing {crc4 | no-crc4} [name]

You use the **framing** {**crc4** | **no-crc4**} [*name*] command to configure the E1 framing format. **crc4** sets framing for CRC type 4. *name* is optional for E1 only. If the trunk will be connected to a device in say, Ireland, enter "Ireland" as the name option.

linecode {ami | hdb3}

You use the **linecode** {**ami** | **hdb3**} command to configure the line encoding format. **ami** specifies alternate mark inversion (AMI) encoding. For E1 only, **hdb3** specifies HDB3 encoding. The **hdb3** setting is required for ATM and Frame Relay traffic.

After configuring these parameters for the E1 controller, enter the following commands to activate the settings and leave configuration mode and return to the privileged prompt and verify the settings:

- Use the **no shutdown** command to activate the T1 controller.
- Use the **end** command to exit configuration mode.
- Use the **show controller E1 0** command if you want to verify your E1 controller configuration settings. (Remember that controller 0 is always serial port 2.)

ccs connect {serial | atm} number [dlci dlci | pvc vci | pvc vcd | pvc vpi/vci | pvc string]

The **ccs connect** {**serial** | **atm**} *number* [**dlci** *dlci* | **pvc** *vci* | **pvc** *vcd* | **pvc** *vpi/vci* | **pvc** *string*] command configures a common channel signaling (CCS) connection on an interface configured to support CCS frame forwarding. Table 7-25 describes the parameters of the **ccs connect** command.

Table 7-25   **ccs connect** *Command Options*

Command Option	Description
serial	Makes a serial CCS connection.
atm	Makes an ATM CCS connection.
number	Specifies the connection number. Choose one of the following options: • **dlci** *dlci*—Specifies the DLCI number. • **pvc** *vci*—Specifies the PVC virtual circuit identifier. • **pvc** *vcd*—Specifies the PVC virtual circuit descriptor. • **pvc** *vpi/vci*—Specifies the PVC virtual path identifier/virtual channel identifier. • **pvc** *string*—Specifies the PVC string.

mode ccs {cross-connect | frame-forwarding}

The **mode ccs {cross-connect | frame-forwarding}** command configures the T1/E1 controller to support CCS cross-connect or CCS frame-forwarding. Table 7-26 describes the parameters of the **mode css** command.

Table 7-26   **mode ccs** *Command Options*

Command Option	Description
cross-connect	Enables CCS cross-connect on the controller.
frame-forwarding	Enables CCS frame-forwarding on the controller

Example 7-15 shows how to enable CCS cross-connect on controller T1 1.

Example 7-15   *Enabling cross-connect*

```
Router(config)#controller T1 1
Router(config-controller)#mode ccs cross-connect
```

ds0-group *ds0-group-no* timeslots *timeslot-list* type *signal-type*

The **ds0-group** *ds0-group-no* **timeslots** *timeslot-list* **type** *signal-type* command specifies the DS0 timeslots that make up a logical voice port on a T1 or E1 controller, and specifies the signaling type.

The **ds0-group** command automatically creates a *logical voice port.* Each DS0 group may contain one or more DS0 channels. Calls are routed to individual DS0 channels in the DS0 group. DS0 groups are specified as follows:

- On Cisco 2600 and 3600 series routers—*slot/port:ds0-group-no*

- On the Cisco MC3810—*slot:ds0-group-no*

On the Cisco MC3810, *slot* is the controller number. Although only one voice port is created for each group, applicable calls are routed to any channel in the group.

Example 7-16 shows how to configure ranges of T1 controller timeslots for FXS ground-start and E&M wink signaling on a Cisco 2600 or 3600 series router.

Example 7-16   *Signaling Example*

```
router(config)# controller T1 1/0
router(config-controller)# framing esf
router(config-controller)# linecode b8zs
router(config-controller)# ds0-group 1 timeslot 1-5 type fxs-ground-start
router(config-controller)# ds0-group 2 timeslot 6-10 type fxo-loop-start
router(config-controller)# ds0-group 2 timeslot 12-24 type e&m-wink-start
```

Example 7-17 shows how to configure DS0 groups 1 and 2 on controller T1 1 on the Cisco MC3810 to support Transparent CCS.

Example 7-17   *CSS Configuration*

```
router(config)# controller T1 1
router(config-controller)# mode ccs cross-connect
router(config-controller)# ds0-group 1 timeslot 1-10 type ext-sig master
router(config-controller)# ds0-group 2 timeslot 11-24 type ext-sig
```

Table 7-27 describes the **ds0-group** command.

Table 7-27   **ds-0 group** *Command Options*

Command Option	Description
ds0-group-no	Specifies a value from 0 to 23 (T1) or 0 to 30 (E1) that identifies the DS0 group.
timeslot-list	timeslot-list is a single timeslot number, a single range of numbers, or multiple ranges of numbers separated by commas. For T1, allowable values are from 1 to 24. Examples are: 2 1–15, 17–24 1–23 2, 4, 6–12

continues

Table 7-27   **ds-0 group** *Command Options (Continued)*

Command Option	Description
type	The signaling method selection for type depends on the connection that you are making. The E&M interface allows connection for PBX trunk lines (tie lines) and telephone equipment. The FXS interface allows connection of basic telephone equipment and PBXs. The FXO interface is for connecting the central office (CO) to a standard PBX interface where permitted by local regulations. The FXO interface is often used for off-premises extensions.
	The options are as follows:
	• **e&m-immediate-start**—no specific off-hook and on-hook signaling
	• **e&m-delay-dial**—the originating endpoint sends an off-hook signal and then waits for an off-hook signal followed by an on-hook signal from the destination
	• **e&m-wink-start**—the originating endpoint sends an off-hook signal and waits for a wink signal from the destination
	• **fxs-ground-start**—Foreign exchange station ground-start signaling support
	• **fxs-loop-start**—Foreign exchange station loop-start signaling support
	• **fxo-ground-start**—Foreign exchange office ground-start signaling support
	• **fxo-loop-start**—Foreign exchange office loop-start signaling support
	The following options are available only on E1 controllers on the Cisco MC3810:
	• **e&m-melcas-immed**—E&M Mercury Exchange Limited Channel Associated Signaling (MELCAS) immediate start signaling support
	• **e&m-melcas-wink**—E&M MELCAS wink-start signaling support
	• **e&m-melcas-delay**—E&M MELCAS delay-start signaling support
	• **fxo-melcas**—MELCAS foreign exchange office signaling support
	• **fxs-melcas**—MELCAS foreign exchange station signaling support
	The **ext-sig** option is available only when the **mode ccs** command is enabled.

Configuring Voice Ports for ISDN PRI

If you are using the AS5300, before you configure your voice port, you can configure ISDN PRI. This section describes the necessary configurations you can use to configure your voice ports as an ISDN PRI group. It also provides a tutorial on how to configure the D channel.

To configure a voice port for ISDN PRI, perform the following steps:

1   From the global configuration mode, enter your telephone company's switch type with the **isdn switch-type** command.

2   Enter controller configuration mode and prepare to configure the controller with the **controller** command.

3   Enter your telephone company's framing type with the **framing** command.

4   Enter your telephone company's line-code type with the **linecode** command.

5   Enter the **clock source** for the line.

6   Configure all channels for ISDN with the **pri-group timeslots** command.

Repeat Steps 2 through 6 for the remaining three controllers. These steps and command options are repeated in the following text.

NOTE   This section provides only the necessary information and commands to configure an AS5300 router's voice port as an ISDN PRI group. For an additional tutorial regarding ISDN, refer to the "Configuring, Monitoring, and Troubleshooting Dialup Services" course.

In global configuration mode, you must specify your telephone company's switch that you are attaching to with the **isdn switch-type [primary-4ess | primary-5ess | primary-dms100 | primary-net5 | primary-ntt | primary-ts014]** command.

You must then enter controller configuration mode and specify the controller port you wish to configure using the **controller [t1 | e1] [0 | 1 | 2 | 3]** configuration command. The controller ports are labeled 0 to 3 on the quad T1/PRI and quad E1/PRI cards.

Find out your telephone company's framing type and line-code type and then configure the framing and line-code types.

First use the **framing {sf | esf | crc4 | no-crc4}** command to specify your telephone company's framing type. Table 7-28 describes the **framing** command options.

Table 7-28   framing *Command Options*

Command Option	Description
sf	Specifies Superframe as the T1 frame type.
esf	Specifies Extended Superframe as the T1 frame type.
crc4	Specifies CRC4 as the E1 frame type.
no-crc4	Specifies no CRC4 as the E1 frame type.

Then use the **linecode {ami | b8zs | hdb3}** command to select the line-code type for your T1 or E1 line. Table 7-29 describes the **linecode** command options.

Table 7-29   linecode *Command Options*

Command Option	Description
ami	Specifies alternate mark inversion (AMI) as the line-code type.
b8zs	Specifies B8ZS as the line-code type. Valid for T1 controllers only.
hdb3	Specifies high-density bipolar 3 as the line-code type. Valid for E1 controllers only.

You must also enter the clock source for the line. One line should be **clock source line primary**. The others should be configured as **clock source line secondary** or **clock source internal**.

NOTE	Only one PRI can be clock source primary and only one PRI can be clock source secondary. Remaining PRIs must be configured as clock source internal.

You must also configure all channels for ISDN. Enter **pri-group timeslots 1-24** for T1. If E1, enter **pri-group timeslots 1-31**.

Once you have completed these steps, you are ready to configure the ISDN D channels, which carry the control and signaling information for ISDN calls, for each ISDN PRI line.

Enter serial interface configuration mode by entering the **interface** command. After you have configured the controller, a corresponding D channel serial interface is created instantly. For example, serial interface 0:23 is the D channel for controller 0. You must configure each serial interface to receive incoming and send outgoing modem signaling.

You must also configure all incoming voice calls to go to the modems by using the **isdn incoming-voice modem** command.

To configure your D channels, perform the following steps:

1 From the global configuration mode, enter interface configuration mode with the **interface** command.

2 Configure incoming voice calls to go to the modem with the **isdn incoming-voice modem** command.

Configuring Dial Peers

Once you have established the configuration for the physical voice ports, you must now establish the relationship to the network services. This relationship consists of mapping the voice ports to the method of voice encapsulation, that is the voice-over method, which is being used. The commands in the following sections outline how to map the POTS ports configured earlier in this chapter to the voice-over encapsulations. This section describes the commands **dial-peer voice**, **destination-pattern**, **num-exp**, and **vad**.

dial-peer voice *tag* {pots | voip | voatm | vofr}

You use the **dial-peer voice** command to enter dial-peer configuration mode and specify the method of voice encapsulation. Options for this command are listed in Table 7-30.

Table 7-30   **dial-peer voice** *Command Options*

Command Option	Description
tag	A number identifying a particular dial peer. Valid entries are 1–2147483647.
pots	(For all platforms.) POTS dial peer using basic telephone service.
voip	(For all platforms.) VoIP dial peer using voice encapsulation on the POTS network.
voatm	(Cisco 3600 and MC3810 only. Soon to be for the 2600 as well.) Voice over ATM dial peer using real-time AAL5 voice encapsulation on the ATM backbone network.
vofr	(For all platforms.) VoFR dial peer using encapsulation on the Frame Relay backbone network.

The **dial-peer voice** commands define the dialing plan within the router. They specify both the remote phone numbers (VoIP, VoFR, VoATM) and the locally connected phone numbers (POTS). The digits in the destination pattern can either be complete numbers or partial numbers with wildcard digits, represented by . (period). Each . represents an individual digit for collection.

Each leg of a call peer must be configured. The **dial-peer voice** command ties each of the dialed digits exit to the encapsulation method. The next few sections describe POTS and voice-over dial peers.

Configuring Local Dial Peers

The POTS dial peer identifies how to configure the basic telephone functions of a voice port, such as the FXS interface. The commands described in the following sections are used to identify the ports and provide information such as destination patterns (that is, dialed numbers).

dial-peer voice *tag-number* pots

Use the **dial-peer voice** *tag-number* **pots** global configuration command to enter the dial-peer configuration mode and specify using basic telephone service or POTS. The destination for a POTS dial peer is a telephony device attached to the router's voice port. Table 7-31 shows the **dial-peer voice** command options.

Table 7-31 **dial-peer voice** *Command Options*

Command Option	Description
tag-number	The tag-number is an arbitrary identifier you assign to uniquely identify the dial peer. Valid entries are from 1–2147483647 (up to 100000 on the MC3810).
pots	Indicates that this is a POTS peer using basic telephone service.

To modify the *tag* configuration after you configure the dial-peer voice tag, enter the dial-peer voice and the tag number and press Enter.

TIP   For documentation purposes, you should make the *tag-number* relevant to a destination pattern for the remote site

destination-pattern *string*

Once in dial peer configuration mode, use the **destination-pattern** command to specify the extension or the full E.164 telephone number (depending on your dial plan) of the destination dial peer.

The **destination-pattern** command specifies either the prefix, a dial string, or the full E.164 telephone number (depending on your dial plan) to be used for a dial peer. This command is used to match dialed digits to a dial peer. The dial peer is then used to complete the call. When a router receives voice data, it compares the called number (the full E.164 telephone number) in the packet header with the number configured as the **destination-pattern** for the voice-telephony peer. The router then strips out the left-justified numbers corresponding to the destination pattern.

If you have configured a prefix, the prefix is appended to the front of the remaining numbers, creating a dial string, which the router then dials. If all numbers in the destination pattern are stripped-out, the user receives a dial tone.

There are certain areas in the world (for example, in certain European countries) where valid telephone numbers can vary in length. Use the optional control character, **t**, to indicate that a particular destination-pattern value is a variable-length dial-string. In this case, the system does not match the dialed numbers until the interdigit timeout value has expired. Table 7-32 shows the **destination-pattern** command options.

NOTE   The Cisco IOS software does not check the validity of the E.164 telephone number; it accepts any series of digits as a valid number.

Table 7-32   **destination-pattern** *Command Options*

Command Option	Description
string	Specifies the E.164 or private dialing plan telephone number. Valid entries are digits 0 through 9, the letters A through D, and the following special characters:
	+ (plus sign), which is optionally used as the first digit to indicate an E.164 standard number.
	* (asterisk) and **#** (pound sign), which appear on standard touch-tone dial pads. On the Cisco 3600 only, these characters cannot be used as leading characters in a string (for example, *650).
	, (comma) inserts a pause between digits.
	. (period) matches any entered digit. (Used as a wildcard.)
	t control character indicating that the destination-pattern value is a variable length dial-string.

port {*slot-number/subunit-number/port* | *slot/port*}

The **port** command is used for calls incoming from a telephony interface to select a dial peer and for VoIP calls to match a port with the selected outgoing dial peer. This command is applicable only to POTS peers on both the Cisco voice-over gateway routers. Table 7-33 shows the options for the **port** command.

Table 7-33   **port** *Command Options*

Command Option	Description
slot-number	Slot in the router where the voice interface card is installed. Valid entries are from 0–3, depending on the slot where it has been installed.
subunit-number	Subunit on the voice interface card in the router where the voice port is located. Valid entry is 0 or 1.
port	Voice port number. Valid entries are 0 or 1.
MC3810 Slot	Slot number where the voice interface card is installed. Valid entries are 1 or 0.
port	Voice port number. Valid entries are: analog voice ports: 1–6; digital voice ports: T1: 1–24; E1: 1–15, and 16–31.

NOTE   When configuring dial peers, you need to understand the relationship between the **destination pattern** and the **port**. The destination pattern is the telephone number or numbers of the voice device attached to the voice port. The port represents the route from the router to the telephony device.

num-exp *extension-number expanded-number*

You use the **num-exp** global configuration command to expand a set of numbers (for example, an extension number) into a destination pattern. With this command, you can map specific extensions and expanded numbers together by explicitly defining each number, or you can define extensions and expanded numbers using variables. You can also use this command to convert seven-digit numbers to numbers containing more or fewer than seven digits.

In most corporate environments, office PBXs are usually configured so the user can dial a local call (within the same PBX) by dialing the extension only. For example, you may call another telephone number, 1 (408) 555-9210, within the PBX with the four-digit extension 9210 or the five-digit extension 5-9210.

In a VoIP network, by using **num-exp**, the router expands a particular sequence of dialed numbers into a complete telephone number (destination pattern). Number expansion takes place before dial-peer matching.

NOTE

When wildcarding digits in a phone number for commands like **num-exp**, use a period (.) as a variable or wild card, representing a single number. Use a separate period for each number you want to represent with a wildcard; if you want to replace four numbers in an extension with wildcards, type in four periods.

In Example 7-18, extension number 55541 will be expanded to 14085555541.

Example 7-18 *Using the **num-exp** Command for 55541*

```
num-exp 55541 14085555541
```

In Example 7-19, all five-digit extensions beginning with 5 will be expanded to 1408555.

Example 7-19 *Using the **num-exp** Command for All Extensions Starting with 5*

```
num-exp 5.... 1408555
```

vad

The **vad** command enables VAD for VoFR, VoATM, and VoIP on the Cisco MC3810 for the calls using this voice port. With VAD, silence is not transmitted over the network, only audible speech. If you enable VAD, the sound quality will be slightly degraded but the connection will monopolize much less bandwidth. If you use the **no** form of this command, VAD is disabled on the voice port.

In a telephone conversation, you are listening to the conversation approximately half the time and speaking the other half. During the silence, voice packets of the silence are continually being generated that can consume much bandwidth in your network. To eliminate the silence and conserve bandwidth, use silence suppression like VAD.

Enable VAD with the **vad** command if bandwidth requirements are an issue. Disable VAD with **no vad** if you are operating in a high-bandwidth network and voice quality is of the highest importance.

NOTE

On the Cisco 3600 series, VAD is assigned to the dial peer using the **vad dial-peer configuration** command. Only the MC3810 allows you to configure the **vad** command on the voice port. The VoIP capable routers require you to configure VAD on the dial peer.

prefix *string*

The **prefix** command is used to optionally specify a prefix for a specific dial peer. When an outgoing call is initiated to this dial peer, the prefix *string* value is sent to the telephony interface first, before the telephone number associated with the dial peer.

The *string* can be integers representing the prefix of the telephone number associated with the specified dial peer. Valid numbers are 0 through 9, and a comma (,). Use the comma to include a pause in the prefix.

If you want to configure different prefixes for dialed numbers on the same interface, you need to configure different dial peers.

forward-digits {*num-digit* | **all** | **extra**}

The **forward-digits** command specifies which digits to forward for voice calls. If the **no** form of this command is entered, any digits not matching the destination pattern are not forwarded. Forwarded digits are always right-justified so that extra leading digits are stripped. The destination pattern includes both explicit digits and wildcards, if present.

Table 7-34 describes the **forward-digits** command.

Table 7-34   **forward-digits** *Command Options*

Command Option	Description
num-digit	The number of digits to be forwarded. If the number of digits is greater than the length of a destination phone number, the length of the destination number is used. The valid range is 0–32. Setting the value to 0 is equivalent to entering no forward-digits.
all	Forward all digits. If all is entered, the full length of the destination pattern is used.
extra	If the length of the dialed digit string is greater than the length of the dial-peer destination pattern, the extra right-justified digits are forwarded. However, if the dial-peer destination pattern is variable length (ending with character "T", for example: T, 123T, 123...T), extra digits are not forwarded.

NOTE   The **forward-digits** command applies only to POTS dial peers.

An Example of Configuring POTS Dial Peers

Figure 7-12 and Example 7-20 illustrate proper POTS dial-peer configuration on a 2600 or 3600 router.

Figure 7-12 *POTS Dial-Peer Example*

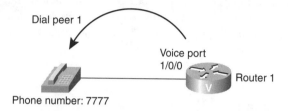

Example 7-20 *POTS Dial-Peer Configuration*

```
Router(config)#dial-peer voice 1 pots
Router(config-dial-peer)#destination-pattern 7777
Router(config-dial-peer)#port 1/0/0
```

By entering **dial-peer voice 1 pots** you are telling Router 1 that dial peer 1 is a POTS dial peer and you are calling it 1. With the **destination-pattern 7777** command, you are telling the router the telephony device's phone number. The **port 1/0/0** command tells the router that the telephony device is plugged into module slot 1, VIC subslot 0, voice port 0.

Configuring VoFR, VoATM, or VoIP Dial Peers

After you set up the POTS port you will need to set up connectivity across the network for the voice traffic. You will set up dial peers with the voice option. The following sections describe the commands you use to set up voice-over technologies.

dial-peer voice *tag-number* {**vofr** | **voatm** | **voip**}

You use the **dial-peer voice** global configuration command to enter the dial peer configuration mode and specify the method of network related encapsulation. Table 7-35 shows the options for the command.

Table 7-35   **dial-peer voice** *Command Options*

Command Option	Description
tag-number	Digit(s), unique to the local router, defining a particular dial peer. Valid entries are from 1–2147483647 (up to 100000 on the MC3810).
vofr	Indicates that this is a VoFR peer using encapsulation on the Frame Relay backbone network.
voatm	Indicates that this is a VoATM peer using the real-time AAL5 voice encapsulation on the ATM backbone network.
voip	Indicates that this is a VoIP dial peer using voice encapsulation on the IP backbone.

NOTE	*tag-number* is an arbitrary identifier you assign to uniquely identify the dial peer. To modify the tag configuration after you configure the dial-peer voice tag, enter the **dial-peer voice** and the *tag-number* and press Enter. This number should have some meaning to the administrator for the sake of manageability.

destination-pattern *string*

The **destination-pattern** command is used once in dial-peer configuration mode. Use the **destination-pattern** command to specify the full E.164 telephone number (depending on your dial plan) of the destination dial peer. This command must be used in both local and remote dial-peer statements. Table 7-36 shows the **destination-pattern** command options.

NOTE	Remember, the destination for a VoFR, VoATM, or VoIP dial peer is another network device (for example, voice-capable router) you are sending a voice call.

Table 7-36   **destination-pattern** *Command Option*

Command Option	Description
string	Digits 0 through 9, letters A through D, or private dialing plan telephone number. Valid entries are: ***** (asterisk) and **#** (pound sign) that appear on standard touch-tone dial pads. On the Cisco 3600 only, these characters cannot be used as leading characters in a string (for example, *650). **+** (plus sign), which is optionally used as the first digit to indicate an E.164 standard number. **,** (comma), which inserts a pause between digits. **.** (period), which matches any entered digit. **t** control character indicating that the destination-pattern value is a variable length dial-string.

When configuring dial peers, you need to understand the relationship between the **destination-pattern** and the **session target**. Destination pattern is the telephone number of the voice device attached to a voice port at the other end of the network connection. Session target represents the route to a port on the other end of the network connection. The **session target** command shows the target for a Frame Relay voice-over connection.

session target *interface dlci* [*cid*]

Use the **session target** command if you are configuring a VoFR, VoATM, or VoIP dial peer to specify a network address for a specified dial peer.

You use the **session target** command to specify the network address of the router you are trying to contact. Table 7-37 shows the options for this command.

Table 7-37   **session target** *VoFR Options*

Command Option	Description
interface	Specifies the serial interface and interface number (slot number/port number) associated with this dial peer.
dlci	Specifies the data link connection identifier for this dial peer. The valid range is from 16 to 1007.
cid	(Optional.) Specifies the DLCI subchannel to be used for data on FRF.11 calls. A CID must be specified only when the session protocol is frf11-trunk. When the session protocol is Cisco switched, the CID is dynamically allocated. For VoFR dial peers, the cid option is not allowed when using the cisco-switched session protocol. The valid range is from 4 to 255. Note that by default, CID 4 is used for data; CID 5 is used for call control. Cisco recommends that you select CID values between 6 and 63 for voice traffic. If the CID is greater than 63, the FRF.11 header will contain an extra byte of data.

Example 7-21 shows how to configure serial interface 1/1, DLCI 200 as the session target for switched VoFR dial peer 400, using the cisco-switched session protocol and starting from global configuration mode on a 2600/3600/7200 voice router.

Example 7-21   *VoFR dial-peer example*

```
router(config)# dial-peer voice 400 vofr
router(config-dial-peer)# destination-pattern 13102221111
router (config-dial-peer)# session target serial 1/1 200
```

Example 7-22 shows how to configure a session target for a VoFR dial peer with a session target on Serial Port 1 and a DLCI of 200 on a MC3810.

Example 7-22   *VoFR dial-peer example for MC3810*

```
router(config)# dial-peer voice 11 vofr
router(config-dial-peer)# destination-pattern 13102221111
router (config-dial-peer)# session target Serial1 200
```

You can also use the session target to map voice over ATM.

session target {*interface* **pvc** *name vpi/vci* | *vci*}

Use the **session target** command to specify a network-specific address or domain name for a dial peer. Whether you select a network-specific address or a domain name depends on the session protocol you select. Table 7-38 shows the options for this command.

Table 7-38   **session target** *VoATM Options*

Command Option	Description
interface	Specifies the interface type and interface number on the router. **For MC3810**: The only valid number is 0.
pvc	Indicates the specific ATM permanent virtual circuit for this dial peer.
name	Specifies the PVC name.
vpi/vci	The ATM network virtual path identifier (VPI) and virtual channel identifier (VCI) of this PVC.
	For 3600: If you have the Multiport T1/E1 ATM Network Module with IMA installed, the valid range for *vpi* is 0–15, and the valid range for *vci* is 1–255.
	If you have the OC3 ATM Network Module installed, the valid range for *vpi* is 0–15, and the valid range for *vci* is 1–1023.
vci	The ATM network VCI of this PVC.

Example 7-23 illustrates a typical ATM connection using an MC3810 as an example. The session target is sent to ATM interface 0 and is for a PVC with a *vci* of 20.

Example 7-23   *VoATM Connection*

```
Router(config)#dial-peer voice 12 voatm
Router(config-dial-peer)#destination-pattern 13102221111
Router(config-dial-peer)#session target atm0 pvc 20
```

Example 7-24 shows an ATM connection using a 2600 series router. The session target is sent to ATM interface 0 and is for a **pvc** with a *vpi/vci* of 1/100.

Example 7-24   *VoATM Connection for 2600 Series*

```
Router(config)#dial-peer voice 12 voatm
Router(config-dial-peer)#destination-pattern 13102221111
Router(config-dial-peer)#session target atm1/0 pvc 1/100
```

session target can also be used to specify a VoIP connection.

session target {ipv4:*destination-address* | dns:[s. | d. | e. | u.] *host-name* | loopback:rtp | loopback:compressed | loopback:uncompressed}

You use the **session target ipv4** command to configure a network-specific address for a dial peer.

The **dns** command can be used with or without the specified wildcards. Using the optional wildcards can reduce the number of VoIP dial-peer session targets you need to configure if you have groups of numbers associated with a particular router. Table 7-39 shows how to use this command.

Table 7-39   **session target** *VoIP Options*

Command Option	Description
ipv4:*destination-address*	The IP address of the dial peer.
dns:*host-name*	Indicates that the domain name server will be used to resolve the name of the IP address. Valid entries for this parameter are characters representing the name of the host device.
	You can optionally use one of the following wildcards with this keyword when defining the session target for VoIP peers:
	• **s**—Indicates that the source destination pattern will be used as part of the domain name.
	• **d**—Indicates that the destination number will be used as part of the domain name.
	• **e**—Indicates that the digits in the called number will be reversed, periods will be added between each digit of the called number, and this string will be used as part of the domain name.
	• **u**—Indicates that the unmatched portion of the destination pattern (such as a defined extension number) will be used as part of the domain name.

For example, you can define a session target for a VoIP dial peer using DNS for a host, voice_router, in the domain cisco.com. Example 7-25 shows how to configure Voice over IP using DNS to specify the IP target location.

Example 7-25   *VoIP using DNS*

```
voip(config)# dial-peer voice 10 voip
voip (config-dial-peer)# session target dns:voice_router.cisco.com
```

You can also define a session target for a VoIP dial peer using DNS with the optional **u** wildcard. In this example, the destination pattern has been configured to allow for any four-digit extension, beginning with the numbers 1310222. The optional wildcard **u** indicates that the router will use the unmatched portion of the dialed number—in this case, the four-digit extension, to identify the dial peer. As in the previous example, the domain is cisco.com. Example 7-26 shows how to use the **u** wildcard for DNS targets.

Example 7-26 *VoIP Using DNS and the u Wildcard*

```
voip(config)# dial-peer voice 10 voip
voip (config-dial-peer)# destination-pattern 1310222....
voip (config-dial-peer)# session target dns:$u$.cisco.com
```

You can also define a session target for a VoIP dial peer using DNS with the optional **d** wildcard. In this example, the destination pattern has been configured for 13102221111. The optional wildcard **d** indicates that the router will use the destination pattern to identify the dial peer in the cisco.com domain. Example 7-27 shows the usage of the **d** wildcard to specify the target.

Example 7-27 *VoIP Using DNS and the d Wildcard*

```
voip(config)# dial-peer voice 10 voip
voip (config-dial-peer)# destination-pattern 13102221111
voip (config-dial-peer)# session target dns:$d$.cisco.com
```

Finally, you can define a session target for a VoIP dial peer using DNS, with the optional **e** wildcard. In this example, the destination pattern has been configured for 12345. The optional wildcard **e** indicates that the router will reverse the digits in the destination pattern, add periods between the digits, and then use this reverse-exploded destination pattern to identify the dial peer in the cisco.com domain. Example 7-28 shows the usage of the **e** wildcard to specify the target..

Example 7-28 *VoIP Using DNS and the e Wildcard*

```
voip(config)# dial-peer voice 10 voip
voip (config-dial-peer)# destination-pattern 12345
voip (config-dial-peer)# session target dns:$e$.cisco.com
```

loopback {loopback:rtp | loopback:compressed | loopback:uncompressed}

You use the **loopback** command for testing the voice transmission path of a call. The loopback point depends on the call origination and the loopback type selected. Table 7-40 describes the **loopback** command.

Table 7-40   **loopback** *Command Options*

Command Option	Description
loopback:rtp	Indicates that all voice data will be looped back to the originating source. This is applicable for VoIP peers.
loopback:compressed	Indicates that all voice data will be looped back in compressed mode to the originating source. This is applicable for POTS peers.
loopback:uncompressed	Indicates that all voice data will be looped back in uncompressed mode to the originating source. This is applicable for POTS peers.

signal-type {cas | cept | ext-signal | transparent}

The **signal-type** command applies to VoFR and VoATM dial peers. It is used with permanent connections only (Cisco trunks and FRF.11 trunks), not with switched calls.

The **signal-type** command is used to inform the local telephony interface of the type of signaling it should expect to receive from the far-end dial peer. To turn signaling off at this dial peer, select the **ext-signal** option. If signaling is turned off and there are no external signaling channels, a permanent trunk exists, enabling this dial peer to connect to anything at the far end.

When you connect an FXS to another FXS, or if you have anything other than an FXS/FXO or E&M/E&M pair, the appropriate signaling type on Cisco 2600 series and 3600 series routers is **ext-signal** (disabled).

If you have a digital E1 connection at the remote end that is running cept/MELCAS signaling and you then trunk that across to an analog port, you should make sure that you configure both ends for the **cept** signal-type.

For a T1 or E1 connection at both ends, and where the T1/E1 is running a signaling protocol that is neither EIA-464 nor cept/MELCAS, you may want to configure the signal-type for the **transparent** option in order to pass through the signaling. Table 7-41 shows the **signal-type** command options.

Table 7-41   **signal-type** *Command Options*

Command Option	Description
cas	North American EIA-464 Channel-Associated Signaling (robbed bit signaling). If the Digital T1 Packet Voice Trunk Network Module is installed, this option might not be available.
cept	Provides a basic E1 ABCD signaling protocol. Used primarily for E&M interfaces. When used with FXS/FXO interfaces, this protocol is equivalent to MELCAS.

continues

Table 7-41   **signal-type** *Command Options (Continued)*

Command Option	Description
ext-signal	External signaling. The DSP does not generate any signaling frames. Use this option when there is an external signaling channel (that is, CCS) or when you need to have a permanent "dumb" voice pipe.
transparent	On the Cisco MC3810, selecting this option produces different results depending on whether you are using a digital voice module (DVM) or an analog voice module (AVM). For DVM: The ABCD signaling bits are copied from or transported through the T1/E1 interface "transparently," without modification or interpretation. This enables the MC3810 to handle arbitrary or unknown signaling protocols. For AVM: It is not possible to provide "transparent" behavior because the MC3810 must interpret the signaling information in order to read/write the correct state to the analog hardware. This option is mapped to be equal to **cas**.

voice class permanent *tag*

The **voice class permanent** command in global configuration mode creates a voice class for a Cisco trunk or FRF.11 trunk to a dial peer. This command can be used for VoFR, VoATM, and VoHDLC trunks. Table 7-42 shows the command options for this command.

Table 7-42   **voice class permanent** *Command Option*

Command Option	Description
tag	Specifies the unique tag number you assign to the permanent voice class. The valid range for this tag is 1–10000. The tag number must be unique on the router.

This command can also be used in dial-peer configuration mode. The **voice-class permanent** *tag* dial-peer command assigns a previously configured voice class for a Cisco trunk or FRF.11 trunk to a dial peer. In interface configuration mode, this command allows you to specify a unique tag number to the permanent voice call. Table 7-43 describes this command.

Table 7-43   **voice-class permanent** *Dial-Peer Command Option*

Command Option	Description
tag	Specifies the unique tag number assigned to the permanent voice class. The valid range for this tag is 1–10000. The tag number maps to the tag number created using the **voice class permanent** global configuration command.

NOTE   The **voice class permanent** command in global configuration mode is entered *without* a hyphen. The **voice-class permanent** command in dial peer configuration mode is entered *with* the hyphen.

Example 7-29 shows how to configure a permanent voice class starting from global configuration mode, configure parameters for that voice class, and then assign the voice class to a dial peer.

Example 7-29   *Assigning Voice Class*

```
router(config)# voice class permanent 10
router(config-class)# signal pattern idle transmit 110
router(config-class)# exit
router(config)# dial-peer voice 100 vofr
router(config-dial-peer)# voice-class permanent 10
```

signal pattern {idle | oos} {transmit | receive} *word*

The **signal pattern** command configures the ABCD bit pattern for Cisco trunks and FRF.11 trunks. Before configuring the signaling pattern, you must use the voice class permanent command in global configuration mode to create a voice class for the Cisco trunk or FRF.11 trunk. The voice class must then be assigned to a dial peer.

This command must be entered twice. When you enter the command to specify the signaling pattern for the idle transmit state, you must reenter the command to specify the signaling pattern for the idle receive state.

The idle state of a call is normally based on both the transmit and receive idle patterns matching the signaling state in the signaling packets. If only one direction is configured (transmit or receive), the idle state will be detected based only on the direction that is configured. The out-of-service (oos) transmit pattern is matched against the signaling state from the PBX (and transmitted to the network). This is used in conjunction with either the suppress-voice timing parameter or the suppress-all parameter.

The oos receive pattern is the pattern sent to the PBX if the *signal timing oos* timeout timer expires during which no signaling packets are received from the network. The oos receive pattern is not used for pattern matching against the signaling packets received from the network. The receive packets directly indicate an oos condition by setting the AIS alarm indication bit in the packet.

To "busy out" a PBX if the network connection fails, set the oos receive pattern to match the seized state (busy), then set the signal timing oos *timeout* value. When the timeout value expires and no signaling packets have been received, the router will send the oos receive pattern to the PBX.

Use the busy seized pattern only if the PBX does not have a special pattern specifically intended to indicate an oos state. If the PBX does have a specific oos pattern, use that pattern instead. Table 7-44 shows the options for this command.

Table 7-44 **signal pattern** *Command Options*

Command Option	Description
idle receive	Specifies that the signal pattern applies to the idle state of the call for receive bits. The receive direction is from the network to the PBX.
idle transmit	Specifies that the signal pattern applies to the idle state of the call for transmit bits. The transmit direction is from the PBX to the network.
oos receive	Specifies that the signal pattern applies to the out-of-service state of the call for receive bits.
oos transmit	Specifies that the signal pattern applies to the out-of-service state of the call for transmit bits.
word	The ABCD bit pattern. Valid values are from 0000–1111.

Example 7-30 shows how to configure the signaling bit pattern for the idle receive and transmit states.

Example 7-30 *Signaling Bit Pattern Configuration for Idle States*

```
router(config)# voice class permanent 10
router(config-class)# signal keepalive 3
router(config-class)# signal pattern idle receive 0101
router(config-class)# signal pattern idle transmit 0101
router(config-class)# exit
router(config)# dial-peer voice 100 vofr
router(config-dial-peer)# voice-class permanent 10
```

Example 7-31 shows how to configure the signaling bit pattern for the out-of-service receive and transmit states.

Example 7-31 *Bit Pattern for Out-of-Service States*

```
router(config)# voice class permanent 10
router(config-class)# signal keepalive 3
router(config-class)# signal pattern oos receive 0001
router(config-class)# signal pattern oos transmit 0001
router(config-class)# exit
router(config)# dial-peer voice 100 vofr
router(config-dial-peer)# voice-class permanent 10
```

Table 7-45 lists additional related signaling commands for voice class configuration.

Table 7-45 *Signaling Commands for Voice Class Configuration*

Command	Description
signal keepalive	Configures the keepalive signaling packet interval for Cisco trunks and FRF.11 trunks.
signal timing idle suppress-voice	Configures the signal timing parameter for the idle state of a call.
signal timing oos	Configures the signal timing parameter for the out-of-service (oos) state of a call.
signal-type	Sets the signaling type to be used when connecting to a dial peer.
voice class permanent (global configuration mode)	Creates a voice class for a Cisco trunk or FRF.11 trunk.

voice-card *slot*

The **voice-card** command configures a voice card and enters voice-card configuration mode. Table 7-46 shows the **voice-card** command options.

Table 7-46 **voice-card** *Command Option*

Slot Option	Valid Options for Listed Device
slot	For the 2600/3600 platforms, a value from 0–6 that identifies the physical slot in the chassis where the voice card is located.
	For an MC3810 concentrators with one or two HCMs installed, 0 only; this applies to the entire chassis.

Example 7-32 shows how to enter voice-card configuration mode for the voice card in slot 1 on a Cisco 2600 or 3600 router.

Example 7-32 *Voice-Card config Mode on the 2600/3600 Series*

```
router(config)# voice-card 1
router(config-voicecard)#
```

For the MC3810 the only slot available is 0, this covers the entire chassis. Example 7-33 enters voice-card configuration mode on a Cisco MC3810.

Example 7-33 *Voice-Card config Mode on an MC3810*

```
router(config)# voice-card 0
router(config-voicecard)#
```

codec complexity {high | medium}

You can configure **codec complexity** only in voice-card configuration mode. On the Cisco 2600 and 3600 platforms, the slot corresponds to the physical slot in the chassis. On the Cisco MC3810, the slot is always 0, and all changes made in voice-card mode apply to the entire Cisco MC3810. On Cisco MC3810 series concentrators, this command is available only if the chassis is equipped with one or two HCMs.

The **codec complexity** command is used to match the DSP complexity packaging to the codec(s) to be supported. Codec complexity refers to the amount of processing required to perform voice compression. Codec complexity affects the call density—the number of calls that can take place on the digital signal processors (DSPs). With higher codec complexity, fewer calls can be handled.

Select a higher codec complexity if that is required in order to support a particular codec or combination of codecs.

Select a lower codec complexity to support the greatest number of voice channels, provided that the lower complexity is compatible with the particular codecs in use.

To change codec complexity, all of the DSP voice channels must be in the idle state and the card must be shutdown.

Table 7-47 describes the **codec complexity** command keywords.

Table 7-47 codec complexity *Command Options*

Command Option	Description
high	With high-complexity packaging, each DSP supports two voice channels encoded in any of the following formats: G.711ulaw, G.711alaw, G.723.1(r5.3), G.723.1 Annex A(r5.3), G.723.1(r6.3), G.723.1 Annex A(r6.3), G.726(r16), G.726(r24), G.726(r32), G.729, G.729 Annex B, G.728, and fax relay.
medium	With medium complexity packaging, each DSP supports four voice channels encoded in any of the following formats: G.711ulaw, G.711alaw, G.726(r16), G.726(r24), G.726(r32), G.729 Annex A, G.729 Annex B with Annex A, and fax relay. This is the default.

NOTE On the Cisco MC3810, the **codec complexity** command is valid only with HCM(s) installed, and you must specify voice card 0 in the command mode. If two HCMs are installed, the **codec complexity** command configures both HCMs at once.

Example 7-34 sets the codec complexity to high on a Cisco MC3810 containing one or two HCMs.

Example 7-34   *Codec Settings on an MC3810*

```
router(config)# voice-card 0
router(config-voicecard)# codec complexity high
```

Example 7-35 sets the codec complexity to high on voice card 1 in a Cisco 2600 or 3600 router.

Example 7-35   *Codec Settings on the 2600/3600 Series Router*

```
router(config)# voice-card 1
router(config-voicecard)# codec complexity high
```

An Example of Configuring VoIP Dial Peers

The voice-over dial peers example shown in Figure 7-13 is an example using VoIP. This figure shows the basic configuration discussed in this chapter.

Figure 7-13   *VoIP Dial-Peers Example*

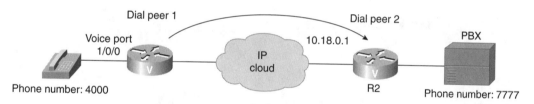

Example 7-36 illustrates proper VoIP dial-peer configuration on a 2600 or 3600 router. By entering **dial-peer voice 2 voip** you are identifying (on Router 1) Router 2 as dial peer 2 and that it is a VoIP dial peer. With the **destination-pattern +1214444….** you are telling Router 1 the phone number of the telephony device you are calling. The **session target ipv:10.18.0.1** command is the IP address of Router 2.

Example 7-36   *VoIP Example Configurations for Router 1*

```
Router1(config)#dial-peer voice 2 voip
Router1(config-dial-peer)#destination-pattern 7777
Router1(config-dial-peer)#session target ipv4:10.18.0.1
```

NOTE	As indicated on the **destination-pattern** command table, Table 7-36, the "...." replacing the last four telephone number digits are "wildcard" placeholders that match any digit 0 through 9. The placeholder "." is commonly used in the VoIP dial-peer configuration mode and means that from router 10.18.0.2, for example, calling any number string that begins with the digits "+1214444" plus the remaining phone number digits will result in a connection to router 10.18.0.1 and implies that router 10.18.0.1 services all numbers beginning with those digits.

An Example of Configuring Dial Peers

Figure 7-14 shows that dial peers 3 and 4 must now be configured on Router 2 so there can be an end-to-end call. The example is a VoIP example using a 2600 or 3600 router.

Figure 7-14 *Dial-Peers Example*

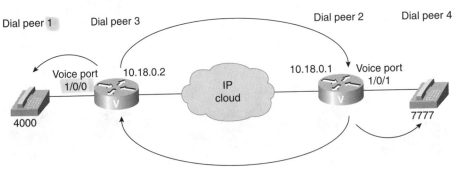

Example 7-37 shows the proper configuration for these peers.

Example 7-37 *Dial-Peer Configuration for VoIP*

```
Router1(config)#dial-peer voice 1 pots
Router1(config-dial-peer)#destination pattern 4000 port 1/0/0
Router1(config-dial-peer)#dial-peer voice 2 voip
Router1(config-dial-peer)#destination-pattern 7777
Router1(config-dial-peer)#session target ipv4:10.18.0.1

Router2(config)#dial-peer voice 4 pots
Router2(config-dial-peer)#destination pattern 7777 port 1/0/1
Router2(config-dial-peer)#dial-peer voice 3 voip
Router2(config-dial-peer)#destination-pattern 4000
Router2(config-dial-peer)session target ipv4:10.18.0.2
```

Verifying Dial-Peer Configuration

You verify that the voice connection is working by doing the following:

1 Pick up the handset on a telephone connected to the configuration and verify that you can get a dial tone.

2 Make a call from the local telephone to a configured dial peer and verify that the call attempt is successful.

The following sections describe the commands involved in verifying dial-peer configuration.

show dial-peer voice

You can check your dial-peer configuration by using the **show dial-peer voice** command to verify that the data configured is correct. Use this command to display a specific dial peer or to display all configured dial peers.

Output for a POTS dial peer should look like the output in Example 7-38. Output for a VoFR, VoATM, or VoIP dial peer should be similar to the output in this example.

Example 7-38 *POTS Dial-Peer Output*

```
Router1#show dial-peer voice
VoiceEncapPeer1
 tag = 1, destination-pattern = '4000',
 answer-address = '',
 group = 1, Admin state is up, Operation state is up
 type = pots, prefix = '',
 session-target = '', voice-port = 1/0/0
 Connect Time = 46214, Charged Units = 0
 Successful Calls = 50, Failed Calls = 0
 Accepted Calls = 65, Refused Calls = 0
 Last Disconnect Cause is "10 "
 Last Disconnect Text is "normal call clearing."
 Last Setup Time = 34485570
VoiceOverIpPeer2
 tag = 2, destination-pattern = '7777',
 answer-address = '',
 group = 2, Admin state is up, Operation state is up
 type = voip, session-target = 'ipv4:10.18.0.1',
 ip precedence: 0 UDP checksum = disabled
 session-protocol = cisco, req-qos = best-effort,
 acc-qos = best-effort,
 fax-rate = voice, codec = g729r8,
 Expect factor = 10, Icpif = 30,
 VAD = enabled, Poor QOV Trap = disabled
 Connect Time = 44739, Charged Units = 0
 Successful Calls = 35, Failed Calls = 0
```

continues

Example 7-38   *POTS Dial-Peer Output (Continued)*

```
Accepted Calls = 35, Refused Calls = 0
Last Disconnect Cause is "10 "
Last Disconnect Text is "normal call clearing."
Last Setup Time = 34461803
```

How to Configure Dial Peers

To configure dial peers on your router, you would perform the following steps:

Step 1   Enter the dial-peer configuration mode and specify the POTS dial peer you wish to configure.

Step 2   Set the destination pattern of the POTS peer.

Step 3   Specify the voice port to which the POTS peer is connected.

Step 4   Exit the dial-peer configuration mode for the specific POTS peer.

Step 5   Repeat for all POTS dial peers.

Step 6   Enter the dial-peer configuration mode and specify the VoFR, VoATM, or VoIP dial peer you wish to configure.

Step 7   Set the destination pattern of the VoFR, VoATM, or VoIP peer.

Step 8   Specify the session target of the peer to which you wish to attach.

Step 9   Exit VoFR, VoATM, or VoIP dial-peer configuration mode.

Step 10   Repeat for all VoFR, VoATM, or VoIP dial peers.

Step 11   Verify proper configuration with the **show dial-peer voice** command.

Monitoring, Testing, and Troubleshooting Commands

You can verify voice port configuration by performing the following steps:

1   Pick up the handset of an attached telephony device and check for dial tone.

2   If you have dial tone, check for DTMF, voice-band tones like touch-tone detection. If the dial tone stops when you dial a digit, then the voice port is most likely configured properly.

3   Use the **show voice port** command to verify that the data configured is correct. The output in Example 7-39 is from a 2600 or 3600 router. You should see output similar to this.

Example 7-39   *Using the* **show voice port** *Command to Verify Data*

```
Router3#show voice port

Foreign Exchange Station 1/0/0 Slot is 1, Sub-unit is 0, Port is 0
 Type of VoicePort is FXS
 Operation State is DORMANT
 Administrative State is UP
 No Interface Down Failure

<Omitted Information>

 Analog Info Follows:
 Region Tone is set for northamerica
 Currently processing Voice
 Maintenance Mode Set to None (not in mtc mode)
 Number of signaling protocol errors are 0
 Impedance is set to 600r Ohm

 Voice card specific Info Follows:
 Signal Type is loopStart
 Ring Frequency is 25 Hz
 Hook Status is On Hook
<Omitted Information>
```

If you are having trouble connecting a call and you suspect the problem is associated with voice-port configuration, you can try to resolve the problem by performing the following steps:

1 Ping the associated IP address to confirm connectivity. If you cannot successfully ping your destination, confirm your network configurations with the **show running-config** command.

2 Use the **show voice port** command to make sure that the port is enabled. If the port is off line, use the **no shutdown** command.

3 If you have configured E&M interfaces, make sure that the values pertaining to your specific PBX setup are correct. Specifically check for 2-wire or 4-wire, wink-start, immediate or delay-dial signal types, and the E&M interface type.

4 Check that the VIC has been correctly installed.

NOTE   When configuring an analog E&M interface, pay particular attention to the wiring and timing parameters of the PBX you are connecting. Analog FXS and FXO defaults work most of the time.

show Commands

You issue the **show** commands from the router prompt to aid in the verification of configurations and troubleshooting. Table 7-48 describes some of the **show** port commands available to help with the configurations discussed in this chapter.

Table 7-48   *Available* **show** *Commands*

Command	Description
show voice port	Shows all voice port configurations in detail.
show voice port *x/y*	Shows one voice port configuration in detail.
show voice port summary	Shows all voice port configurations in brief.
show voice call *output modifiers*	Shows all or one voice port call status. The *output modifiers* options are: • **1-1**—Voice interface slot number • **summary**—Summary of all voice calls

Example 7-40 displays the output from the **show voice call** command.

Example 7-40   *Using the* **show voice call** *Command*

```
Router#show voice call 1/1
DS0#1(orig):
src_ds0=1/1, dst_ds0=1/2, state=ST_ACTIVE
CODEC=G729AR8, dst_id=522, is_voice_ready=1
connect_type=local, DSP-PCM-Port={(-1,-1), 23, (1,1)}
talk_duration=277, seize-talk-at(136682, 136955).
LFXS= connected,  CPD= connected
lss_voice = PASS, cps_voice = PASS, digit = PASS
00:22:52: no DSP assigned to this call, can't show levels
```

Table 7-49 lists the available of the **show voice** commands.

Table 7-49   **show voice** *Commands*

Command	Description
show voice busyout	Shows all configured as busyout.
show voice dsp	Shows all dsp status.
show dial-peer voice {*tag*}	Shows specific dial-peer information.

Example 7-41 gives a summary of the dial peers.

Example 7-41   *Using the* **show dial-peer voice summary** *command*

```
rtr2dflt#show dial-peer voice summary
 TAG TYPE    ADMIN OPER PREFIX    DEST-PATTERN    FWD  PREF SESS-TARGET PORT
 6 pots    up up    ... 3 3    1/6
 3 pots    up up    3.. 3 3    1/3
 1 pots    up up    201 0 0    1/1
 2 pots    up up    202 0 0    1/2
   12 vofr    up up    1.. 0 0    Serial0 122
   23 vofr    up up    3.. 0 0    Serial1 123
```

Other useful commands include the T1- and E1-specific **show** commands. These functions allow you to check to status of the controllers. Table 7-50 outlines the **show** commands associated with the controllers.

Table 7-50   *T1/E1-specific* **show** *Commands*

Command	Description
**show controllers T1	E1 0 or 1**	Provides the operational status of the controller.
show interface serial1:x	Provides the operational status of the controller T1	E1 1 common or D channel.
show interface serial0:y	Provides the operational status of the controller T1	E1 0 serial interface opened by the channel-group command.
show frame-relay pvc	Provides status on active PVCs.

You may also have a need to check the status of ISDN connectivity for support of voice services. Table 7-51 lists the ISDN specific commands.

Table 7-51   *ISDN-specific* **show** *Commands*

Command	Description
show isdn active	Shows ISDN active calls.
show isdn history	Shows ISDN call history.
show isdn memory	Shows ISDN memory information.
show isdn nfas	Displays ISDN NFAS information for the AS5300.
show isdn service	Shows ISDN service information.
show isdn status	Shows ISDN line status.
show isdn timers	Shows ISDN timer values.

debug Commands

The most powerful troubleshooting tools available are the **debug** commands. The **debug** commands are issued from the router prompt and provide feedback to logging facilities like the console. While these commands are extremely useful they can also severely impact router performance and should be used with great care. Table 7-52 lists some useful debug commands when dealing with voice.

Table 7-52 *Useful **debug** Commands*

Command	Description
debug voice all	Debugs all calls in process or a particular call if a specific voice port is selected.
debug voice cp	Turns on voice call processing state machine debug for all calls in process or for a particular call if a specific voice port is selected.
debug voice eecm	Turns on end-to-end CallManager debug for all calls in process or for a particular call if a specific voice port is selected.
debug voice protocol	Turns on line protocol state machine debug for all APM calls in process or for a particular call if a specific voice port is selected.
debug voice signaling	Turns on voice signaling interface debug for all APM calls in process or for a particular call if a specific voice port is selected.
debug DSX1	Debugs T1	E1 controllers.
debug Interface	Debugs an interface.
debug isdn events	Displays ISDN events.
debug isdn q921	Displays ISDN Q921 packet history.
debug isdn q931	Displays ISDN Q931 packet history.

test Commands

With release 12.07XK of the Cisco IOS, several voice **test** commands were added. Table 7-53 describes the command syntax and usage of these commands.

Table 7-53 *Voice* **test** *Commands*

Command	Description
test voice port detector {*option*}	Forces a detector into specific states for testing. For each signaling type (E&M, FXO, FXS), only the applicable keywords are displayed. The options are: {**m-lead** I **battery-reversal** I **ring** I **tip-ground** I **ring-ground** I **ring-trip**} {**on** I **off** I **disable**}
test voice port inject-tone {*option*}	Injects a test tone into a voice port. A call must be established on the voice port under test. When you are finished testing, be sure to enter the disable command to end the test tone. The options are: {**local** I **network**} {**1000hz** I **2000hz** I **200hz** I **3000hz** I **300hz** I **3200hz** I **3400hz** I **500hz** I **quiet** I **disable**}
test voice port loopback {*option*}	Performs loopback testing on a voice port. A call must be established on the voice port under test. When you are finished testing, be sure to enter the disable command to end the forced loopback. The options are: {**local** I **network** I **disable**}
test voice port relay {*option*}	Tests relay-related functions on a voice port. The options that can be tested are: {**e-lead** I **loop** I **ring-ground** I **battery-reversal** I **power-denial** I **ring** I **tip-ground**} {**on** I **off** I **disable**}
test voice port switch {*option*}	Forces a voice port into fax or voice mode for testing. If no fax data is detected by the voice port, the voice port remains in fax mode for 30 seconds and then reverts automatically to voice mode. After you enter the test voice port switch fax command, you can use the show voice call or show voice call summary command to check whether the voice port is able to operate in fax mode. The options are: {**fax** I **disable**}

The commands listed in Table 7-53 are **test** commands introduced in Cisco IOS Release 12.0(7)XK that expand the capabilities to analyze and troubleshoot voice ports on the Cisco 2600 and 3600 series routers and MC3810 series concentrators.

These voice-port testing commands allow you to force voice ports into specific states for testing. Refer to the *Voice Port Testing Enhancements in Cisco 2600 and 3600 Series Routers and MC3810 Series Concentrators* document for further information.

Summary

In this chapter you have learned the basics to configuring voice-over services using Cisco voice-capable routers. You have seen how dial legs are established and how to configure both the voice ports and the dial peers. The voice ports, like FXS and E&M, connect to telephony devices or POTS services and dial peers establish the logical network connections to provide the call legs services. This chapter also discussed in detail many of the commands used to provide call legs services along with example configurations. While this chapter discusses basics and builds a foundation for voice-over services, the next few chapters expand on these functions.

Review Questions

The following questions should help you gauge your understanding of this chapter. You can find the answers in Appendix A, "Answers to Review Questions."

1 Which voice port standard allows connectivity to a standard PBX interface or the PSTN?

2 What is a call leg?

3 Which global configuration command allows you to configure properties on an FXS port?

4 What is the European codec standard for ITU-T Pulse Code Modulation?

5 True or False: Voice transmissions are like data transmissions and are not seriously affected by delay of packet delivery.

_____

_____

_____

6 How would you configure the router to expand the dialed digits 55512 to include a 1 and the area code 859?

_____

_____

_____

7 What is the first thing you should check when troubleshooting a voice port?

_____

_____

_____

8 What powerful troubleshooting tool can severely impact a router's performance?

_____

_____

_____

Upon completion of this chapter, you will be able to perform the following tasks:

- Describe Voice over Frame Relay functionality including call setup signaling, addressing, routing, and delay characteristics.

- Explain the features of Cisco voice-capable routers that support VoFR.

- Configure Frame Relay voice ports, dial peers, and QoS to successfully complete a VoFR voice call.

- Given quality of service (QoS) considerations for a VoFR network, select the optimum solution using Cisco's recommendations.

CHAPTER 8

Configuring Cisco Routers for VoFR

This chapter introduces you to Voice over Frame Relay (VoFR), gives a brief overview of how Frame Relay operates, and provides a description of relevant VoFR hardware, cable interface connectivity, and software features. The chapter also provides a summary of configuration concepts and commands and an example of a VoFR configuration.

This chapter discusses the following:

- Understanding VoFR basics
- An introduction to Cisco's implementation of VoFR
- Applying QoS for VoFR
- Using VoFR configuration commands

VoFR Basics

Frame Relay is a networking technology that allows multiple logical paths to be connected via a single physical line to form virtual meshed networks. This technology provides a way of sharing network bandwidth that provides flexibility and higher effective throughputs through oversubscription and instantaneous dynamic bandwidth allocation.

Frame Relay supersedes X.25 and older public networking systems and is extremely well-suited for both public and private networking implementations due to its configuration flexibility, high-speed interfaces, and ability to integrate traditional Systems Network Architecture (SNA) and packet data, bursty LAN data, Internet connectivity, and both voice and video communications.

Time-division multiplexing (TDM)-based solutions reserve a fixed time slot for a device whether the device has anything to send or not. Frame Relay, conversely, statistically multiplexes traffic over a single physical transmission link, thus allocating any unused bandwidth "on demand" or on an as-needed basis.

To configure Frame Relay for voice, you must specify the serial interface you wish to use for transporting voice through a network service cloud in the form of Frame Relay data transmission packets. You must also specify the data link connection identifier (DLCI) you wish to use to carry the voice traffic in the dial-peer statement, as described in Chapter 7, "Configuring Voice Ports and Dial Peers for Voice." You may be required to specify the channel ID for data and call control, depending on if you configure VoFR services to use FRF.11. This subject will be covered in greater depth in the section "Using VoFR Configuration Commands" in this chapter. You must also configure the QoS features to provide proper voice quality.

PVCs, SVCs, and DLCIs

It is common to hear the terms and abbreviations for permanent virtual circuits (PVCs), switched virtual circuits (SVCs), and data link connection identifiers (DLCIs) when talking Frame Relay. A PVC is a connection that is available at all times through the Frame Relay cloud from one endpoint to another endpoint. An SVC is a connection through the Frame Relay cloud from one endpoint to another endpoint that is switched into and out of service as needed. The concept behind SVCs is that the bandwidth is not available when not needed, so service providers can bill at a lower rate. In practice, SVCs are not common— yet. They may become common someday. So what is a DLCI? The DLCI is simply the local identifier by which either a PVC or an SVC is known. The terms are frequently used interchangeably, but it is technically incorrect to refer to the number of a PVC; the number is the DLCI and has only local significance.

VoFR Call Setup

When you pick up a phone connected to a Cisco voice-enabled router and dial a number, the following events occur:

1   Going off-hook. (A signal is sent from the telephone hardware interface to the CPU; the CPU allocates resources.)

2   Dialing digits. (The DSP interprets digits; dial mapper interprets a number based on dial-peer commands.)

3   Finding the route. (The end-to-end call manager (EECM) finds the route and requests and completes the connection.)

4   Digital signal processor (DSP) takes over. (The call is "released" to DSP.)

Frame Relay then performs the four steps listed in Figure 8-1:

1   Dialed digits translated to the PVC

2   Call setup frame sent

3   Confirm frame (that is, call accepted, call rejected, or congestion on the link)

4   Telephone rings (assuming call accepted frame is returned)

Figure 8-1 *VoFR Call Setup*

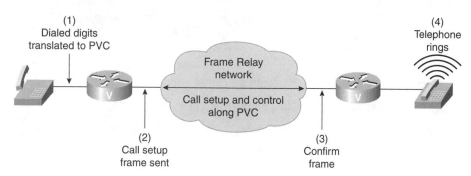

VoFR Addressing and Routing

The basic component of Frame Relay addressing is the data link connection identifier (DLCI), which is typically a value between 16 and 1007. The DLCI has local significance only, meaning the same DLCI can be used multiple times in different locations throughout a service provider's network. Mapping of the telephone numbers (ITU-T standard E.164 addresses) is handled through static tables, and the dialed digits are mapped to specific PVCs on a specific interface, not subinterface. See Chapter 7 for specifics on how to configure the dial-peer statements.

DLCI Numbering

DLCI addressing can be a real time-saver in troubleshooting your network. Some service providers will let you assign the DLCIs in their network, providing that the same DLCI is not in use on the switch that will connect to your site. One of the authors worked as a system engineer at a service provider that did just this, and by using DLCIs that reflected the IP network address of each site—as well as the IPX network at each site—it was possible to identify the troubled site immediately. All addresses were "synchronized" so that the last three digits were the same. This probably works best if you use private IP addressing (see RFC 1918). If your service provider will let you request DLCIs, do it.

VoFR routing is dependent on the configuration of the Frame Relay network. PVC endpoints are determined by the customer, but the service provider determines the internal configuration of the Frame Relay network. The service provider's network or pricing scheme may limit your network design options. It is possible to carry multiple calls and data per PVC, assuming proper sizing guidelines are followed.

Options in Designing a VoFR Network

Frame Relay provides economical transport for voice traffic compared to other transport mechanisms. Some specific design criteria should be kept in mind when designing a VoFR network. Using VoIP over Frame Relay is also a popular option, but the configuration of the voice design criteria needs to include both network layer and data link layer considerations. Both of these subjects will be covered in this section.

VoFR should be considered as a design option for the following reasons:

- Frame Relay is quite common and inexpensive in many parts of the world.

- Frame Relay customer premise equipment (CPE) supports voice now.

- Many branch-to-headquarters access networks have been implemented:

  — Delivering significant cost savings

  — Using fragmentation, signaling, compression

- VoFR standards provide for better voice support:

  — FRF.12 frame fragmentation

  — FRF.11 voice signaling carried in Frame Relay frames

- CS-ACELP (G.729) is becoming the de facto standard for voice compression in Frame Relay.

As a protocol specification, Frame Relay is used solely as a transport mechanism. However, many vendors have implemented proprietary VoFR solutions, and this serves to validate the VoFR market independently of the standards-based products Cisco delivers.

A full mesh of voice and data PVCs minimizes the number of network transit hops and maximizes the ability to establish different qualities of service. A network designed in this fashion minimizes delay and improves voice quality, but this design option usually represents the highest public network cost.

To reduce costs, both data and voice segments can be configured to use the same PVC, which reduces the number of PVCs required. Using a star topology is frequently the most economical implementation of Frame Relay. It also generally introduces the most delay.

In a star topology design, the central site switch reroutes voice calls. This network design has the potential problem of creating a transit hop when voice needs to go from one remote office to another remote office. However, it avoids the compression-decompression common when using a tandem PBX.

NOTE   Separate voice and data PVCs maximize quality of service if no call admission control is used, but this may increase the cost. Combining voice and data on one PVC generally minimizes the recurring costs. Based on your traffic patterns and analysis, you may want to use some combination of these two design options.

There are two components to delay in any network—fixed delay and variable delay—as discussed in the following sections.

Fixed Delay

Fixed delay is composed of:

- Propagation delay
- Serialization delay
- Processing delay

These components are illustrated in Figure 8-2 and are described in the following sections.

Figure 8-2   *Fixed Delay Components*

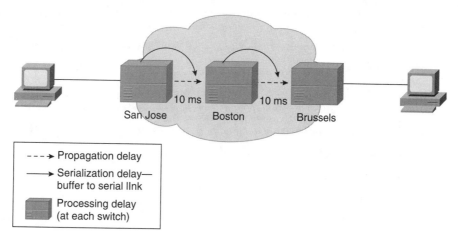

Propagation Delay

Propagation delay is the delay caused by the distance the electrons need to travel—it takes 6 microseconds per kilometer. You can only affect propagation delay by avoiding long-distance spans, such as those of satellite links, when there are shorter alternative links available. We need to consider propagation delay in our calculations, especially if the sites in our design are separated by a large geographic distance.

Serialization Delay

Serialization delay is the delay caused by clocking the bits on to the line. Serialization delay can be improved in our delay budget if we are willing to spend the extra dollars for higher line speeds. However, there may be trade-offs of cost versus bandwidth and in-country versus international spans.

Another consideration for serialization delay is that if voice is queued behind large data packets on low-speed links, we may exceed our delay budget on just one link. A QoS feature called *fragmentation* can mitigate this effect.

Processing Delay

Processing delay can be affected by:

- Algorithmic factors such as codec selection/compression/decompression
- Packetization, or the number of samples per packet

To overcome processing delay, you might need to switch to a different codec and perhaps thereby change voice quality. This option depends on how much processing delay you need to save. QoS features called *silence suppression* or *voice activity detection (VAD)* may be enough to free up the required bandwidth. Increasing the bandwidth may mitigate some packetization delay.

Variable Delay

Variable delays are caused by:

- De-jitter buffers
- Queuing delays
- WAN delays

These components are illustrated in Figure 8-3 and are described in the following sections.

Figure 8-3   *Variable Delay Components*

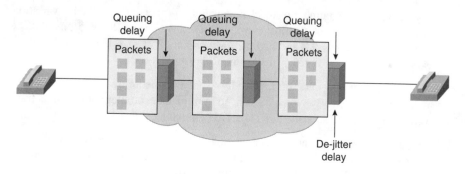

De-jitter Buffers

De-jitter buffers are time-delay buffers that delay the beginning of a "talk spurt" from being played to the user. Think for a moment: What would happen if the first packet of a talk spurt is played to the end user but the next packet hasn't arrived and been queued? There are correction methods in some algorithms, but after they run out, the user hears cracking and popping sounds. The de-jitter buffer delays the play-out of the first packet in the talk spurt, so the arriving packets are more likely to be in the queue when the first packet has been played.

Queuing Delay

Queuing delay results from network congestion and the need to shape the traffic to a contract on the egress port. In Frame Relay networks, to overcome queuing delay, we need to provide queuing *priority*.

WAN/Service Provider Delay

WAN/service provider delays are delays in the network. You may or may not own the links that these delays occur on. You need to understand what the service provider will do with your packets under congestion conditions. In public networks, we may not be able to affect network congestion. Even in some private networks, the backbone may not be under our control. In some cases, we can request priority PVCs from the network provider and service-level agreements (SLAs) for delay maximums.

Minimizing Delay and Delay Variation

Frame Relay has a number of mechanisms to minimize delay and delay variation in a network. The ones that are most commonly used are Frame Relay fragmentation (FRF.11 Annex C for VoFR PVCs and FRF.12 for data-only PVCs), prioritization, and traffic shaping as solutions to delay in a Frame Relay network.

The presence of long data frames on a low-speed Frame Relay link can cause unacceptable delays for time-sensitive voice frames. To minimize the problem of long data frames, some vendors implement smaller frame sizes to help reduce delay and delay variation.

FRF.12 is an industry standard for products from different vendors to be able to interoperate. FRF.12 fragments large frames and interleaves small frames among the fragments. The Frame Relay Forum's FRF.11 Annex C provides for the fragmentation of data packets when doing VoFR. FRF.12 is a data-only fragmentation specification, while FRF.11 Annex C provides for the fragmentation of data when the PVC is servicing both voice and data queues.

Prioritization sends time-sensitive voice frames before data frames. Methods for prioritizing voice frames over data frames also help reduce delay and delay variation. Prioritization and the use of smaller frame sizes are vendor-specific implementations.

To ensure voice quality, the *committed information rate* (CIR) on each PVC should be set properly to ensure that voice frames are not discarded:

- Each virtual circuit is assigned a minimum average service threshold; this is the CIR. A user can transmit traffic at a rate exceeding the CIR, but excess traffic might be discarded in the event of congestion.

- The minimum guaranteed service is the minCIR. Service may fall below the CIR during periods of congestion but will never fall below the minCIR.

- Backward explicit congestion notification (BECN) informs the source device that the network is experiencing congestion.

- Forward explicit congestion notification (FECN) informs the next device that the network is experiencing congestion.

Frame Relay traffic shaping (FRTS) is the mechanism used to make your traffic conform to a traffic contact agreed upon with a service provider and therefore to avoid packet drops in the service provider cloud. Traffic shaping is relevant when transporting VoFR and VoIP over Frame Relay service. Frame Relay traffic shaping shapes total PVC traffic to conform to the CIR, committed burst (Bc), and excessive burst (Be). Figure 8-4 illustrates reasons to shape traffic:

1 An ingress traffic policer tags frames that are not in compliance with the SLA with DE=1 and may discard them.

2 In case of congestions, tagged frames are dropped.

3 Bad voice quality is experienced due to packet loss.

Figure 8-4   *Reasons for Frame Relay Traffic Shaping*

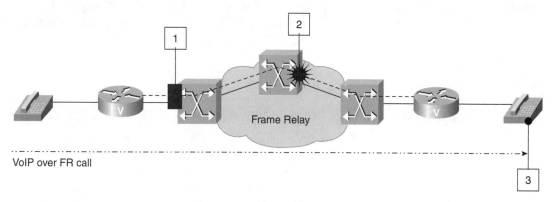

VoIP over FR call

Frame Relay traffic shaped to the service provider contracted rate avoids frame tagging and hence avoids packet drops. The secondary effect is that a source router can react to possible congestion by changing its transmission rate dynamically. The traffic contract, or SLA, says how much data the service provider agrees to transport over its Frame Relay service *with* guarantees and how much *without* guarantees. Traffic without guarantees can be dropped at any time (for example, due to congestion).

On a PVC transporting voice, it is recommended to do Frame Relay traffic shaping (not generic traffic shaping) to the CIR set up with the service provider, where CIR reflects a bandwidth required to transport required VoX calls plus a little more for overhead, thus ensuring reasonable voice quality. An example of a map-class Frame Relay group that provides traffic shaping is listed in Example 8-1.

Example 8-1   *Configuring Frame Relay Traffic Shaping on a Subinterface*

```
p1r3#config t
Enter configuration commands, one per line. End with CNTL/Z.
p1r3(config)#interface serial 0/0
p1r3(config-if)#encapsulation frame-relay
p1r3(config-if)#frame-relay traffic-shaping
p1r3(config-if)#interface serial 0/0.121 point-to-point
p1r3(config-subif)#frame-relay interface-dlci 121
p1r3(config-fr-dlci)#class voice2
p1r3(config-fr-dlci)#vofr cisco
p1r3(config-fr-dlci)#exit
p1r3(config-subif)#exit
p1r3(config-if)#exit
p1r3(config)#map-class frame-relay voice2
p1r3(config-map-class)#frame-relay adaptive-shaping becn
p1r3#conf t
p1r3(config)#map-class frame-relay voice
p1r3(config-map-class)#frame-relay cir 128000
p1r3(config-map-class)#frame-relay bc 1280
p1r3(config-map-class)#frame-relay-be 0
p1r3(config-map-class)#end
p1r3#
```

NOTE   The commands in Example 8-1 are explained in detail in the section "Using VoFR Configuration Commands" later in this chapter. The values for **cir** and **bc** in Example 8-1 are per PVC contracted values. These values are discussed in more detail later in this chapter.

An Introduction to Cisco's Implementation of VoFR

Cisco's implementation of VoFR is significant in that it supports purely standards-based VoFR and provides a solution for Cisco-based networks. The former is significant in that it permits a customer to implement Cisco equipment into a heterogeneous network and interoperate with equipment from other vendors. The latter provides a more robust call setup than using FRF.11. Customers are free to select which of the two implementations they prefer in their environment.

In this section, we will discuss the background of Cisco's implementation of VoFR and both standards-based and Cisco-enhanced solutions. We also will highlight the following:

- The VoFR functions these routers perform and the standards Cisco supports for VoFR
- Benefits of the latest releases of Cisco IOS
- Restrictions of Cisco IOS in performing VoFR functions

Background on Cisco's Implementation of VoFR

The Cisco voice-capable routers integrate voice, LAN, synchronous data, video, and fax traffic for transport over a public or private Frame Relay network. Cisco voice-enabled routers from multiple remote sites can be multiplexed into a central site router through Frame Relay links. Cisco optimizes network bandwidth over these links by multiplexing voice and data on the same circuit or physical interface.

Cisco voice-capable routers perform the following functions:

- Enable real-time, delay-sensitive voice traffic to be carried over slow Frame Relay links
- Allow dedicated 64-kbps time-division multiplexing (TDM) telephony circuits to be replaced by more economical Frame Relay permanent virtual circuits (PVCs)
- Utilize voice compression technology that conforms to International Telecommunication Union Telecommunication Standardization Sector (ITU-T) specifications
- Provide support for Frame Relay fragmentation

- Allow intelligent setup of proprietary switched VoFR connections between two VoFR endpoints, saving the extensive configuration overhead associated with pure FRF.11 implementations

- Support standards-based FRF.11 and FRF.12 functionality, allowing Cisco routers to interconnect with other equipment supporting these specifications

The following Frame Relay Forum specifications are supported and are considered significant:

- FRF.1: User-Network Interface
- FRF.3.1: Multiprotocol Encapsulation (RFC 1490)
- FRF.5: Frame Relay/ATM Network Interworking
- FRF.8: Frame Relay/ATM PVC Service Interworking
- FRF.9: Data Compression over Frame Relay
- FRF.11: Standard for VoFR
- FRF.12: Frame Fragmentation

For more information on these standards, you can go to Frame Relay Forum's Web site (www.frforum.com), ANSI's Web site (www.ansi.org), and IETF's Web site (www.ietf.org).

Benefits of Cisco's VoFR

Cisco IOS Release 12.0(7)XK and later provide the following VoFR benefits:

- Configuration consistency across all VoFR routers. (Cisco 2600 and 3600 series routers, the Cisco 7200 Series routers, and the Cisco MC3810 multiservice access concentrator).

- Digital voice support on the Cisco 2600 and 3600 series routers. (Cisco 3600 series routers support digital voice calls for VoFR.)

- Interoperability with Cisco 7200.

NOTE   In Cisco IOS releases prior to Release 12.0(5)XE, the Cisco 7200 could only act as a tandem router in VoFR topologies and could not terminate VoFR calls. Beginning in Cisco IOS Release 12.0(5)XE2 and 12.1(1)T, the Cisco 7200 can terminate VoFR calls. Remember to check the release notes for the version of Cisco IOS you are using to ascertain current release capabilities and restrictions.

Restrictions of Cisco's VoFR

There are some restrictions related to using VoFR as implemented by Cisco IOS:

- The Cisco 2600 series and 3600 series routers cannot terminate calls initiated by a Cisco MC3810 using VoFR implementations prior to Cisco IOS Release 12.0(3)XG or 12.0(4)T.

- Cisco MC3810 concentrators running Cisco IOS versions prior to Release 12.0(3)XG or 12.0(4)T cannot tandem VoFR calls from Cisco 2600 series, 3600 series, and 7200 series routers.

- It is currently not possible to translate from the VoIP transport protocol to VoFR. A call coming in on a VoIP connection will not be (tandem) switched to a VoFR connection.

Applying QoS to VoFR

QoS covers four distinct functional areas, each with its own specialization and its own terminology. Below are the functional areas and the terms that apply to those functions in their transport arenas:

- **(Traffic) policing**—The process used to measure the actual traffic flow across a given connection and compare it to the total admissible traffic flow for that connection. Traffic outside of the agreed upon flow can be tagged (where the CLP bit in ATM is set to 1) and can be discarded en route if congestion develops. Traffic policing is not done with VoFR.

- **Traffic shaping**—Uses queues to limit surges that can congest a network. Data is buffered and then sent into the network in regulated amounts to ensure that the traffic will fit within the promised traffic envelope for the particular connection. Traffic shaping is used in Frame Relay, ATM, and other types of networks. It performs the following tasks:

  — Limiting the packet rate

  — Buffering to smooth traffic flow

  — Output mechanism

  — Generic traffic shaping (GTS), Frame Relay traffic shaping (FRTS), ATM shaping

- **Queuing/scheduling**—Queuing involves organizing packets waiting to go out on an interface. Scheduling involves, when interface is free, deciding which of the waiting packets to send next.

- **Tagging/marking/coloring**—Terms used to describe setting bits in the packet header to guide priority and queuing mechanisms. These can be changed/adjusted by any network node. They covers such things as:

    — IP Precedence—Using the IP Type of Service (ToS) field to prioritize the traffic

    — DiffServ Code Point (DSCP)—Using the redefined ToS field to provide differentiated service

    — Discard Eligible bit (DE)—Using the Frame Relay DE bit to mark a frame as eligible for discard during congestion

NOTE   Two of the mechanisms listed (IP Precedence and DiffServ Code Point) are used by Layer 3—IP-based—traffic. They are listed only to provide the reader with a complete picture of tagging/marking/coloring schemes.

Because of the importance of QoS, we need to look at the specific considerations necessary in designing and implementing a VoFR network and the tools to overcome these issues. The following sections address these areas.

QoS Considerations

You need to consider several factors when designing and implementing the QoS tools you use in a network. Factors to consider based on network location are illustrated in Figure 8-5 and discussed below:

Figure 8-5   *QoS Considerations*

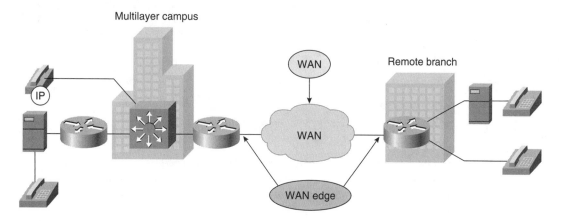

- **Campus**: Bandwidth minimizes QoS issues to some extent, but buffers must be kept in mind where speeds change. A high-speed pipe supplying a slower link may result in buffer overflow, causing lost data or sequencing issues.

- **WAN edge**: Often slow access links connect to high-speed LAN links. If WAN links are less than 2 MB, QoS techniques are a must to attain acceptable voice quality.

- **WAN**: Often QoS techniques are overlooked or misunderstood. Keep these points in mind:

  — Speed mismatches—Speed mismatches between campus and remote sites may cause buffer overflow problems.

  — Oversubscription—Too many users on one side of a WAN link may cause dropped packets if the WAN link is oversubscribed.

  — Lack of control over a service provider (SP) network—Despite your best engineering efforts, the SP may have policies in place that defeat your QoS techniques.

There are several symptoms of poor QoS:

- **Loss**: Packet drops when congestion occurs. There are either inadequate buffers to hold the voice packets, or the packets are held too long in a buffer and time out.

- **Delay**: Delay includes fixed (codec, serialization, processing, WAN propagation) and variable (that is, jitter; including queuing, SP WAN, de-jitter buffer, traffic shaping) delay.

- **Bandwidth**: Contention induces delay (traffic shaping and queuing may alleviate this problem).

- **Echo:** Can be aggravated by delay.

QoS Tools

To solve the problem of poor QoS as identified by the preceding symptoms, you can use various QoS tools. The following apply to Frame Relay in particular:

- **Prioritization**. Gives priority treatment to real-time sensitive traffic.

- **Queuing/scheduling.** PQ-WFQ or LLQ.

- **Link/bandwidth efficiency**. Limits delay on slow links.

Why is there a need to fragment? You need to fragment to contain serialization delay of large data packets on slow links.

When large data frames precede voice frames on a low-speed link, the residency or occupancy of the large data frame will cause a significant penalty in the voice delay budget. Fragmentation of large data frames allows voice frames to be interleaved with the data fragments.

The chart in Figure 8-6 is a guide to calculating the fragment size, based on the maximum delay that the low-speed link may add to the delay budget.

Figure 8-6   *Fragmentation Frame Size Matrix*

Real-time packet interval

Link speed	10 ms	20 ms	30 ms	40 ms	50 ms	100 ms	200 ms
56 kbps	70 bytes	140 bytes	210 bytes	280 bytes	350 bytes	700 bytes	1400 bytes
64 kbps	80 bytes	160 bytes	240 bytes	320 bytes	400 bytes	800 bytes	1600 bytes
128 kbps	160 bytes	320 bytes	480 bytes	640 bytes	800 bytes	1600 bytes	3200 bytes
256 kbps	320 bytes	640 bytes	960 bytes	1280 bytes	1600 bytes	3200 bytes	6400 bytes
512 kbps	640 bytes	1280 bytes	1920 bytes	2560 bytes	3200 bytes	6400 bytes	12,800 bytes
768 kbps	1000 bytes	2000 bytes	3000 bytes	4000 bytes	5000 bytes	10,000 bytes	20,000 bytes
1536 kbps	2000 bytes	4000 bytes	6000 bytes	8000 bytes	10,000 bytes	20,000 bytes	40,000 bytes

▢ = fragmentation not an issue due to BW + interval combination

Example of Fragmentation Size Matrix Use

If the maximum acceptable delay that the low-speed link may add to the budget is 20 ms and the link speed is 64 kbps, then the maximum fragment size may be no more than 160 bytes. The blocks in Figure 8-6 that are shaded are generally of sufficient speed that fragmentation will not be required.

The chart illustrated in Figure 8-6, in conjunction with Figure 8-7, is a powerful tool to begin designing a VoFR network.

For Frame Relay PVCs, the fragment size should be set according to the speed of the PVC (for example, a 128-kbps PVC on a T1 would have the fragment size on that PVC set to 160 bytes).

Figure 8-7   *Fragmentation Recommendations*

Link speed	Frame size						
	1 byte	64 bytes	128 bytes	256 bytes	512 bytes	1024 bytes	1500 bytes
56 kbps	143 μs	9 ms	18 ms	36 ms	72 ms	144 ms	214 ms
64 kbps	125 μs	8 ms	16 ms	32 ms	64 ms	128 ms	187 ms
128 kbps	62.5 μs	4 ms	8 ms	16 ms	32 ms	64 ms	93 ms
256 kbps	31 μs	2 ms	4 ms	8 ms	16 ms	32 ms	46 ms
512 kbps	15.5 μs	1 ms	2 ms	4 ms	8 ms	16 ms	23 ms
768 kbps	10 μs	640 μs	1.28 ms	2.56 ms	5.1 ms	10.2 ms	15 ms
1536 kbps	5 μs	320 μs	640 μs	1.28 ms	2.56 ms	5.12 ms	7.5 ms

Here is the formula for determining fragmentation size. Assume a 10-ms blocking delay fragment:

$$\frac{10 \text{ ms}}{\text{time for 1 byte at bandwidth}} = \text{fragment size}$$

Now consider four G.729 calls on 128-kbps circuit. From the chart in Figure 8-6, we know that a 128-kbps circuit with a 10-ms blocking delay permits a fragment size of 160 bytes.

Here is the general formula for finding the worst-case delay based on actual value calculations for frame size, number of calls, bandwidth, and blocking delay:

$$Q = (Pv \times N/C) + LFI$$

This is the basic formula we will use to determine the fragment size. The terms of the equation are defined as follows:

- **Q** = Worst-case delay of voice packet (in ms). In our example, we have calculated this to be 25 ms.
- **Pv** = Size of a voice packet in bits (at Layer 1). So we include the G.729 frame and the Frame Relay header for a total of 60 bytes.
- **N** = Number of calls (for this example, 4).
- **C** = The link capacity in bps (for this example, 128 kbps).
- **LFI** = Fragment size serialization delay in ms (for this example, 10 ms).

Table 8-1 illustrates the recommended fragment size for various link speeds based on a 10-ms blocking delay. The fragment size should be as large as possible to increase throughput and decrease overall delay.

Table 8-1   *Recommended Fragment Sizes*

Link Speed	Fragment Size
56 kbps	70 bytes
64 kbps	80 bytes
128 kbps	160 bytes
256 kbps	320 bytes
512 kbps	640 bytes
768 kbps	1000 bytes
1536 kbps	2000 bytes

Applying this formula to the case of 4 calls on a 128-kbps circuit with a 10-ms blocking delay yields the following formula:

$$Q = (480 \text{ bits} \times 4/128{,}000) + 10 \text{ ms} = 25 \text{ ms}$$

Worst case queuing delay $= 25$ ms

From this, we can see that the worst-case queuing delay we should expect for voice over this circuit is 25 ms. We can now use that figure in calculating our overall delay to ensure that we are within the recommended range (less than 150 ms total delay one way).

Fragmentation generally is not needed above 768 kbps. Some other means of dealing with symptoms of poor QoS are:

- **Send fewer packets:** Use voice activity detection (VAD) so that silence is not sampled and forwarded to the person at the other end of the phone call.

- **Call admission control.** Use of Cisco IOS features that check to ensure that there is adequate bandwidth for a call before sending a call setup frame prevents admitting calls that will not be completed.

- **Traffic shaping**. Smoothes out speed mismatches. Frame Relay traffic shaping (FRTS) conforms output to PVC values as negotiated with the SP.

- **Bandwidth management**. Check/reserve/restrict bandwidth for certain flows and the number of ingress ports. Use of the **voice bandwidth** command sets the bandwidth permitted for voice calls.

Using VoFR Configuration Commands

You can configure VoFR calls on Cisco 2600, 3600, and 7200 series routers and on Cisco MC3810 concentrators. To configure calls on routers that support VoFR, use the commands described in the following sections from interface configuration mode.

The encapsulation frame-relay Command

The **encapsulation frame-relay** command enables Frame Relay on the interface. Note that Frame Relay traffic shaping should be enabled on the interface if sending voice and data traffic over a single Frame Relay PVC over a public Frame Relay network. Also, be sure to use Frame Relay traffic shaping only; do not use generic traffic shaping.

The frame-relay interface-dlci Command

The **frame-relay interface-dlci** *dlci* [**ieft** | **cisco**] command assigns a DLCI to a specified Frame Relay subinterface on the interface or subinterface.

Subinterfaces are logical interfaces associated with a physical interface. This command is required for all point-to-point subinterfaces; it is also required for multipoint subinterfaces for which dynamic address resolution is enabled. It is not required for multipoint subinterfaces configured with static address mappings. You must specify both the interface and subinterface before you can use this command to assign any DLCIs and any encapsulation or broadcast options.

On subinterfaces, using the **frame-relay interface-dlci** command will enable the use of routing protocols on interfaces that use Inverse ARP. It is also valuable for assigning a specific class to a single PVC where special characteristics are desired. Use of the command on an interface (rather than on a subinterface) prevents the device from forwarding packets intended for that DLCI.

The options for the **frame-relay interface-dlci** command are described in Table 8-2.

Table 8-2   **frame-relay interface-dlci** *Command Parameters*

Command Parameters	Description
dlci	Specifies the DLCI number to be used on the specified subinterface
broadcast	Sends broadcasts on this DLCI
ietf	**cisco**	(Optional.) Specifies the encapsulation type: Internet Engineering Task Force (IETF) Frame Relay encapsulation or Cisco Frame Relay encapsulation

NOTE	The **voice-encap** *size* option of the **frame-relay interface-dlci** command on the Cisco MC3810 is no longer supported beginning with Release (12.0(7)XK).

The class Command

You use the **class** *map-class-name* command to associate a map class with a specified DLCI. The **class** *map-class-name* command permits you to assign the same traffic shaping parameters to multiple interfaces with a single command.

The vofr Command

The **vofr** [[**cisco**]|[[**data** *cid*] [**call-control** [*cid*]]]] command enables VoFR on a specific DLCI and optionally configures specific subchannels on that DLCI. Use the **vofr** command from Frame Relay DLCI configuration mode. Use the **no** form of the command to disable VoFR on a specific DLCI.

The options for this command include whether to use Cisco's proprietary encapsulation for backward compatibility or to specifically define the subchannel ID for the data channel and the subchannel ID for the voice channel. Table 8-3 lists the range of options for this command and their purposes.

Table 8-3   **vofr** *Command Options*

Command Parameters	Description
cisco	(Optional.) Cisco proprietary voice encapsulation for VoFR with data carried on CID 4 and call-control on CID 5.
	Note: This option is required when configuring switched calls or Cisco trunks to Cisco MC3810 concentrators running Cisco IOS releases prior to 12.0(7)XK.
	If configuring switched calls or Cisco trunks to Cisco MC3810 concentrators running Cisco IOS Release 12.0(7)XK or 12.1(2)T and later, do not use this option.
data	(Required for Cisco-trunk permanent calls. Optional for switched calls.) Selects a subchannel (CID) for data other than the default subchannel, which is 4.
cid	(Optional.) Specifies the subchannel to be used for data. Valid values are from 4 to 255; the default is 4. If data is specified, a valid CID must be entered.

continues

Table 8-3   **vofr** *Command Options (Continued)*

Command Parameters	Description
call-control	(Optional.) Used to specify that a subchannel will be reserved for call-control signaling. Note: This option is not supported on the Cisco MC3810.
cid	(Optional.) Specifies the subchannel to be used for call-control signaling. Valid values are from 4 to 255; the default is 5. If call-control is specified and a CID is not entered, the default CID will be used.

Combinations of the **vofr** Command

The **vofr** [*option*] command has several options, depending on the type of call being made (see Table 8-4).

Table 8-4   **vofr** [*option*] *Command Options for Various Types of Calls*

Type of Call	vofr Command Combination to Use
Switched call (user dialed or auto-ringdown) to other routers supporting VoFR	**vofr** [**data** *cid*] [**call-control** *cid*]
Switched call (user dialed or auto-ringdown) to a Cisco MC3810 running IOS Releases prior to 12.0(7)XK	**vofr cisco**
Cisco-trunk permanent call to other routers supporting VoFR	**vofr data** *cid* **call-control** *cid*
Cisco-trunk permanent call (private-line) to a Cisco MC3810 running IOS Releases prior to 12.0(7)XK	**vofr cisco**

The recommended use of this command is **vofr data 4 call-control 5** on new installations when backward compatibility is not an issue. It is standards-compliant on fragmentation and is the best current practice for new networks.

vofr cisco is backward compatible in fragmentation with proprietary Cisco VoFR installations. There is no need to upgrade your network if this is what you've already got. You must use this option if you have an older release of Cisco IOS running on the MC3810s in your network. This command consumes data CID 4 and call-control CID 5.

Special Conditions and Restrictions for Using **vofr**

There are some special considerations that you should be aware of when using the **vofr** [*option*] command:

- The **voice-encap** option of the **frame-relay interface-dlci** command on the Cisco MC3810 is no longer supported as of Cisco IOS Release 12.0(7)XK.

- For switched VoFR calls, use **vofr cisco** on the Cisco 2600, 3600, and 7200 series routers.

- When the **vofr** command is used *without* the **cisco** option, all subchannels on the DLCI are configured for FRF.11 encapsulation. If the **vofr** command is entered without any keywords or arguments, the data subchannel will be CID 4, and there will be no call-control subchannel. Both of these encapsulations are FRF.11 encapsulations.

- The **session protocol cisco-switched** dial-peer configuration command option is the default setting. If you do not enter this command, the setting will still apply.

- Cisco proprietary fragmentation is based on an early draft of FRF.11 Annex C and is compatible with Cisco MC3810 concentrators running software versions prior to Cisco IOS Release 12.0(3)XG or 12.0(4)T.

Use of Cisco trunks for permanent calls (private-line) is recommended over FRF.11-trunk calls unless FRF.11-compliant, standards-based interworking is required with non-Cisco devices. The Cisco trunk protocol is a superset of the FRF.11 protocol and contains Cisco proprietary extensions designed to support switched call routing and other advanced features.

Using the **vofr** Command with the MC3810

Use of the **cisco** option was modified beginning in release 12.0(7)XK. You now use the **cisco** option only when configuring connections to Cisco MC3810 concentrators running Cisco IOS releases before 12.0(7)XK.

There are some conditions and restrictions for using the **vofr** command on an MC3810 router for releases prior to 12.0(7)XK.

- Cisco trunks must use the **vofr cisco** command.

- Switched VoFR must use the **vofr cisco** command.

If the **data** option is selected, a numeric value must be entered to complete the command. If the **call-control** option is selected, you need not enter a numeric value if you wish to accept the default **call-control** subchannel.

When the **vofr** command is used on a Cisco MC3810 without the **cisco** option, switched calls are not permitted. Only permanent FRF.11-trunk calls can be made.

Example 8-2 shows how to enable VoFR on Serial 1/1, DLCI 100 on a Cisco 2600, 3600, or 7200 series router or on an MC3810 concentrator, starting from global configuration mode.

Example 8-2   *Enabling the* **vofr** [**cisco**] *Command on the Cisco 2600, 3600, and 7200*

```
router(config)#interface serial 1/1.100 point-to-point
router(config-subif)#frame-relay interface-dlci 100
router(config-fr-dlci)#vofr
```

To configure a Cisco router or MC3810 for a VoFR application with an older release of the MC3810 (prior to Release 12.0(3)XG), enter the command shown in Example 8-3.

Example 8-3   *Enabling* **vofr cisco** *on an MC3810*

```
router#conf t
Enter configuration commands, one per line. End with CNTL/Z.
router(config)#int s0.100 point-to-point
router(config-subif)#frame-relay interface-dlci 100
router(config-fr-dlci)#vofr cisco
```

The frame-relay traffic-shaping Command

The **frame-relay traffic-shaping** interface configuration command is used to enable both traffic shaping and per-virtual circuit queuing for all PVCs on a Frame Relay interface. To disable traffic shaping and per-virtual circuit queuing, use the **no** form of this command.

NOTE   To configure weighted fair queuing, use the **frame-relay fair-queue** command in the **map-class frame-relay** *map-class-name* command. See the section "The **frame-relay fair-queue** Command" for more information.

For virtual circuits for which no specific traffic shaping or queuing parameters are specified, a set of default values is used. The default queuing is performed on a first-come, first-served basis. You should always use traffic shaping when servicing voice traffic on a PVC.

The map-class frame-relay Command

Use the **map-class frame-relay** *map-class-name* command in interface configuration mode to configure the Frame Relay map-class. In this configuration mode, you can use the following commands to provide traffic shaping:

- **frame-relay cir [in | out]** *bps*
- **frame-relay mincir [in | out]** *bps*
- **frame-relay bc [in | out]** *bits*
- **frame-relay be [in | out]** *bits*
- **frame-relay fair-queue** [*congestive_discard_threshold* [*number_dynamic_conversation_queues* [*number_reservable_conversation_queues* [*max_buffer_size_for_fair_queues*]]]]
- **frame-relay voice bandwidth** *bps_reserved*
- **frame-relay adaptive-shaping**

Many of these commands are covered in the following sections. Example 8-4 shows the use of some of these commands.

Example 8-4   **map-class frame relay** *Example*

```
router#conf t
Enter configuration commands, one per line. End with CNTL/Z.
router(config)#map-class frame vofr
router(config-map-class)frame-relay cir 64000
router(config-map-class)frame-relay bc 640
router(config-map-class)frame-relay fragment 80
router(config-map-class)frame-relay fair-queue
router(config-map-class)frame-relay voice bandwidth 48000
```

The frame-relay voice bandwidth Command

The **frame-relay voice bandwidth** *bps_reserved* [**queue** *depth*] command specifies how much bandwidth should be reserved for voice traffic on a specific DLCI and the number of voice packets to hold in the queue. The default value for *bps_reserved* is 0, which disables voice calls; the valid range is 8000 to 45,000,000. The default value for **queue** *depth* is 100, with a valid range of 30 to 1000.

To use this command, you must first associate a Frame Relay map-class with a specific DLCI and then enter map-class configuration mode and set the amount of bandwidth to be reserved for voice traffic for that map-class.

If a call is attempted and there is not enough remaining bandwidth reserved for voice to handle the additional call, the call will be rejected.

For example, if 64 kbps is reserved for voice traffic and a codec and payload size is being used that requires 10 kbps of bandwidth for each call, then the first 6 calls attempted will be accepted, but the 7th call will be rejected.

Cisco strongly recommends that you set **voice bandwidth** *bps_reserved* [**queue** *depth*] to a value less than the CIR if Frame Relay traffic shaping is configured. Cisco also strongly recommends that you set the minimum CIR (using the **frame-relay mincir** command) to be at least equal to or greater than the voice bandwidth.

The voice bandwidth Command

The **voice bandwidth** [*kbps*] command is applicable only for VoFR. When calculating voice bandwidth, use aggregate voice BW; don't include call setup overhead:

- G.729 VoFR calls need 10.4 kbps.
- Allowing 5 calls would need a **voice bandwidth** of 52 kbps.
- Voice BW defaults to 0; if not specified, no voice calls will be allowed.
- Calls attempted, but exceeding the BW, will be rejected:
  - If no other dial peers match, then an overflow tone is sent to the caller.
  - If another dial peer matches, then the call is rerouted as appropriate.
- Frame Relay traffic shaping (FRTS) is mandatory in order to support fragmentation.
- Use fragmentation with WFQ for non-voice data traffic.

As of Cisco IOS Release 12.0(5)T, the **voice bandwidth** command now triggers priority queue-weighted fair queue (PQ-WFQ).

NOTE   For a full discussion of the **frame-relay fair-queue** command see the section "The **frame-relay fair-queue** Command."

Bandwidth Calculations for VoFR

In this section, we will look at the calculations necessary to determine the amount of bandwidth used by each call. Voice bandwidth used by one voice call, at the physical interface, depends on:

- Media used (Frame Relay, Ethernet, serial/MLPPP, ATM)
- Codec: Payload size, packets-per-second rate (G.729, G.711, and so on)
- VAD

In addition, for Frame Relay (and ATM) networks, the overhead of the voice packet size fitting into fast packets or ATM cells must be taken into account.

The bandwidth required for a voice call depends on the bandwidth of the codec, the voice packetization overhead, and the voice frame payload size. The smaller the voice frame payload size, the higher the bandwidth required for the call.

To make the calculation, use the following formula:

required_bandwidth = codec_bandwidth × (1 + overhead / payload_size)

The overhead for VoFR voice frame is between 6 and 8 bytes. The frame is illustrated in Figure 8-8, and the fields are described below:

- 2-byte Frame Relay header.
- 1- or 2-byte FRF.11 header (depending on the CID value).
- 2-byte CRC.
- 1-byte trailing flag. (If voice sequence numbers are enabled in the voice packets, there is an additional 1-byte sequence number.)

For example, using a G.729 codec with a bandwidth of 8 kbps, an overhead of 6 bytes, and a payload size of 80 bytes, the formula would be:

required_bandwidth = 8000 × (1 + 6/30)

required_bandwidth = 8000 × (1.2)

required_bandwidth = 9600, or 9.6 kbps

Figure 8-8   *VoFR Frame*

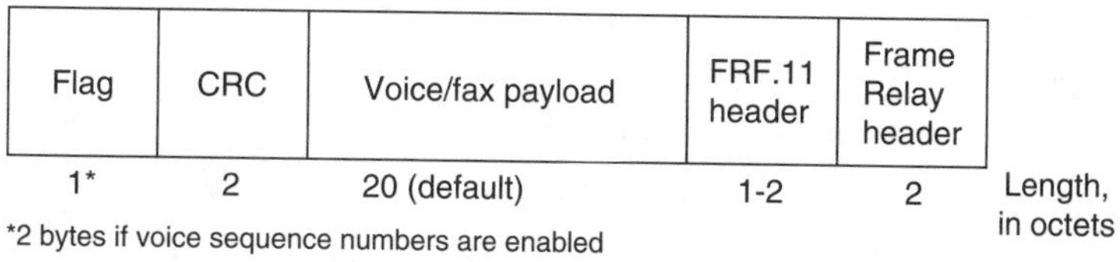

Flag	CRC	Voice/fax payload	FRF.11 header	Frame Relay header
1*	2	20 (default)	1-2	2

*2 bytes if voice sequence numbers are enabled

Length, in octets

Table 8-5 shows the required voice bandwidth for the G.729 8000-bps speech codec for various payload sizes.

Table 8-5   *Voice Bandwidth Required Per Codec*

Codec	Codec Bandwidth	Voice Frame Payload Size	Required Bandwidth Per Call (6 byte OH)	Required Bandwidth Per Call (8 byte OH)
G.729	8000 bps	120 bytes	8400 bps	8534 bps
G.729	8000 bps	80 bytes	8600 bps	8800 bps
G.729	8000 bps	40 bytes	9200 bps	9600 bps
G.729	8000 bps	30 bytes	9600 bps	10134 bps
G.729	8000 bps	20 bytes (Default)	10400 bps	11200 bps

NOTE	Voice payload sizes above 40 bytes with G.729 may cause unacceptable delay.

To configure the payload size for the voice frames, use the **codec** command from dial-peer configuration mode.

Example 8-5 shows how to reserve 64 kbps for voice traffic for the vofr Frame Relay map-class on a Cisco 2600, 3600, or 7200 series router or on an MC3810 concentrator, starting from global configuration mode.

Example 8-5   *Reserving 64 kbps for VoFR Traffic*

```
router(config)#interface serial 1/1
router(config-if)# frame-relay interface-dlci 100
router(config-fr-dlci)# class vofr
router(config-fr-dlci)# exit
router(config)# map-class frame-relay vofr
router(config-map-class)# frame-relay voice bandwidth 64000
```

The frame-relay fair-queue Command

The **frame-relay fair-queue** [*congestive_discard_threshold* [*number_dynamic_conversation_queues* [*number_reservable_conversation_queues* [*max_buffer_size_for_fair_queues*]]]] command enables weighted fair queuing for one or more Frame Relay PVCs. Use this command in conjunction with the **map-class frame-relay** *map-class-name* command. The options for the **frame-relay fair-queue** command are listed in Table 8-6.

Table 8-6   *Options for the* **frame-relay fair-queue** *Command*

Command Parameters	Description
congestive_discard_threshold	(Optional.) Specifies the number of messages allowed in each queue. The range is from 1 to 4096 messages; the default is 64.
number_dynamic_conversation_queues	(Optional.) Specifies the number of dynamic queues to be used for best-effort conversations—normal conversations not requiring any special network services. Valid values are 16, 32, 64, 128, 256, 512, 1024, 2048, and 4096; the default is 16.
number_reservable_conversation_queues	(Optional.) Specifies the number of reserved queues to be used for carrying voice traffic. The range is from 0 to 100; the default is 2. (The CLI will not allow a value less than 2 if fragmentation is configured on the Frame Relay map-class.)
max_buffer_size_for_fair_queues	(Optional.) Specifies the maximum buffer size in bytes for all of the fair queues. The range is from 0 to 4096 bytes; the default is 600.

To use this command, you must first associate a Frame Relay map-class with a specific DLCI and then enter map-class configuration mode and enable (or disable) weighted fair queuing for that map class.

When Frame Relay fragmentation is enabled, weighted fair queuing is the only queuing strategy allowed.

If the **frame-relay fair-queue** command is entered without any accompanying options, the default values for each of the four parameters will be set. See Table 8-6 for a description of the options and their default values. If you alter only the value of the first parameter (*congestive_discard_threshold*), you only need to enter the desired value for that parameter. If you alter only the value of the second, third, or fourth parameters, you must enter values for the preceding parameters as well as for the parameter you wish to change.

Example 8-6 shows how to enable WFQ and set the default parameter values for the **vofr** Frame Relay map-class on a Cisco 2600, 3600, or 7200 series router or on an MC3810 concentrator, starting from global configuration mode.

Example 8-6   *Enabling WFQ and Setting Default Values on a Map-Class Group*

```
router(config)# interface serial 1/1
router(config-if)# frame-relay interface-dlci 100
router(config-fr-dlci)# class vofr
router(config-fr-dlci)# exit
router(config)# map-class frame-relay vofr
router(config-map-class)# frame-relay fair-queue
router(config-map-class)#
```

The frame-relay fragment Command

The **frame-relay fragment** *fragment-size* command enables fragmentation of Frame Relay frames for a Frame Relay map-class (see Table 8-7).

Table 8-7   **frame-relay fragment** *fragment-size Options*

Command Parameter	Description
fragment-size	Specifies the number of payload bytes from the original Frame Relay frame that will go into each fragment. This number excludes the Frame Relay header of the original frame.
	All the fragments of a Frame Relay frame except the last will have a payload size equal to *fragment_size*; the last fragment will have a payload less than or equal to *fragment_size*. Valid values are from 16 to 1600 bytes; the default is 53.

Frame Relay fragmentation is enabled on a per-PVC basis. Before enabling Frame Relay fragmentation, you must first associate a Frame Relay map-class with a specific DLCI and then enter map-class configuration mode and enable or disable fragmentation for that map-class. In addition, you must enable Frame Relay traffic shaping on the interface in order for fragmentation to work.

For Frame Relay PVC's, the fragment size should be set according to the speed of the PVC and the desired delay (for example, a 128 kbps PVC on a T1 would have the fragment size on that PVC set to 160 bytes).

Fragmenting Frame Relay

Frame Relay frames are fragmented using one of the following formats, depending on how the PVC is configured:

- Cisco-proprietary format
- Pure end-to-end FRF.12 format (used for data and VoIPoFR traffic)
- FRF.11 Annex C format

End-to-end FRF.12 fragmentation should be used on PVCs that are carrying VoIP packets and on PVCs that are sharing the link with other PVCs carrying VoFR traffic. In pure end-to-end FRF.12 fragmentation, Frame Relay frames with a payload less than the fragment size configured for that PVC are transmitted without the fragmentation header.

FRF.11 Annex C and Cisco proprietary fragmentation are used when a PVC is configured for VoFR. When fragmentation is enabled on a PVC, FRF.11 Annex C format is triggered when **vofr** is configured on that PVC; Cisco proprietary format is triggered when **vofr cisco** is configured.

In FRF.11 Annex C and Cisco proprietary fragmentation, VoFR frames are never fragmented, and all data packets (including VoIP packets) contain the fragmentation header regardless of the payload size.

The frame-relay cir Command

The **frame-relay cir** {**in** | **out**} *bps* command configures the incoming and outgoing CIR for the PVC/SVC. Use this command to specify a CIR.

The frame-relay mincir Command

The **frame-relay mincir** {**in** | **out**} *bps* command configures the incoming and outgoing minimum CIR for the PVC/SVC. The network uses the mincir value when allocating resources for the SVC. If the mincir value cannot be supported, the call is cleared.

The frame-relay bc Command

The **frame-relay bc** {**in** | **out**} *bits* command configures the incoming and outgoing committed burst size (Bc). The Frame Relay committed burst size is specified within a map class to request a certain burst rate for the circuit. Although it is specified in bits, an implicit time factor is the sampling interval Tc on the switch, which is defined as the burst size Bc divided by the CIR. Configure the bits value to a minimum of 1000 for the voice traffic. The default is 7000 bits.

The frame-relay be Command

The **frame-relay be** {**in** | **out**} *bits* command configures the incoming and outgoing excess burst size (Be). The Frame Relay committed burst size is specified within a map-class to request a certain burst rate for the circuit. Although it is specified in bits, an implicit time factor is the sampling interval Tc on the switch, which is defined as the burst size Bc divided by the CIR. Configure the bits value to a minimum of 1000 for the voice traffic. The default is 7000 bits.

The frame-relay adaptive-shaping Command

You use the **frame-relay adaptive-shaping** {**becn** | **foresight**} command to configure the adaptive traffic rate adjustment to support BECN or ForeSight backward congestion notification messages.

Configuration Examples

In this section, you can review several examples of router configurations for various implementations. The examples included are:

- PQ-WFQ
- WFQ
- End-to-end FRF.12 fragmentation
- FRF.11 Annex C
- Cisco Frame Relay
- VoFR CIR
- Routers using a VoFR CIR
- Cisco-trunk (private-line) Call
- FRF.11 trunk
- Tandem configuration for switched calls

PQ-WFQ Configuration Example

Example 8-7 is a partial example of PQ-WFQ for VoFR. It illustrates a configuration that provides CIR of 32 kbps, committed burst of 400 bits, fragmentation size of 50 bytes, PQ-WFQ, and a priority queue of 24 kbps for voice traffic.

Example 8-7   *PQ-WFQ Configuration Example*

```
p1r3#config t
Enter configuration commands, one per line. End with CNTL/Z.
p1r3(config)#interface serial0
p1r3(config-if)#encapsulation frame-relay
p1r3(config-if)#frame-relay traffic-shaping
p1r3(config-if)#interface s0.200 point-to-point
p1r3(config-subif)#frame-relay interface-dlci 200
p1r3(config-fr-dlci)#class vofr
p1r3(config-fr-dlci)#vofr cisco
p1r3(config-fr-dlci)#map-class frame vofr
p1r3(config-map-class)#frame-relay cir 64000
p1r3(config-map-class)#frame-relay bc 640
p1r3(config-map-class)#frame-relay be 0
p1r3(config-map-class)#frame-relay fragment 80
p1r3(config-map-class)#frame-relay fair-queue
p1r3(config-map-class)#frame-relay voice bandwidth 24000
p1r3(config-map-class)#exit
```

The command **frame-relay class vofr** applies the parameters listed in Example 8-7 to the subinterface s0.200. The **frame-relay interface-dlci 200** command is used to associate DLCI 200 with this subinterface. The **vofr cisco** command enables Cisco-proprietary voice encapsulation for VoFR with data carried on CID 4 and call-control on CID 5.

WFQ Example

Example 8-8 shows how to enable weighted fair queuing and set the number of reservable conversation queues to a value of 25 for the **vofr** Frame Relay map class on a Cisco 2600, 3600, or 7200 series router or on an MC3810 concentrator, starting from global configuration mode.

Example 8-8   *Reserving 25 Conversation Queues with WFQ*

```
router(config)# interface serial 1/1
router#conf t
Enter configuration commands, one per line. End with CNTL/Z.
router(config)#interface s0
p1r3(config-if)#encapsulation frame-relay
p1r3(config-if)#frame-relay traffic-shaping
router(config-if)#frame-relay interface-dlci 100
router(config-fr-dlci)#class vofr
router(config-fr-dlci)#exit
router(config)#map-class frame-relay vofr
router(config-map-class)#frame-relay fair-queue 64 256 25
router(config-map-class)#end
```

End-to-End FRF.12 Fragmentation Example

Example 8-9 shows how to enable pure end-to-end FRF.12 fragmentation for the frag map class on a Cisco 2600, 3600, or 7200 series router, starting from global configuration mode. The fragment payload size is set to 40 bytes. Frame Relay traffic shaping is required on the PVC; the only queuing type supported on the PVC when fragmentation is configured is WFQ.

Example 8-9   *Configuring End-to-End FRF.12 Fragmentation*

```
router(config)#interface serial 1/0/0
router(config-if)#frame-relay traffic-shaping
router(config-if)#frame-relay interface-dlci 100
router(config-fr-dlci)#class frag
router(config-fr-dlci)#exit

router(config)#map-class frame-relay frag
router(config-map-class)#frame-relay cir 128000
router(config-map-class)#frame-relay bc 1280
router(config-map-class)#frame-relay fragment 160
router(config-map-class)#frame-relay fair-queue
router(config-map-class)#
```

FRF.11 Annex C Fragmentation for Data

Example 8-10 shows how to enable FRF.11 Annex C fragmentation for data on a Cisco MC3810 PVC configured for VoFR. Note that fragmentation must be configured if a VoFR PVC is to carry data. The fragment payload size is set to 40 bytes. Frame Relay traffic shaping is required on the PVC; the only queuing type supported on the PVC when fragmentation is configured is WFQ.

Example 8-10   *Enabling FRF.11 Annex C Fragmentation*

```
router(config)#interface serial 1/1
router(config-if)#frame-relay traffic-shaping
router(config-if)#frame-relay interface-dlci 101
router(config-fr-dlci)#vofr
router(config-fr-dlci)#class frag
router(config-fr-dlci)#exit

router(config)#map-class frame-relay frag
router(config-map-class)#frame-relay cir 64000
router(config-map-class)#frame-relay bc 640
router(config-map-class)#frame-relay fragment 80
router(config-map-class)#frame-relay fair-queue
router(config-map-class)#frame-relay voice bandwidth 44000
```

Cisco Frame Relay Fragmentation Example

Example 8-11 shows how to enable Cisco proprietary Frame Relay fragmentation for the "frag" Frame Relay map-class on a Cisco 2600, 3600, or 7200 series router, starting from global configuration mode. The fragment payload size is set to 40 bytes. Frame Relay traffic shaping is required on the PVC; the only queuing type supported on the PVC when fragmentation is configured is WFQ.

Example 8-11   *Enabling Cisco Proprietary Fragmentation*

```
router(config)#interface serial 2/0/0
router(config-if)#frame-relay traffic-shaping
router(config-if)#frame-relay interface-dlci 102
router(config-fr-dlci)#vofr cisco
router(config-fr-dlci)#class frag
router(config-fr-dlci)#exit

router(config)#map-class frame-relay frag
router(config-map-class)#frame-relay cir 128000
router(config-map-class)#frame-relay bc 1000
router(config-map-class)#frame-relay fragment 40
router(config-map-class)#frame-relay fair-queue
router(config-map-class)#
```

VoFR CIR Example

Figure 8-9 illustrates an example network using specific subchannel IDs for data. Note that, in this example, all of the configurable parameters between the two devices that share the same abbreviation *must* be the same. You cannot, for example, use different fragment sizes on opposite ends of a link.

Figure 8-9   *VoFR CIR Example: Using FRF.11 Annex C Fragmentation*

Routers Using a VoFR PVC

Examples 8-12 and 8-13 illustrate partial configurations for two Cisco 3600 model routers connected via an SP. In the examples are several letters used to denote values. The values in one configuration—as denoted by a letter—must match the value in the other configuration as represented by the same letter. The commands and their symbols are:

vofr data a

frame-relay voice-bandwidth vb

frame-relay min-cir m

frame-relay cir c

frame-relay bc bc

frame-relay fragment f

frame-relay voice-bandwidth vb

frame-relay min-cir m

frame-relay cir c

frame-relay bc bc

frame-relay fragment f

Example 8-12   *VoFR PVC Example—Router 1*

```
Router1#config t
Enter configuration commands, one per line. End with CNTL/Z.
Router1(config)#interface serial 0/0
Router1(config-if)#encapsulation frame-relay
Router1(config-if)#frame-relay traffic-shaping
Router1(config-if)#frame-relay interface-dlci 200
Router1(config-fr-dlci)#vofr data a
Router1(config-fr-dlci)#class lizard
Router1(config-fr-dlci)#map-class frame-relay lizard
Router1(config-map-class)#frame-relay voice-bandwidth vb
Router1(config-map-class)#frame-relay min-cir m
Router1(config-map-class)#frame-relay cir c
Router1(config-map-class)#frame-relay bc bc
Router1(config-map-class)#frame-relay fragment f
```

Example 8-13   *VoFR PVC Example—Router 2*

```
Router2#config t
Enter configuration commands, one per line. End with CNTL/Z.
Router2(config)#interface serial 0/0
Router2(config-if)#encapsulation frame-relay
Router2(config-if)#frame-relay traffic-shaping
Router2(config-if)#frame-relay interface-dlci 200
Router2(config-fr-dlci)#vofr data a
Router2(config-fr-dlci)#class lizard
Router2(config-fr-dlci)#map-class frame-relay lizard
Router2(config-map-class)#frame-relay voice-bandwidth vb
Router2(config-map-class)#frame-relay min-cir m
Router2(config-map-class)#frame-relay cir c
Router2(config-map-class)#frame-relay bc bc
Router2(config-map-class)#frame-relay fragment f
```

Cisco-Trunk (Private-Line) Call Example

The router configurations for the network illustrated in Figure 8-10 are shown in Examples 8-14 and 8-15. Again, as in the preceding example, those items that share the same abbreviation must be the same on both devices. Note that in this configuration we have indicated in the **dial-peer** statement which protocol we will use for the call. As in Examples 8-12 and 8-13, the values represented by a letter in one configuration must match the value represented by the same letter in the other configuration. The commands and their symbols are:

frame relay cir c

frame relay bc bc

frame-relay voice bandwidth vb

frame-relay min-cir m

frame-relay fragment f

codec x bytes y

Figure 8-10 *Cisco-Trunk (Private-Line) Call Example*

Example 8-14 *Cisco-Trunk (Private-Line) Example—Router 1*

```
Router2#config t
Enter configuration commands, one per line. End with CNTL/Z.
Router1(config)#interface serial 0/0
Router1(config)#ip address 192.168.1.1  255.255.255.0
Router1(config)#encapsulation frame-relay
Router1(config)#frame-relay traffic shaping
Router1(config)#frame-relay interface-dlci 100
Router1(config-fr-dlci)#class voice
Router1(config-fr-dlci)#vofr data 4 call-control 5
Router1(config-fr-dlci)#map-class frame-relay voice
Router1(config-map-class)#frame relay cir c
Router1(config-map-class)#frame relay bc bc
Router1(config-map-class)#frame-relay voice bandwidth vb
Router1(config-map-class)#frame-relay min-cir m
Router1(config-map-class)#frame-relay fragment f
Router1(config-map-class)#voice-port 2/0/0
Router1(config-)#connection trunk 6001 answer-mode
Router1(config-)#dial-peer voice 1 pots
Router1(config-)#destination-pattern 7001
```

Example 8-14 *Cisco-Trunk (Private-Line) Example—Router 1 (Continued)*

```
Router1(config-voiceport)#port 2/0/0
Router1(config-voiceport)#dial-peer voice 2 vofr
Router1(config-dial-peer)#codec x bytes y
Router1(config- dial-peer)#destination-pattern 6001
Router1(config- dial-peer)#session protocol cisco-switched
Router1(config- dial-peer)#session target S0/0 100
```

Example 8-15 *Cisco-Trunk (Private-Line) Example—Router 2*

```
Router2#config t
Enter configuration commands, one per line. End with CNTL/Z.
Router2(config)#interface serial 0/0
Router2(config-if)#ip address 192.168.1.2  255.255.255.0
Router2(config-if)#encapsulation frame-relay
Router2(config-if)#frame-relay traffic shaping
Router2(config-if)#frame-relay interface-dlci 100
Router2(config-fr-dlci)#class voice
Router2(config-fr-dlci)#vofr data 4 call-control 5
Router2(config-fr-dlci)#map-class frame-relay voice
Router2(config-map-class)#frame relay cir c
Router2(config-map-class)#frame relay bc bc
Router2(config-map-class)#frame-relay voice bandwidth vb
Router2(config-map-class)#frame-relay min-cir m
Router2(config-map-class)#frame-relay fragment f
Router2(config-map-class)#voice-port 1/5
Router2(config-voiceport)#connection trunk 7001
Router2(config-voiceport)#dial-peer voice 2 pots
Router2(config-dial-peer)#destination-pattern 6001
Router2(config-dial-peer)#port 1/5
Router2(config-dial-peer)#dial-peer voice 4 vofr
Router2(config-dial-peer)#codec x bytes y
Router2(config-dial-peer)#destination-pattern 7001
Router2(config-dial-peer)#session protocol cisco-switched
Router2(config-dial-peer)#session target S0/0 100
```

FRF.11-Trunk Call Example

The sample configuration shown in Example 8-16 of the network illustrated in Figure 8-11 demonstrates an FRF.11-trunk call from a Cisco voice-capable router to another FRF.11-capable device. Like example pair Examples 8-12 and 8-13 and example pair Examples 8-14 and 8-15, the letters here represent values that must be matched by the device on the other end of the link for proper operation. The commands that must be matched and their symbols are:

frame relay cir c

frame-relay min-cir in m

frame relay bc bc

frame-relay voice bandwidth vb

frame-relay fragment f

codec x bytes y

Example 8-16   *FRF.11-Trunk Example—Router 1*

```
Router2#config t
Enter configuration commands, one per line. End with CNTL/Z.
Router2(config)#interface serial 0
Router2(config-if)#ip address 192.168.2.1 255.255.255.0
Router2(config-if)#encapsulation frame-relay
Router2(config-if)#frame-relay traffic shaping
Router2(config-if)#frame-relay interface-dlci 150
Router2(config-fr-dlci)#class voice
Router2(config-fr-dlci)#vofr data 4
Router2(config-fr-dlci)#map-class frame-relay voice
Router2(config-map-class)#frame relay cir c
Router2(config-map-class)#frame-relay min-cir in m
Router2(config-map-class)#frame relay bc bc
Router2(config-map-class)#frame-relay voice bandwidth vb
Router2(config-map-class)#frame-relay fragment f
Router2(config-map-class)#voice-port 1/5
Router2(config-voiceport)#connection trunk 7001
Router2(config-voiceport)#voice-port 1/5
Router2(config-voiceport)#connection trunk 7001
Router2(config-voiceport)#dial-peer voice 2 pots
Router2(config-dial-peer)#destination-pattern 6001
Router2(config-dial-peer)#port 1/5
Router2(config-dial-peer)#dial-peer voice 4 vofr
Router2(config-dial-peer)#codec x bytes y
Router2(config-dial-peer)#destination-pattern 7001
Router2(config-dial-peer)#session protocol frf11-trunk
Router2(config-dial-peer)#session target S0 100
Router2(config-dial-peer)#dtmf-relay
Router2(config-dial-peer)#vad
Router2(config-dial-peer)#end
Router2#
```

Figure 8-11   *FRF.11-Trunk Call Example*

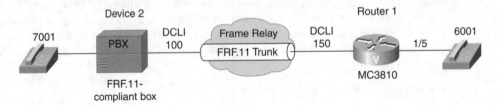

Tandem Configuration for Switched Calls Example

Tandeming is switching incoming VoFR calls on a Frame Relay DLCI to an outgoing VoFR-enabled DLCI. Tandeming works for switched calls and Cisco-trunk permanent calls only. *You cannot tandem FRF.11-trunk calls over a multihop network.*

Tandeming is supported on the Cisco MC3810 and on Cisco 2600, 3600, and 7200 series routers. When you configure a tandem node, you must configure two VoFR dial peers, one for each tandem connection.

Depending on which router is used as the end node and which router is used as the tandem node, you must use the correct Frame Relay PVC type when configuring your connections.

Table 8-8 shows the different combinations of routers that can serve as end nodes and tandem nodes and the Frame Relay PVC type required.

Table 8-8   *VoFR End Node and Tandem Node Combinations Supported*

End Node(s)	Tandem Node	vofr Command to Enter for the Frame Relay DLCI
Cisco 2600/3600, Cisco MC3810, or Cisco 7200	Cisco 2600, Cisco 3600, Cisco MC3810, or Cisco 7200	**vofr call-control**
Cisco 2600/3600, Cisco MC3810	Cisco MC3810 running Cisco IOS releases before 12.0(7)XK	**vofr cisco**
Cisco MC3810 running Cisco IOS releases before 12.0(7)XK	Cisco 2600, Cisco 3600, or Cisco 7200	**vofr cisco**

When creating voice networks with a mixture of router types, the Cisco MC3810 must be running Cisco IOS Release 12.0(3)XG, 12.0(4)T, or later releases to act as a tandem node.

Figure 8-12 illustrates the network configuration that the sample configurations in Examples 8-17, 8-18, and 8-19 are based on. Once again, the letters here represent values that must be matched by the device on the other end of the link for proper operation. The commands that must be matched and their symbols are:

frame relay cir c

frame-relay min-cir in m

frame relay bc bc

frame-relay voice bandwidth vb

frame-relay fragment d

Figure 8-12 *Tandem Configuration for Switched Calls Example*

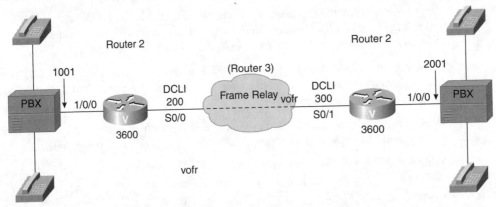

Example 8-17 *Tandem Configuration Example—Router 1*

```
Router1#config t
Enter configuration commands, one per line. End with CNTL/Z.
Router1(config)#interface serial 0/0
Router1(config-if)#encapsulation frame-relay
Router1(config-if)#frame-relay traffic-shaping
Router1(config-if)#frame-relay interface-dlci 200
Router1(config-fr-dlci)#class voice
Router1(config-fr-dlci)#vofr data 4 call-control 5
Router1(config-fr-dlci)#map-class frame-relay voice
Router1(config-map-class)#frame-relay cir c
Router1(config-map-class)#frame-relay min-cir m
Router1(config-map-class)#frame-relay bc bc
Router1(config-map-class)#frame-relay voice bandwidth vb
Router1(config-map-class)#frame-relay fragment d
Router1(config-map-class)#dial-peer voice 1 pots
Router1(config-dial-peer)#destination-pattern 1001
Router1(config-map-class)#port 1/0/0
Router1(config-map-class)#dial-peer voice 2 vofr
Router1(config-map-class)#destination-pattern 2...
Router1(config-map-class)#session target serial 0/0 300
Router1(config-map-class)#voice-port 1/0/0
```

Example 8-18 *Tandem Configuration Example—Router 2*

```
Router2(config)#interface serial 0
Router2(config-if)#encapsulation frame-relay
Router2(config-if)#frame-relay interface-dlci 200
Router2(config-fr-dlci)#class voice
Router2(config-fr-dlci)#vofr data 4 call-control 5
Router2(config-fr-dlci)#interface serial 1
```

Example 8-18   *Tandem Configuration Example—Router 2 (Continued)*

```
Router2(config-if)#encapsulation frame-relay
Router2(config-if)#frame-relay interface-dlci 300
Router2(config-fr-dlci)#class voice
Router2(config-fr-dlci)#vofr data 4 call-control 5
Router2(config-fr-dlci)#map-class frame-relay voice
Router2(config-map-class)#frame-relay cir c
Router2(config-map-class)#frame-relay min-cir m
Router2(config-map-class)#frame-relay bc bc
Router2(config-map-class)#frame-relay voice bandwidth vb
Router2(config-map-class)#frame-relay fragment d
Router2(config-map-class)#dial-peer voice 1 vofr
Router2(config-dial-peer)#destination-pattern 1...
Router2(config-dial-peer)#session target serial 0/0 200
Router2(config-dial-peer)#dial-peer voice 2 vofr
Router2(config-dial-peer)#destination-pattern 2...
Router2(config-dial-peer)#session target serial 0/1 300
```

Example 8-19   *Tandem Configuration Example—Router 3*

```
Router3(config)#interface serial 0/0
Router3(config-if)#encapsulation frame-relay
Router3(config-if)#frame-relay traffic-shaping
Router3(config-if)#frame-relay interface-dlci 300
Router3(config-fr-dlci)#class voice
Router3(config-fr-dlci)#vofr data 4 call-control 5
Router3(config-fr-dlci)#map-class frame-relay voice
Router3(config-map-class)#frame-relay cir c
Router3(config-map-class)#frame-relay min-cir m
Router3(config-map-class)#frame-relay bc bc
Router3(config-map-class)#frame-relay voice bandwidth vb
Router3(config-map-class)#frame-relay fragment d
Router3(config-map-class)#dial-peer voice 1 pots
Router3(config-dial-peer)#destination-pattern 2001
Router3(config-dial-peer)#port 1/0/0
Router3(config-dial-peer)#dial-peer voice 2 vofr
Router3(config-dial-peer)#destination-pattern 1...
Router3(config-dial-peer)#session target serial 0/0 200
Router3(config-dial-peer)#voice-port 1/0/0
```

Monitoring and Troubleshooting Commands

Several **show** and **debug** commands can be used to help monitor and troubleshoot VoFR configurations. Many of these commands have general applicability and are not covered here. Only the commands that pertain to items specific to VoFR are covered in detail in the following sections.

show Commands

You are probably aware of the several commands that will help you monitor a Frame Relay network. Table 8-9 is a quick summary of several with which you should already be familiar.

Table 8-9   show *Command Summary*

Command	Results
show interface *interface number*	Displays detailed reporting of the serial interface.
show controller [*interface* | *controller*] *number*	Displays detailed reporting of the controller. Provides per-DS0 reporting on integrated CSU/DSU.
show frame-relay pvc	Displays statistics for each configured PVC.
show frame-relay map	Displays Frame Relay mappings.
show frame-relay lmi	Displays LMI Statistics

Some other **show** commands that deal expressly with VoFR are listed in the following sections.

The **show frame-relay vofr** Command

The **show frame-relay vofr** [*interface* [*dlci* [*cid*]]] command is used to display information about the FRF.11 subchannels being used on VoFR DLCIs; use the **show frame-relay vofr** command from privileged EXEC mode. Command options for the **show frame-relay vofr** command are listed in Table 8-10.

Table 8-10   show frame-relay vofr *Command Options*

Command Parameter	Description
interface	(Optional.) The specific interface type and number for which you wish to display FRF.11 subchannel information.
dlci	(Optional.) The specific DLCI for which you wish to display FRF.11 subchannel information.
cid	(Optional.) The specific subchannel for which you wish to display information.

If this command is entered without specifying an interface, FRF.11 subchannel information will be displayed for all VoFR interfaces and DLCIs configured on the router.

This command is currently not supported on the Cisco MC3810 for PVCs configured with the **vofr cisco** command or the **frame-relay interface-dlci voice-encap** command.

Example 8-20 is sample output from the **show frame-relay vofr** command when an interface is not specified.

Example 8-20   **show frame-relay vofr** *Command Output with No Interface Specified*

```
3640_vofr#show frame-relay vofr
interface          vofr-type   dlci   cid   cid-type
Serial0/0.1        VoFR        16     4     data
Serial0/0.1        VoFR        16     5     call-control
Serial0/0.1        VoFR        16     10    voice
Serial0/1.1        VoFR cisco  17     4     data
```

Example 8-21 is sample output from the **show frame-relay vofr** command when an interface is specified.

Example 8-21   **show frame-relay vofr** *Command Output with an Interface Specified*

```
3640_vofr#show frame-relay vofr serial0
interface          vofr-type   dlci   cid   cid-type
Serial0             VoFR        16     4     data
Serial0             VoFR        16     5     call-control
Serial0             VoFR        16     10    voice
```

Example 8-22 is sample output from the **show frame-relay vofr** command when an interface and a DLCI are specified:

Example 8-22   **show frame-relay vofr** *Command Output with Both an Interface and a DLCI Specified*

```
3640_vofr#show frame-relay vofr serial0 16
VoFR Configuration for interface Serial0

dlci vofr-type  cid cid-type        input-pkts     output-pkts    dropped-pkts
16   VoFR       4   data            0              0              0
16   VoFR       5   call-control    85982          86099          0
16   VoFR       10  voice           2172293        6370815        0
```

Example 8-23 is sample output from the **show frame-relay vofr** command when an interface, a DLCI, and a CID are specified.

Example 8-23   **show frame-relay vofr** *Command Output with an Interface, a DLCI, and a CID Specified*

```
3640_vofr#show frame-relay vofr serial0 16 10
VoFR Configuration for interface Serial0 dlci 16

 vofr-type  VoFR      cid 10      cid-type voice
  input-pkts 2172293   output-pkts 6370815   dropped-pkts 0
```

Table 8-11 describes the fields in the output.

Table 8-11 *The Fields in the* **show frame-relay vofr** *Command Output (Shown in Examples 8-20 Through 8-23)*

Field	Description
interface	The number of the interface that has been selected for observation of FRF.11 subchannels.
vofr-type	The type of the VoFR DLCI being observed.
cid	The portion of the specified DLCI that is carrying the designated traffic type. A DLCI can be subdivided into 255 subchannels.
cid-type	The type of traffic carried on this subchannel.
input-pkts	The number of packets received by this subchannel.
output-pkts	The number of packets transmitted on this subchannel.
dropped-pkts	The total number of packets discarded by this subchannel.

The **show frame-relay fragment** Command

You can use the **show frame-relay fragment** [**interface** *interface* [*dlci*]] command to display information about the Frame Relay fragmentation taking place in your Cisco router from privileged EXEC mode (see Table 8-12).

Table 8-12 **show frame-relay fragment** *Command Parameters*

Parameter	Description
interface	(Optional.) Indicates a specific interface for which Frame Relay fragmentation information will be displayed.
interface	(Optional.) Interface number containing the DLCI(s) for which you wish to display fragmentation information.
dlci	(Optional.) Specific DLCI for which you wish to display fragmentation information.

When no parameters are specified with the **show frame-relay fragment** command, the output displays a summary of each DLCI configured for fragmentation. The information displayed includes the fragmentation type, the configured fragment size, and the number of fragments transmitted, received, and dropped.

When a specific interface and DLCI are specified, additional details are displayed.

NOTE   The **show frame-relay fragment** command will not produce any output for Cisco MC3810s configured with the **frame-relay interface-dlci voice-encap** command.

Example 8-24 is sample output for the **show frame-relay fragment** command without any parameters specified.

Example 8-24   **show frame-relay fragment** *Command with No Parameters*

```
router#show frame-relay fragment
interface      dlci  frag-type   frag-size  in-frag  out-frag  dropped-frag
Serial0        108   VoFR-cisco  100        1261     1298      0
Serial0        109   VoFR        100        0        243       0
Serial0        110   end-to-end  100        0        0         0
```

Example 8-25 is sample output for the **show frame-relay fragment** command when an interface and DLCI are specified.

Example 8-25   **show frame-relay fragment** *Command When an Interface and DLCI Are Specified*

```
router# show frame-relay fragment interface Serial1/0 16
  fragment-size 45                  fragment type end-to-end
  in fragmented pkts 0              out fragmented pkts 0
  in fragmented bytes 0             out fragmented bytes 0
  in un-fragmented pkts 0           out un-fragmented pkts 0
  in un-fragmented bytes 0          out un-fragmented bytes 0
  in assembled pkts 0               out pre-fragmented pkts 0
  in assembled bytes 0              out pre-fragmented bytes
  in dropped reassembling pkts 0    out dropped fragmenting pkts 0
  in timeouts 0
  in out-of-sequence fragments 0
  in fragments with unexpected B bit set 0
  out interleaved packets 0
```

Table 8-13 describes the significant fields in the outputs of Examples 8-24 and 8-25.

Table 8-13   **show frame-relay fragment** *Field Descriptions*

Field	Description
interface	The subinterface containing the DLCI to which the fragmentation information pertains.
dlci	The DLCI to which the displayed fragmentation information applies.
frag-type	The type of fragmentation configured on the designated DLCI. Supported types are end-to-end, VoFR, and VoFR-cisco.
frag-size	The configured fragment size in bytes.
in-frag	The total number of fragments received by the designated DLCI.
out-frag	The total number of fragments transmitted by the designated DLCI.

continues

Table 8-13 **show frame-relay fragment** *Field Descriptions (Continued)*

Field	Description
dropped-frag	The total number of fragments dropped by the designated DLCI.
in/out fragmented pkts	The total number of frames received/transmitted by this DLCI that have a fragmentation header.
in/out fragmented bytes	The total number of bytes, including those in the Frame Relay headers, that have been received/transmitted by this DLCI.
in/out un-fragmented pkts	The number of frames received/transmitted by this DLCI that do not require reassembly and therefore do not contain the FRF.12 header. These counters can be incremented only when the end-to-end fragmentation type is set.
in/out un-fragmented bytes	The number of bytes received/transmitted by this DLCI that do not require reassembly and therefore do not contain the FRF.12 header. These counters can be incremented only when the end-to-end fragmentation type is set.
in assembled pkts	The total number of fully reassembled frames received by this DLCI, including the frames received without a Frame Relay fragmentation header (in unfragmented packets). This counter corresponds to the frames viewed by the upper-layer protocols.
out pre-fragmented pkts	The total number of fully reassembled frames transmitted by this DLCI, including the frames transmitted without a Frame Relay fragmentation header (out unfragmented packets).
in assembled bytes	The number of bytes in the fully reassembled frames received by this DLCI, including the frames received without a Frame Relay fragmentation header (in unfragmented bytes). This counter corresponds to the total number of bytes viewed by the upper-layer protocols.
out pre-fragmented bytes	The number of bytes in the fully reassembled frames transmitted by this DLCI, including the frames transmitted without a Frame Relay fragmentation header (out unfragmented bytes). This counter corresponds to the total number of bytes viewed by the upper-layer protocols.
in dropped reassembling pkts	The number of fragments received by this DLCI that are dropped for reasons such as running out of memory, receiving segments out of sequence, receiving an unexpected frame with a B bit set, or timing out on a reassembling frame.

Table 8-13   **show frame-relay fragment** *Field Descriptions (Continued)*

Field	Description
out dropped fragmenting pkts	The number of fragments that are dropped by this DLCI during transmission because of running out of memory.
in timeouts	The number of reassembly timeouts that have occurred on incoming frames to this DLCI. (A frame that does not fully reassemble within two minutes is dropped, and the timeout counter is incremented.)
in out-of-sequence fragments	The number of fragments received by this DLCI that have an unexpected sequence number.
in fragments with unexpected B bit set	The number of fragments received by this DLCI that have an unexpected B bit set. When this occurs, all fragments being reassembled are dropped, and a new frame is begun with this fragment.
out interleaved packets	The number of packets leaving this DLCI that have been interleaved between segments.

debug Commands

As with the **show** commands, there are many **debug** commands with which you should be familiar. Three of them are listed in Table 8-14.

Table 8-14   *Useful **debug** Commands for Troubleshooting*

Command	Results
debug frame-relay events	Displays important Frame Relay packet events
debug frame-relay lmi	Displays LMI packet exchanges with service provider
debug frame-relay packets	Displays summary Frame Relay frame decode

Five other **debug** commands may be useful in troubleshooting VoFR sessions. These are listed in the following sections.

The **debug ccfrf11 session** Command

To display the ccfrf11 function calls during call setup and teardown, use the **debug ccfrf11 session** command from privileged EXEC mode. Use the **no** form of this command to turn off the debug function.

This command does not apply to the Cisco MC3810. A Cisco 7500 series router requires a versatile interface processor (VIP) to utilize this command.

This command can be used to display debug information about the various FRF.11 VoFR service provider interface (SPI) functions. Note that this debug command does not display any information regarding the proprietary Cisco switched-VoFR SPI.

This debug is only useful when the session protocol is "frf11-trunk."

Example 8-26 shows sample output from the **debug ccfrf11 session** command. Note that some of these outputs have been modified to fit the printed page.

Example 8-26   **debug ccfrf11 session** *Command Output*

```
router# debug ccfrf11 session
INCOMING CALL SETUP (port setup for answer-mode):
*Mar  6 18:04:07.693:ccfrf11_process_timers:scb (0x60EB6040) timer (0x60EB6098)
  expired
*Mar  6 18:04:07.693:Setting accept_incoming to TRUE
*Mar  6 18:04:11.213:ccfrf11_incoming_request:peer tag 800:callingNumber=
  +2602100, calledNumber=+3622110
*Mar  6 18:04:11.213:ccfrf11_initialize_ccb:preffered_codec set(-1)(0)
*Mar  6 18:04:11.213:ccfrf11_evhandle_incoming_call_setup_request:calling
  +2602100, called +3622110 Incoming Tag 800
*Mar  6 18:04:11.217:ccfrf11_caps_ind:PeerTag = 800
*Mar  6 18:04:11.217:    codec(preferred) = 4, fax_rate = 2, vad = 2
*Mar  6 18:04:11.217:    cid = 30, config_bitmask = 0, codec_bytes = 20,
  signal_type=2
*Mar  6 18:04:11.217:    required_bandwidth 8192
*Mar  6 18:04:11.217:ccfrf11_caps_ind:Bandwidth reservation of 8192 bytes
  succeeded.
*Mar  6 18:04:11.221:ccfrf11_evhandle_call_connect:Entered

CALL SETUP (MASTER):
5d22h:ccfrf11_call_setup_request:Entered
5d22h:ccfrf11_evhandle_call_setup_request:Entered
5d22h:ccfrf11_initialize_ccb:preffered_codec set(-1)(0)
5d22h:ccfrf11_evhandle_call_setup_request:preffered_codec set(9)(24)
5d22h:ccfrf11_call_setup_trunk:subchannel linking successful
5d22h:ccfrf11_caps_ind:PeerTag = 810
5d22h:    codec(preferred) = 512, fax_rate = 2, vad = 2
5d22h:    cid = 30, config_bitmask = 1, codec_bytes = 24, signal_type=2
5d22h:    required_bandwidth 6500
5d22h:ccfrf11_caps_ind:Bandwidth reservation of 6500 bytes succeeded.

CALL TEARDOWN:
*Mar  6 18:09:14.805:ccfrf11_call_disconnect:peer tag 0
*Mar  6 18:09:14.805:ccfrf11_evhandle_call_disconnect:Entered
*Mar  6 18:09:14.805:ccfrf11_call_cleanup:freeccb 1, call_disconnected 1
*Mar  6 18:09:14.805:ccfrf11_call_cleanup:Setting accept_incoming to FALSE and
  starting incoming timer
*Mar  6 18:09:14.809:timer 2:(0x60EB6098)starts - delay (70000)
*Mar  6 18:09:14.809:ccfrf11_call_cleanup:Alive timer stopped
*Mar  6 18:09:14.809:timer 1:(0x60F64104) stops
*Mar  6 18:09:14.809:ccfrf11_call_cleanup:Generating Call record
*Mar  6 18:09:14.809:cause=10 tcause=10    cause_text="normal call clearing."
*Mar  6 18:09:14.809:ccfrf11_call_cleanup:Releasing 8192 bytes of reserved
```

Example 8-26   **debug ccfrf11 session** *Command Output (Continued)*

```
   bandwidth
*Mar  6 18:09:14.809:ccfrf11_call_cleanup:ccb 0x60F6404C, vdbPtr 0x610DB7A4
        freeccb_flag=1, call_disconnected_flag=1
```

The **debug ccswvoice vofr-debug** Command

To display the ccswvoice function calls during call setup and teardown, use the **debug ccswvoice vofr-debug** command from privileged EXEC mode. Use the **no** form of this command to turn off the debug function.

This command does not apply to the Cisco MC3810. A Cisco 7500 series router requires a VIP to utilize this command.

This command should be used when attempting to troubleshoot a VoFR call that uses the "cisco-switched" session protocol. It provides the same information as the **debug ccswvoice vofr-session** command but includes additional debugging information relating to the calls.

Example 8-27 shows sample output from the **debug ccswvoice vofr-debug** command.

Example 8-27   **debug ccswvoice vofr-debug** *Command Output*

```
router# debug ccswvoice vofr-debug
CALL TEARDOWN:
3640_vofr(config-voiceport)#
*Mar  1 03:02:08.719:ccswvofr_bridge_drop:dropping bridge calls src 17 dst 16
  dlci 100
      cid 9 state ACTIVE
*Mar  1 03:02:08.727:ccswvofr:callID 17 dlci 100 cid 9 state ACTIVE event O/G REL
*Mar  1 03:02:08.735:ccswvofr:callID 17 dlci 100 cid 9 state RELEASE event I/C
  RELCOMP
*Mar  1 03:02:08.735:ccswvofr_store_call_history_entry:cause=22 tcause=22
      cause_text=no circuit.
3640_vofr(config-voiceport)#

CALL SETUP (outgoing):
*Mar  1 03:03:22.651:ccswvofr:callID 23 dlci -1 cid -1 state NULL event O/G SETUP
*Mar  1 03:03:22.651:ccswvofr_out_callinit_setup:callID 23 using dlci 100 cid 10
*Mar  1 03:03:22.659:ccswvofr:callID 23 dlci 100 cid 10 state O/G INIT event I/C
  PROC
*Mar  1 03:03:22.667:ccswvofr:callID 23 dlci 100 cid 10 state O/G PROC event I/C
  CONN
ccfrf11_caps_ind:codec(preferred) = 0
```

The **debug ccswvoice vofr-session** Command

To display the ccswvoice function calls during call setup and teardown, use the **debug ccswvoice vofr-session** command from privileged EXEC mode. Use the **no** form of this command to turn off the debug function.

This command does not apply to the Cisco MC3810. A Cisco 7500 series router requires a VIP to utilize this command.

This command can be used to show the state transitions of the cisco-switched-vofr state machine as a call is processed. It should be used when attempting to troubleshoot a VoFR call that uses the "cisco-switched" session protocol.

Example 8-28 shows sample output from the **debug ccswvoice vofr-session** command.

Example 8-28   **debug ccswvoice vofr-session** *Command Output*

```
router# debug ccswvoice vofr-session
CALL TEARDOWN:
3640_vofr(config-voiceport)#
*Mar  1 02:58:13.203:ccswvofr:callID 14 dlci 100 cid 8 state ACTIVE event O/G REL
*Mar  1 02:58:13.215:ccswvofr:callID 14 dlci 100 cid 8 state RELEASE event I/C
  RELCOMP
3640_vofr(config-voiceport)#

CALL SETUP (outgoing):
*Mar  1 02:59:46.551:ccswvofr:callID 17 dlci -1 cid -1 state NULL event O/G SETUP
*Mar  1 02:59:46.559:ccswvofr:callID 17 dlci 100 cid 9 state O/G INIT event I/C
  PROC
*Mar  1 02:59:46.567:ccswvofr:callID 17 dlci 100 cid 9 state O/G PROC event I/C
  CONN
3640_vofr(config-voiceport)#
```

The **debug frame-relay fragment** Command

To display information related to Frame Relay fragmentation on a PVC, use the **debug frame-relay fragment** [**event** | **interface** *type number dlci*] command in privileged EXEC mode. Use the **no** form of this command to turn off the debug function.

This command will display event or error messages related to Frame Relay fragmentation; it is only enabled at the PVC level on the selected interface.

This command is not supported on the Cisco MC3810 for fragments received by a PVC configured via the **voice-encap** command.

Example 8-29 shows sample output from the **debug frame-relay fragment interface** command.

Example 8-29   **debug frame-relay fragment interface** *Command Output*

```
router#debug frame-relay fragment interface serial 0/0 109
This may severely impact network performance.
You are advised to enable 'no logging console debug'. Continue?[confirm]
Frame Relay fragment/packet debugging is on
Displaying fragments/packets on interface Serial0/0  dlci 109 only
Serial0/0(i): dlci 109, rx-seq-num 126, exp_seq-num 126, BE bits set, frag_hdr
  04 C0 7E
Serial0/0(o): dlci 109, tx-seq-num 82, BE bits set, frag_hdr 04 C0 52
```

Example 8-30 shows sample output from the **debug frame-relay fragment event** command.

Example 8-30   **debug frame-relay fragment event** *Command Output*

```
router#debug frame-relay fragment event
This may severely impact network performance.
You are advised to enable 'no logging console debug'. Continue?[confirm]
Frame Relay fragment event/errors debugging is on
Frame-relay reassembled packet is greater than MTU size, packet dropped on
    serial 0/0 dlci 109
Unexpected B bit  frame rx on serial0/0 dlci 109, dropping pending segments
Rx an out-of-sequence packet on serial 0/0 dlci 109, seq_num_received 17
        seq_num_expected 19
```

The **debug voice vofr** Command

To show Cisco trunk and FRF.11-trunk call setup attempts and to show which dial peer is used in the call setup, you use the **debug voice vofr** command from privileged EXEC mode. You use the **no** form of this command to turn off the debug function.

This command applies to Cisco trunks and FRF.11 trunks only; it does not apply to switched calls. A Cisco 7500 series router requires a VIP to utilize this command.

This command applies to VoFR, VoATM, and VoHDLC dial peers on the Cisco MC3810.

Example 8-31 shows sample output from the **debug voice vofr** command for a Cisco trunk.

Example 8-31   **debug voice vofr** *Command Output on a Cisco Trunk*

```
router# debug voice vofr
1d05h: 1/1:VOFR, unconf ==> pending_start
1d05h: 1/1:VOFR,create VOFR
1d05h: 1/1:VOFR,search dial-peer 7100 preference 0
1d05h: 1/1:VOFR, pending_start ==> start
1d05h: 1/1:VOFR,
1d05h:voice_configure_perm_svc:
1d05h:dial-peer 7100 codec = G729A payload size = 30 vad = off dtmf relay = on
    seq num = off
1d05h:voice-port 1/1 codec = G729A payload size = 30 vad = off dtmf relay = on
    seq num = off
1d05h: 1/1:VOFR,SIGNAL-TYPE = cept
1d05h:init_frf11 tcid 0 master 0 signaltype 2
1d05h:Going Out Of Service on tcid 0  with sig state 0001
1d05h: 1/1:VOFR, start get event idle
1d05h: 1/1:VOFR, start get event
1d05h: 1/1:VOFR, start get event set up
1d05h: 1/1:VOFR, start ==> pending_connect
1d05h: 1/1:VOFR, pending_connect get event connect
1d05h: 1/1:VOFR, pending_connect ==> connect
1d05h: 1/1:VOFR,SIGNAL-TYPE = cept
1d05h:init_frf11 tcid 0 master 1 signaltype 2
1d05h:start_vofr_polling on port 0 signaltype 2
```

Example 8-32 shows sample output from the **debug voice vofr** command for an FRF.11 trunk.

Example 8-32   **debug voice vofr** *Command Output on an FRF.11 Trunk*

```
router# debug voice vofr
1d05h: 1/1:VOFR,search dial-peer 7200 preference 2
1d05h: 1/1:VOFR,SIGNAL-TYPE = cept
1d05h:Launch Voice Trunk:signal-type 2
1d05h:calculated bandwidth = 10, coding = 6, size = 30
1d05h:%Voice-port 1/1 is down.
1d05h: 1/1:VOFR, pending_start get event idle
1d05h:Codec Type = 6 Payload Size = 30 Seq# off
1d05h:%Voice-port 1/1 is up.
1d05h:init_frf11 tcid 0 master 1 signaltype 2
1d05h:status OK :cid = 100
1d05h: 1/1:VOFR,
1d05h:start FRF11
1d05h: 1/1:VOFR, pending_start ==> frf11
1d05h: 1/1:VOFR,SIGNAL-TYPE = cept
```

Summary

In this chapter, we have examined several aspects of Frame Relay operation, especially as it pertains to providing the necessary quality of service for VoFR traffic. Fixed delay components, such as propagation delay, serialization delay, and processing delay, were described and discussed. QoS tools to overcome or mitigate these forms of delay are additional bandwidth, fragmentation, and voice activity detection. The variable delay components of de-jitter buffers, queuing delay, and WAN/service provider delays were also described and discussed. The role of the codec choice in bandwidth requirements is another important feature highlighted in this chapter. The need for Frame Relay traffic shaping, means of shaping traffic, and various queuing mechanisms provided by Cisco IOS were all described to help you in building a good integrated network. Well over half of the chapter discussed various commands that you can use in Cisco IOS to configure Frame Relay for servicing voice traffic or for monitoring and troubleshooting configured voice-over services on Cisco.

Review Questions

The following questions should help you gauge your understanding of this chapter. You can find the answers in Appendix A, "Answers to Review Questions."

1 What kind of Frame Relay network minimizes transit hops and maximizes the ability to establish different qualities of service?

   _____

   _____

   _____

2 Name the two general categories of delay and three specific types of delay within each.

   _____

   _____

   _____

3 What is the standards-based mechanism for providing fragmentation of data over Frame Relay?

   _____

   _____

   _____

4 What is the standards-based mechanism for providing fragmentation of data in a PVC used for VoFR?

   _____

   _____

   _____

5 What command is used to configure PQ-WFQ?

   _____

   _____

   _____

6 What is the preferred queuing method for VoFR?

   _____

   _____

   _____

7 How do you configure the CIR for traffic shaping?

_____

_____

_____

8 How do you configure per-virtual circuit queuing and traffic shaping for all PVCs on a Frame Relay interface?

_____

_____

_____

9 What is the maximum one-way delay acceptable when designing a VoFR network?

_____

_____

_____

10 The overhead in a VoFR frame is how many bytes?

_____

_____

_____

Upon completion of this chapter you will be able to perform the following tasks:

- Describe the basic operation of an ATM network, including the contents of an ATM cell, service classes, bandwidth allocation, and QoS.

- Explain the features of Cisco voice-capable routers that support VoATM.

- Configure WAN services, voice port interfaces, and dial peers on the Cisco 3600 and MC3810 to support VoATM using the MFT WAN trunk and IMA card.

CHAPTER 9

Configuring Cisco Routers for VoATM

This chapter provides an introduction to the theory and operation of the Asynchronous Transfer Mode (ATM) protocol. We will look at the various types of traffic in an ATM network, the methods we can use in an ATM network to minimize delay and delay variation (also known as *jitter*), and how to configure and troubleshoot Voice over ATM (VoATM) in more depth.

Introduction to ATM

ATM is a switching method for transmitting information in fixed-length cells, based on application demand and priority. The following are some of the basic characteristics of ATM:

- It uses small, fixed-sized cells (53 octets).
- It is a connection-oriented protocol.
- It supports multiple service types.
- It is applicable to LAN and WAN traffic.
- ATM virtual circuits (VCs) emulate PSTN circuits.
- ATM minimizes delay and delay variation.

The aspects of ATM that permit it to be successful as a voice transport mechanism include:

- Various service classes, which permit assigning different data types to different service classes for best traffic management
- Bandwidth allocation techniques that, when used in conjunction with service classes, permit more granularity in quality of service (QoS) than in other transport mechanisms such as Frame Relay and X.25
- Architectural design that minimizes delay and delay variation

Each of these is covered in more detail in the following sections.

ATM Service Classes: VBR, CBR, UBR, and ABR

ATM provides different classes of service, and the throughput desired for each traffic type must be matched to the appropriate service class. Some traffic classes provide only a best-effort transport, and others guarantee a minimum amount of bandwidth to a specific traffic class.

Constant bit rate (CBR) and variable bit rate (VBR) classes have provisions for passing real-time traffic, such as voice or video—including videoconferencing—and are suitable for guaranteeing a certain level of service. CBR in particular allows the amount of bandwidth, end-to-end delay, and delay variation to be specified during the call setup. VBR has both a real-time and a non-real-time equivalent, VBR-nrt. VBR-rt is used for connections that require a fixed timing relationship between samples, while VBR-nrt is used for connections in which there is no fixed timing relationship between samples but that still need a guaranteed QoS.

Unspecified bit rate (UBR) and available bit rate (ABR) were designed with bursty traffic in mind and are more suitable for data applications. UBR in particular makes no guarantees about the delivery of the data traffic.

Each class is given a guaranteed minimum bandwidth, which ensures deterministic behavior under load (for example, voice will not degrade when data bursts). If additional bandwidth is available, each class can access a fair share, which allows ATM to support multiple, complex classes of service (CoSs).

The first architectural key requirement of ATM is the necessity to implement internal dedicated network (that is, trunk) queues for every service type that needs to be supported.

A simple example is voice versus data: Voice has stringent requirements when it comes to delay behavior. The delay needs to be low and constant to avoid starvation or blocking at the egress point. Voice tolerates a certain degree of cell loss but it should be minimized. Application data, on the other hand, typically does not tolerate cell loss at all, but use of a reliable transport will retransmit lost packets. Thus, upon facing overload, an architecture that handles these traffic types on dedicated queues can, with intelligent queue handling algorithms, decide that, if one cell needs to be dropped, it should be one out of the data queue.

The more queues, the more granularity the queuing algorithms have in order to "outplay" the different service types against each other and guarantee QoS requirements even when available resources become constricted.

ATM is a design that allocates dedicated network queues based on QoS subsets. This means that ATM can handle these queues intelligently and deliver deterministic QoS in an efficient way that does not require overengineering of resources.

NOTE   To be able to guarantee CoS and QoS, dedicated network queues are required for different service types. There are several Cisco IOS images with different core routing features. The differences between the images are the varying types of support for VoATM. Check the Release Notes to ensure that the features you require are included in the Cisco IOS Release you plan to run. Refer to the Cisco Web site to obtain the latest feature sets and Cisco IOS images available.

ATM Bandwidth Allocation

Unlike traditional LAN data, voice has strict requirements in terms of delay and jitter in order to remain intelligible. You need to balance these requirements with the need to use available bandwidth as efficiently as possible.

Figure 9-1 illustrates how ATM can allocate minimum amounts of bandwidth to each application. When one application does not need all the bandwidth, other applications can use it.

Figure 9-1   *ATM Bandwidth Allocation*

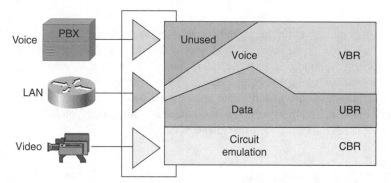

Cisco VoATM devices use queuing and prioritization techniques that handle bandwidth on a per-VC basis to ensure that each separate application has the bandwidth and QoS it requires.

Techniques to Minimize Delay and Variation

There are many means of minimizing delay and variation in a networking environment. ATM starts with a 53-byte cell. Because the cells are small, there is less delay between payloads. The ATM cell is made up of a 5-byte header and a 48-byte payload. When a data packet that is larger than 48 bytes is received for transmission via ATM, that data packet is sliced into multiple 48-byte ATM payloads for reassembly at the receiving end. This is referred to as segmentation and reassembly (SAR). Figure 9-2 illustrates the segmentation of an outbound packet larger than 144 bytes, so it requires 4 ATM cells to carry the payload (144/48 = 3 full ATM frames). The fixed cell size and classes of service permit ATM to provide a more deterministic throughput than other Layer 2 transports.

Figure 9-2   *Creating ATM Cells*

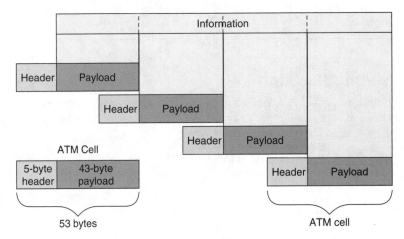

Another means of minimizing delay is by using the various service classes discussed earlier in the chapter, in the section "ATM Service Classes: VBR, CBR, UBR, and ABR." By selecting the appropriate service class, you can ensure that voice, video, and other traffic types that are intolerant of high delay or jitter are provided the appropriate QoS. Each traffic type that is assigned to a different service class travels in a separate VC. Merely by being in a different VC, that traffic has its own queue. This means you can classify traffic so that voice is not queued in the same queue as data.

Cisco VoATM Hardware

Cisco Systems makes a variety of equipment that supports ATM transmission. The following sections discuss only those modules that are used for VoATM in a chassis that supports directly connected voice devices such as analog telephones, key systems, PBXs, and connections to the PSTN. This equipment is capable of multiplexing voice, video, and data onto ATM trunks running at either T1 or E1.

There are two specific hardware options discussed below that provide ATM capability. They are the multiflex trunk (MFT) and the network module-inverse multiplex ATM (NM-IMA) products. Check the Cisco Web site for specific part numbers.

Multiflex Trunk (MFT)

The MFT is used as the trunk interface for many applications on the Cisco MC3810. The MFT provides support for either ANSI T1.403 (T1) or ITU G.703 (E1) and also provides a built-in CSU/DSU.

There are two operating modes for the MFT:

- Multiflex mode
- ATM mode

In multiflex services the interface may be used for data, digital voice trunking, and TDM circuits, but not for ATM services. In ATM mode the full T1/E1 is (and must be) devoted to ATM. Voice connections are made via compressed voice using AAL5 VBR services. Similar to the multiflex mode, multiple services, including LAN and video, may be passed over the ATM trunk—except they all use ATM encapsulation.

Network Module-Inverse Multiplex ATM (NM-IMA)

The multiport T1/E1 ATM network modules with IMA allow users to take advantage of the higher bandwidth and multiservice capabilities available from ATM services.

IMA is an ATM Forum specification that provides a cost-effective and scalable alternative to T3/E3 services by allowing carriers to leverage widely available T1 services over ATM and interoperate with other vendors' standards-based equipment. Multiport T1/E1 IMA network module functionality allows access to ATM WAN uplinks with bandwidth requirements between DS1 and DS3 rates. Each T1 link provides 1.544 megabits per second (Mbps) and each E1 link provides 2.048 Mbps connectivity. With IMA's aggregation of multiple T1/E1s, bandwidth may be increased inexpensively to allow wide-area network (WAN) uplinks at speeds ranging from 1.544 Mbps to 16 Mbps full duplex on a single card.

All four multiport ATM network modules support the ATM Forum Adaptation Layer 5 (AAL5) with ATM QoS categories UBR, VBR-rt and VBR-nrt, and ABR. VoIP and H.323 video over ATM can be supported using VBR-rt CoS parameters with the currently available analog voice/fax network modules with voice interface cards (VICs), or the new T1 digital packet voice trunk network modules with multiflex trunk cards.

The multiport T1/E1 IMA modules' support for VoIP and H.323 video applications provides low-latency, jitter-free voice support to enhance throughput, and high-quality, end-to-end voice and video performance. When combined with high-density packet telephony solutions already available on the Cisco 2600 and 3600 series, the multiport ATM network modules provide a seamless end-to-end multiservice solution, interoperable with other ATM Forum IMA version 1.0 products available from Cisco.

At the time of writing there are both 4-port and 8-port IMA cards available.

ATM VC Traffic Shaping

Traffic shaping is an important component in planning a VoATM network. You must be familiar with the traffic types and their delay budgets and sensitivity to jitter. The following sections discuss the voice sample characteristics, formulas necessary to calculate the bandwidth, and details of traffic-shaping ATM traffic using Cisco IOS.

The codec sample size is used in calculating the delay budget. To help you visualize the differences in codec frame sizes and how they are segmented, the ATM cell structures for CS-ACELP (G.729), and PCM (G.711) voice calls are shown in Figure 9-3. Note that PCM voice sample requires multiple ATM cells to transport a single voice sample.

Figure 9-3 *Voice Sample Sizes*

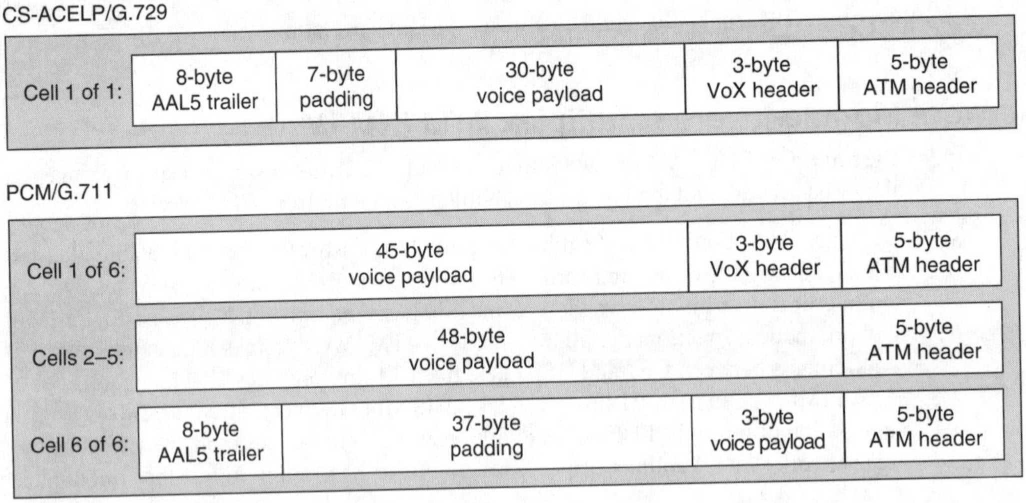

CS-ACELP/G.729

	8-byte AAL5 trailer	7-byte padding	30-byte voice payload	3-byte VoX header	5-byte ATM header
Cell 1 of 1:					

PCM/G.711

	45-byte voice payload	3-byte VoX header	5-byte ATM header
Cell 1 of 6:			

	48-byte voice payload	5-byte ATM header
Cells 2–5:		

	8-byte AAL5 trailer	37-byte padding	3-byte voice payload	5-byte ATM header
Cell 6 of 6:				

ATM Bandwidth Formulas

To calculate the bandwidth for a particular codec, use the following formulas:

ATM bandwidth = ([cells per sample × 53] × 8) / sample rate in milliseconds (ms)
ATM cells per sample = (sample size[bytes]+3[bytes vofr hdr]+8[bytes aal5 OH]) / 48 (round up to whole number)

The following is the default for G.729 (round to next even kilobits per second (kbps); for example, 14.1 would be rounded to 15):

G.729 and G.729a—30 m/sec default = ([1 × 53] × 8) / .030 = 15 kbps

| NOTE | Cisco MC3810s running Cisco IOS 12.03T or earlier will default to 15 kbps (G.729 and G.729a) and 85 kbps (G711) calls. |

Table 9-1 describes how you would set the Cisco traffic type (AAL1, AAL5snap, and so forth) to get one of the following desired service types:

- CBR
- UBR
- VBR

Table 9-1 *1ATM VC Traffic Shaping*

IOS Traffic Type	peak-rate (kbps)	avg-rate (kbps)	burst (cells)	Service Type
aal1	must set	must set	must set	CBR
aal5snap	must set	must set	must set	VBR-rt
aal5snap	may set	must set (UBR+)	—	UBR
aal5muxframe-relay	must set	must set	must set	VBR-nrt
aal5muxframe-relay	may set	must set (UBR+)	—	UBR
aal5muxvoice	must set	must set	must set	VBT-rt

The column heading IOS Traffic Type is used by the **encapsulation** command, while the column headings *peak-rate*, *avg-rate*, and *burst* are all commands used in the ATM virtual circuit configuration mode. See the section titled "Using VoATM Configuration Commands" for further descriptions and examples.

| NOTE | If the parameters are explicitly set, you will get the type of service indicated in the far right column in Table 9-1. However, if you do not set parameters, you will only get UBR services. |

Key parameters to know when dealing with T1 or E1 ATM are as follows:

- There are 424 (that is, 53×8) bits in a cell and 384 (that is, 48×8) bits in payload.
- For E1 there are 4529 cells/sec (= 1.920 Mbps/[424 bits/cell]), 221 microsec/cell.
- For T1 there are 3627 cells/sec (=1.536 Mbps/[424 bits/cell]), 276 microsec/cell.

The ATM interface follows the E1 ATM specification and does not use timeslots 0 and 16.

To avoid loss of data, you must set proper policing parameters—*peak-rate*, *average-rate*, and *burst*—for all virtual channels. To find the allowable burst size in cells, call your carrier and ask for the CDVT (cell delay variation tolerance) that is usually specified in microseconds (µs). If that information is not available, ask, "How many cells or for how long a time period at our access rate can we transmit above peak before being discarded?" The burst size in cells is calculated by dividing the CDVT by the cell size. For example, on an E1 circuit, burst = ([CDVT] / [221 microsec/cell]).

The peak cell rate (PCR) corresponds to the peak-rate parameter, and the average cell rate (ACR) corresponds to average-rate parameter.

The following ATM conversion factors apply:

- *peak-rate* in kbps = peak cell rate (PCR) in cells per second times 424 bits per cell
- *average-rate* in kbps = average or sustained cell rate (ACR or SCR) in cells per second times 424 bits per cell

---

NOTE   The MC3810 uses kbps as the units in the *peak-rate* and *average-rate* parameters.

Traffic Shaping for Data or Video

When setting up ATM traffic shaping, the shaping engine does not take into account ATM overhead for the peak and average rates you enter. Therefore, you *must* account for the overhead when setting the peak and average cell (that is, PCR, ACR) rates, as follows:

- Use 1.13 percent of the application's required data rate as a basis for AAL5snap peak or average bit rate calculations.
- Use 1.14 percent of the application's required data rate as a basis for AAL1 peak or average bit rate calculations.

NOTE   An extra byte in the AAL1 header accounts for the slight increase in the overhead relative to AAL5 VCs.

MC3810 Fixed Prioritization

The ability to distinguish voice and video data types, and service them before non-real-time data types, is a feature of the Cisco MC3810. To do this function, priority queues are used on the trunk, each supporting one of two service classes.

The MC3810 supports two service classes: real time and non-real-time. The real-time class is used for voice and video, and the non-real-time class is used for data. Real-time virtual channel queues are serviced in a round-robin fashion, and they are completely emptied before any non-real-time channel queues are serviced. Non-real-time virtual channel queues will be serviced only when there are no real-time cells in the queues. Data channel queues are also serviced in a round-robin fashion.

Non-real-time may be prioritized within its own domain by the core router. Using Cisco IOS features such as priority queuing and weighted fair queuing, different non-real-time data streams may be prioritized relative to each other before they are presented to the trunk.

Assuming that the Cisco IOS data streams are in priority order by alphabet, the priority of all the data streams in Figure 9-4 is as follows:

- CES (video)
- VBR-rt (voice)
- Data A
- Data B
- Data C

Figure 9-4   *Fixed Prioritization*

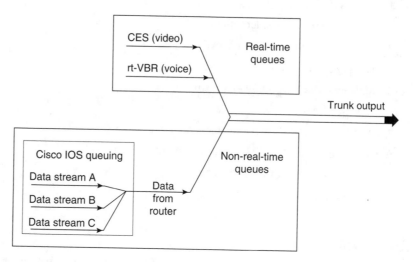

Voice and video traffic always have top priority. There is no way to prioritize non-real-time data over real-time data. Voice always has top priority to prevent telephone users from being locked out. However, non-real-time data streams will never be completely locked out by real-time data if the unit is correctly configured.

For example, out of an entire T1/E1 ATM trunk, if 24 compressed voice channels are present, they will use approximately 240 kbps of bandwidth (about 15 percent of a T1 and

11 percent of an E1). Even if a video service takes up another 384 kbps, plenty of bandwidth would be available for data.

NOTE To ensure that critical data gets through a network, the design should include bandwidth for data over and above the greatest amount of voice bandwidth possible. The total bandwidth includes locally terminated voice connections as well as tandem connections.

Using VoATM Configuration Commands

Several commands are used to configure VoATM using Cisco IOS. This section covers those commands and—if the command is intended to run on only one series of platforms—notes which platform series the command is applicable to. In this section we cover the following commands and subjects:

- **mode** {**atm** | **cas**}
- **interface atm**
- **pvc** *name vpi/vci*
- **encapsulation** {*aal-encap*}
- Virtual template considerations
- An example of configuring VoATM
- **vbr-rt** *peak-rate average-rate burst*
- **dial-peer voice** *tag-number* **voatm**
- **session target**
- Additional ATM PVC parameters and defaults
- An example of a VoATM across a network
- Creating a peer configuration table
- Configuring dial peers

mode {atm | cas}

The **mode** {**atm** | **cas**} command sets the mode of the T1/E1 controller and enters specific configuration commands for each mode type. This command applies to the Cisco MC3810 with the MFT installed. When no mode is selected, channel groups, voice-signaling, and clear channels (data mode) can be created using the **channel group**, **ds0-group**, and **tdm-group** commands, respectively.

On the Cisco MC3810, some DS0s are used exclusively for different signaling modes. The DS0 channels have the following limitations when mixing different applications (such as voice and data) on the same network trunk:

- On E1 controllers, DS0 16 is used exclusively for either channel-associated signaling (CAS) or common channel signaling (CCS), depending on which mode is configured.

- On T1 controllers, DS0 24 is used exclusively for CCS.

Table 9-2 describes the options for the **mode** {**atm** | **cas**} command.

Table 9-2   **mode** {**atm** | **cas**} *Command*

Command Option	Description
atm	Sets the controller into ATM mode and creates an ATM interface (ATM 0) on the Cisco MC3810. When ATM mode is enabled, no channel groups, CAS groups, CCS groups, or clear channels are allowed because ATM occupies all the DS0s on the T1/E1 trunk.
	When you set the controller to ATM mode, the controller framing is automatically set to ESF for T1 or CRC4 for E1. The linecode is automatically set to B8ZS for T1 or HDBC for E1. When you remove ATM mode by entering the **no mode atm** command, ATM interface 0 is deleted.
	ATM mode is supported only on controller 0 (T1 or E1 0).
cas	Sets the controller into CAS mode, which allows you to create channel groups, CAS groups, and clear channels (both data and CAS modes). CAS mode is supported on both controllers 0 and 1.

CAS extraction and insertion is used for T1/E1 voice ports or the trunk transmitting and receiving digital voice traffic from a PBX. In this case, DS0 channels can be configured to directly connect to the T1/E1 network trunk in TDM mode, or the channels can be connected to the Cisco MC3810's digital signaling processor for voice compression. During this process, the CAS bits are extracted from the defined T1/E1 channels and passed to the voice-signaling handler.

Use the **mode cas** command to configure the T1/E1 line to support CAS. This command is also supported on the digital voice module (DVM).

NOTE   The **melcas** options are supported only on E1 and apply to the Mercury Exchange Limited (MELCAS) interpretation of the CEPT standard, used primarily in the United Kingdom.

Example 9-1 configures ATM mode on controller T1 0. This step is required for VoATM.

Example 9-1 *Configuring* **mode atm** *on an MC3810 Controller*

```
p1r3#config t
Enter configuration commands, one per line. End with CNTL/Z.
p1r3(config)#controller T1 0
p1r3(config)#mode atm
```

Example 9-2 configures CAS mode on controller T1 1.

Example 9-2 *Configuring* **mode cas** *on an MC3810 Controller*

```
p1r3#config t
Enter configuration commands, one per line. End with CNTL/Z.
p1r3(config)#controller T1 1
p1r3(config)#mode cas
```

ATM 0 is a logical interface that is opened when the **mode atm** command is issued on the T1/E1 0 controller. The **mode atm** command opens all available timeslots on the T1/E1 facilities.

The ATM 0 interface is configured using the commands and syntax in Example 9-3.

Example 9-3 *Configuring an ATM Interface on an MC3810*

```
rx.y(config)#controller t1 0
rx.y(config-controller)#mode atm
rx.y(config-controller)#exit
rx.y(config)#interface atm0
rx.y(config-if)#no shutdown
rx.y(config-if)#exit
rx.y(config)#end
```

Multiple ATM PVCs may be configured under the primary interface. Multiple subinterfaces with multiple PVCs may be opened using the command string example shown in Example 9-4.

Example 9-4 *Configuring Multiple ATM Subinterfaces on an MC3810*

```
rx.y(config)#interface atm0.1 ?
  multipoint       Treat as a multipoint link
  point-to-point   Treat as a point-to-point link
rx.y(config)#interface atm0.1 point-to-point
rx.y(config-subif)#no shutdown
rx.y(config-subif)#exit
rx.y(config)#end
```

NOTE   ATM 0 replaces the logical serial 2 interface used in the MC3810 11.3MAx Cisco IOS images.

VoATM enables a Cisco MC3810 or 3600 to carry live voice traffic over an ATM network.

interface atm

The **interface atm** command configures an ATM interface type and enters interface configuration mode. Use *subinterface-number* to configure an ATM subinterface.

Table 9-3 describes the options for the **interface atm** *slot*/**0** {*options*} command.

Table 9-3   *Options for the* **interface atm** *slot*/**0** {*options*} *Command*

Command Option	Description
slot	Specifies the backplane slot number of the router you are using. Refer to the specific manual of that equipment.
/0	ATM port number. Because the AIP and all ATM port adapters have a single ATM interface, the port number is always 0.
subinterface-number	Subinterface number in the range 1 to 4294967293.
**multipoint	point-to-point**	Specifies a multipoint or point-to-point subinterface. There is no default.

pvc *name vpi/vci*

Use the **pvc** *name vpi/vci* interface configuration command to do one or more of the following:

- Create an ATM PVC on a main interface or subinterface.
- Assign a name to an ATM PVC.
- Specify ILMI, QSAAL, or SMDS as the encapsulation type on an ATM PVC. (To configure other encapsulation types, see the **encapsulation** command.)
- Enter interface-ATM-VC configuration mode.

The Cisco IOS software dynamically creates rate queues as necessary to satisfy the requests of the **pvc** commands.

The **pvc** command creates a PVC and attaches it to the *vpi* and *vci* specified.

When configuring an SVC, use the **pvc** command to configure the PVC that handles SVC call setup and termination. In this case, specify the **qsaal** keyword.

Once you specify a *name* for a PVC, you can reenter the interface-ATM-VC configuration mode by simply entering **pvc** *name*. You can remove a PVC and any associated parameters by entering **no pvc** *name* or **no pvc** *vpi/vci*.

Table 9-4 describes the options for the **pvc** *name vpi/vci* command.

Table 9-4   *Options for the **pvc** name vpi/vci Command*

Command Option	Description
name	(Optional.) The name of the PVC or map. The name can be up to 16 characters.
vpi/	ATM network virtual path identifier (VPI) for this PVC. The absence of the "/" and a *vpi* value defaults the *vpi* value to 0. The arguments *vpi* and *vci* cannot both be set to 0; if one is 0, the other cannot be 0.
vci	ATM network virtual channel identifier (VCI) for this PVC. This value ranges from 0 to 1 less than the maximum value set for this interface by the **atm vc-per-vp** command. Typically, lower values 0 to 31 are reserved for specific traffic (for example, F4 OAM, SVC signaling, ILMI, and so on) and should not be used. The VCI is a 16-bit field in the header of the ATM cell. The VCI value is unique only on a single link, not throughout the ATM network, because it has local significance only. The arguments *vpi* and *vci* cannot both be set to 0; if one is 0, the other cannot be 0.

After configuring the parameters for an ATM PVC, you must exit the interface-ATM-VC configuration mode in order to create the PVC and enable the settings.

encapsulation *aal-encap*

The **encapsulation** *aal-encap* command configures the ATM adaptation layer (AAL) and encapsulation type for an ATM PVC, SVC, or VC class.

Use one of the **aal5mux** encapsulation options to dedicate the specified PVC to a single protocol; use the **aal5snap** encapsulation option to multiplex two or more protocols over the same PVC.

Selecting **aal5mux** or **aal5snap** encapsulation depends on practical choices, such as the type of network and the pricing offered by the network:

- If the network's pricing depends on the number of PVCs set up, **aal5snap** might be the appropriate choice.

- If pricing depends on the number of bytes transmitted, **aal5mux** might be the appropriate choice because it has slightly less overhead.

Table 9-5 describes the options for the **encapsulation** *aal-encap* command.

Table 9-5   *Options for the* **encapsulation** *aal-encap Command*

Command Option	Description
aal-encap	ATM adaptation layer (AAL) and encapsulation type.
	ATM adaptation layer 5—One of four AALs recommended by the ITU-T. AAL5 supports connection-oriented VBR services and is used predominantly for the transfer of classical IP over ATM and LANE traffic. AAL5 uses SEAL and is the least complex of the current AAL recommendations. It offers low bandwidth overhead and simpler processing requirements in exchange for reduced bandwidth capacity and error-recovery capability.
	When **aal5mux** is specified, a protocol is required. Possible values for *aal-encap* are as follows:
	• **aal5mux frame**—For a MUX-type virtual circuit for Frame Relay–ATM Interworking.
	• **aal5mux voice**—For a MUX-type virtual circuit for Voice over ATM (VoATM).
	• **aal5snap**—The only encapsulation supported for Inverse ARP. Logical Link Control/Subnetwork Access Protocol (LLC/SNAP) precedes the protocol datagram.

The following are issues regarding AAL1 and AAL5:

- No CBR support with VoATM (AAL5)
- No ATM-25 support
- AAL1 support on 3810 is for CES only, not for VoATM

Virtual Template Consideration

If you specify *virtual template* parameters after the ATM PVC is configured, you should issue a **shutdown** command followed by a **no shutdown** command on the ATM subinterface to restart the interface, causing the newly configured parameters (such as an IP address) to take effect.

If the **encapsulation** command is not explicitly configured on an ATM PVC, the VC inherits the following default configuration of the **encapsulation** command in a class assigned to:

- PVC itself
- PVC ATM subinterface
- PVC ATM main interface
- Global default: *aal-encap* = **aal5snap**

An Example of Configuring VoATM

Example 9-5 configures a PVC to support encapsulation for Voice over ATM (VoATM) using aal5mux encapsulation. On the MC3810 the encapsulation type for native VoATM must be aal5mux.

Example 9-5 *Configuring aal5mux voice on ATM*

```
rtr1(config-if)#pvc 5/70
rtr1(config-if-atm-vc)#encapsulation aal5mux voice
rtr1(config-if-atm-vc)#?
ATM virtual circuit configuration commands:
  cbr           Enter Average Cell Rate in Kbps.
  ubr           Enter Unspecified Peak Cell Rate (pcr) in Kbps.
  ubr+          Enter Peak Cell Rate(pcr)Minimum Cell Rate(mcr) in Kbps.
  vbr-nrt       Enter Variable Bit Rate (pcr)(scr)(bcs)
  vbr-rt        Enter Variable Bit Rate (pcr)(average)
rtr1(config-if-atm-vc)#vbr-rt ?
  <56-10000>  Peak Cell Rate(PCR) in Kbps
rtr1(config-if-atm-vc)#vbr-rt 190 ?
  <1-190>  Average Cell Rate in Kbps
rtr1(config-if-atm-vc)#vbr-rt 190 80 ?
  <0-65536>  Burst cell size in number of cells
rtr1(config-if-atm-vc)#vbr-rt 190 80 16
rtr1(config-if-atm-vc)#exit
rtr1(config)#end
rtr1#
```

NOTE	PVCs are configured in the same manner when commands are issued under a subinterface.

The **vbr-rt** *peak-rate average-rate burst* command configures the real-time VBR for VoATM connections.

The **vbr-rt** command configures traffic shaping between voice and data PVCs. Traffic shaping is required so that the carrier does not discard calls.

The vbr-rt *peak-rate average-rate burst* Command

The objective of using ATM as a transport is to guarantee QoS for the voice traffic. To do that, it is necessary to use vbr-rt traffic. This section lists the options to the **vbr-rt** *peak-rate average-rate burst* command and discusses proper sizing of the PVC for voice traffic. The options for the **vbr-rt** *peak-rate average-rate burst* command are listed in Table 9-6.

Table 9-6 *Options for the* **vbr-rt** *peak-rate average-rate burst Command*

Command Option	Description
peak-rate	The peak information rate (PIR) of the voice connection in kbps. The range is 56 to 10000.
average-rate	The average information rate (AIR) of the voice connection in kbps. The range is 1 to 56.
burst	Burst size in number of cells. The range is 0 to 65536.

Calculating the Peak, Average, and Burst Options

To configure voice and data traffic shaping, you must configure the peak, average, and burst options for voice traffic. Configure the burst value if the PVC will be carrying bursty traffic. The peak, average, and burst values are needed so the PVC can effectively handle the bandwidth for the number of voice calls. To calculate the minimum peak, average, and burst values for the number of voice calls, use the following calculations:

- Peak value: $(2 \times$ the maximum number of calls$) \times 16$

- Average value: $(1 \times$ the maximum number of calls$) \times 16$

- Burst value: $(4 \times$ the maximum number of calls$)$

NOTE When you configure data PVCs that will be traffic-shaped with voice PVCs, use the **aalsnap** encapsulation and calculate the overhead as 1.13 times the voice rate.

The peak (optional) value is the maximum rate (in kbps) at which this virtual circuit can transmit. The valid range is from 56 to 10000. If configuring Voice over ATM, you must configure the peak, average, and burst values. To calculate the peak rate for the number of voice calls, use the following calculation:

$(2 \times$ the number of calls$) \times 16$

The average (optional) value is the rate (in kbps) at which this virtual circuit transmits. Valid values are platform dependent. If configuring Voice over ATM, you must configure the peak, average, and burst values. To calculate the average rate for the number of voice calls, use the following calculation:

$(1 \times$ the number of calls$) \times 16$

The burst (optional) value relates to the maximum number of ATM cells the virtual circuit can transmit to the network at the peak rate of the PVC for bursty traffic. If configuring

Voice over ATM, you must configure the peak, average, and burst values. To calculate the burst rate for the number of voice calls, use the following calculation:

$4 \times$ the number of calls

Example 9-6 configures the traffic-shaping rate for ATM PVC 20 on a Cisco 3600. In the example, the peak, average, and burst rates are calculated based on a maximum of 20 calls on the PVC.

Example 9-6 *VoATM Traffic Shaping Example*

```
router(config-if)# pvc 20
router(config-if-atm-pvc)# encapsulation aal5mux voice
router(config-if-atm-pvc)# vbr-rt 640 320 80
```

NOTE   Do not configure the **inarp** or **compress** values for a voice PVC.

dial-peer voice *tag-number* voatm

The **dial-peer voice** *tag-number* **voatm** command enters the dial peer configuration mode and specifies the method of network-related encapsulation. Table 9-7 lists the options for the **dial-peer voice** *tag-number* **voatm** command.

Table 9-7   *Options for the* **dial-peer voice** *tag-number* **voatm** *Command*

Command Option	Description
tag-number	Digit(s), unique to the local router, defining a particular dial peer. Valid entries are from 1 to 2147483647 (up to 100000 on the MC3810). The *tag-number* is an arbitrary identifier you assign to uniquely identify the dial peer.
voatm	Indicates that this is a VoATM peer using the real-time AAL5 voice encapsulation on the ATM backbone network.

NOTE   To modify the tag configuration after you configure the dial peer voice tag, enter the **dial-peer voice** and the *tag-number* and press Enter.

The session target Command

You use the **session target** command to specify a network-specific address or domain name for a dial peer. Whether you select a network-specific address or a domain name depends on the session protocol you select. Table 9-8 lists the options for the **session target** command.

Table 9-8   *Options for the* **session target** *Command*

Command Option	Description
interface	Specifies the interface type and interface number on the router. **For MC3810**: The only valid number is 0.
pvc	Indicates the specific ATM permanent virtual circuit for this dial peer.
name	The PVC name.
vpi/vci	The ATM network VPI and VCI of this PVC. **For 3600**: If you have the Multiport T1/E1 ATM network module with IMA installed, the valid range for *vpi* is 0–15, and the valid range for *vci* is 1–255. If you have the OC3 ATM Network Module installed, the valid range for *vpi* is 0–15, and the valid range for *vci* is 1–1023.
vci	The ATM network virtual channel identifier (VCI) of this PVC.

The Cisco IOS configuration in Example 9-7 illustrates a session target set to ATM interface 0 and the PVC set to a VPI/VCI of 1/100.

Example 9-7   *Example 3600* **dial-peer** *Configuration (Including Subcommands)*

```
p3r2#config t
Enter configuration commands, one per line. End with CNTL/Z.
p3r2(config)#dial-peer voice 12 voatm
p3r2(config-dial-peer)#destination-pattern 13102221111
p3r2(config-dial-peer)#session target atm1/0 pvc 1/100
```

The Cisco IOS configuration in Example 9-8 illustrates a session target set to ATM interface 0 and the PVC configured with a VCI of 20.

Example 9-8   *Example MC3810* **dial-peer** *Configuration (Including Subcommands)*

```
voatm(config)# dial-peer voice 12 voatm
voatm(config-dial-peer)# destination-pattern 13102221111
voatm (config-dial-peer)# session target atm0 pvc 20
```

Additional ATM PVC Parameters and Defaults

The **inarp** *minutes* (optional) (set for the aal5snap encapsulation only) parameter specifies how often Inverse ARP datagrams are sent on this virtual circuit. The valid range is from 1 to 60. The default value is 15 minutes.

If the **inarp** keyword is omitted, Inverse ARPs are not generated. If the **inarp** keyword is present but the timeout value is not given, then Inverse ARPs are generated every 15 minutes.

If peak and average rate values are omitted, the PVC defaults to peak and average rates equal to the link rate. The peak and average rates are then equal. By default, the virtual circuit is configured to run as fast as possible.

NOTE   The order of command options is important. The **inarp** keyword can be specified either separately or before the **oam** keyword has been enabled. The peak, average, and burst values, if specified, cannot be specified after either the **inarp** or the **oam** keywords.

The Cisco IOS software dynamically creates rate queues as necessary to satisfy the requests of **atm pvc** commands. The software dynamically creates a rate queue when an **atm pvc** command specifies a peak or average rate that does not match any user-configured rate queue.

The **atm pvc** command creates a PVC and attaches it to the *vpi* and *vci* specified. Both *vpi* and *vci* cannot be specified as 0; if one is 0, the other cannot be 0. The *aal-encap* argument determines the AAL mode and the encapsulation method used. The *peak-rate* and *average-rate* arguments determine the rate queue used.

NOTE   If you choose to specify *peak-rate* or *average-rate* values, you must specify both.

An Example of VoATM Across a Network

Figure 9-5 is a diagram of a small voice network in which router 1, with ATM virtual circuit 0/20, connects a small sales branch office to the main office through router 2. There are only two devices in the sales branch office that need to be established as dial peers: a basic telephone and a fax machine. Router 2, with an ATM virtual circuit of 0/40, is the primary gateway to the main office; as such, it needs to be connected to the company's PBX. There are three devices that need to be established as dial peers in the main office, all of which are basic telephones connected to the PBX. The numbers in the figure refer to the dial peer numbers.

Figure 9-5   *An Example of a VoATM Network*

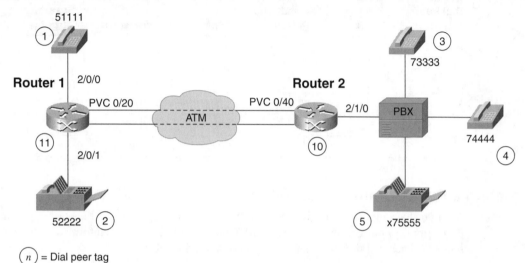

n = Dial peer tag

Table 9-9 shows the peer configuration table for the example VoATM network illustrated in the figure above. The columns are:

- Dial peer tag—Corresponds with the circled number in Figure 9-5
- Prefix—The number to prefix to the dialed digits
- Destination pattern—The dialed digits to match for this dial-peer statement
- Type—The *type* at the end of the **dial-peer voice** *tag type* command
- Voice port—The voice port to which this call is directed (used only on **dial-peer voice** *tag* **pots** statements)
- Session target—The ATM interface and VPI/VCI pair to which this call is directed (used only on **dial-peer voice** *tag* **voatm** statements)

Table 9-9   *Peer Configuration Table for the VoATM Network Example*

Dial Peer Tag	Prefix	Destination Pattern	Type	Voice Port	Session Target
Router 1					
1		51111	pots	2/0/0	
2		52222	pots	2/0/1	
10		7...	voatm		atm0/0 pvc 0/20

continues

Table 9-9 *Peer Configuration Table for the VoATM Network Example (Continued)*

Dial Peer Tag	Prefix	Destination Pattern	Type	Voice Port	Session Target
Router 2					
11		5...	voatm		atm1/0 pvc 0/40
3	7	733...	pots	2/1/0	
4	7	74...	pots	2/1/0	
5	7	75...	pots	2/1/0	

Creating a Peer Configuration Table

After you have merged your telephony and WAN networks together, there are tasks you can do to simplify configuring Voice over ATM. One is to collect all of the information directly related to each dial peer by creating a peer configuration table.

There is specific information relative to each dial peer that needs to be identified before you can configure Voice over ATM. One way to identify this specific information is to create a peer configuration table.

Configuring Dial Peers

Dial peers describe the entities to or from which a call is established. Dial peer configuration tasks define the address or set of addresses serviced by that dial peer and the call parameters required to establish a call to or from that dial peer.

There are two different kinds of dial peers:

- POTS—Dial peer connected via a traditional telephony network. POTS (plain old telephone service) dial peers point to a particular voice port on a voice-network device.
- VoATM—Dial peer connected via an ATM WAN backbone. Voice over ATM dial peers point to specific voice-network devices.

POTS dial peers associate a telephone number with a particular voice port so that incoming calls for that telephone number can be received and outgoing calls can be placed. VoATM dial peers point to specific voice-network devices (by associating destination telephone numbers with a specific ATM VC) so that incoming calls can be received and outgoing calls can be placed. Both POTS and VoATM dial peers are required if you want to both send and receive calls using VoATM.

Establishing two-way communication using VoATM requires establishing a specific voice connection between two defined endpoints. As shown in Figure 9-5, for outgoing calls (from the perspective of the POTS dial peer 1), the POTS dial peer establishes the source (the originating telephone number and voice port) of the call. The VoATM dial peer establishes the destination by associating the destination phone number with a specific ATM virtual circuit.

Monitoring and Troubleshooting Commands

This section covers the monitoring and troubleshooting commands you will find helpful in managing a Cisco VoATM network. Keep in mind that some commands, especially the **debug** command, may require intensive CPU cycles, reducing the percent of available time to basic routing and packet-forwarding functions.

show Commands

There are several **show** commands associated with ATM services. We will cover only a few of them here; for further information see the *Wide-Area Networking Command Reference* for the current release of Cisco IOS.

show atm pvc

The **show atm pvc** command provides a summary of the PVCs that you have configured on your router and their VPI/VCI as well as several other fields. Sample output is displayed in Example 9-9.

Example 9-9   **show atm pvc** *Command*

```
p1r2#show atm pvc
            VCD /                                Peak  Avg/Min Burst
 Interface  Name VPI   VCI  Type  Encaps   SC   Kbps   Kbps  Cells  Sts
 0           1    1    101  PVC   SNAP    VBR    128     64     0   UP (   64)
 0           2    1    102  PVC   VOICE   VBR    192     96    48   UP (   96)
```

show atm pvc {*vpi/vci*}

The **show atm pvc** {*vpi/vci*} command provides detailed information about a specific virtual circuit. A sample output is displayed in Example 9-10.

Example 9-10 **show atm pvc** {*vpi/vci*} *Command*

```
p1r2#show atm pvc 1/102
ATM0: VCD: 2, VPI: 1, VCI: 102
VBR-RT, PeakRate: 192, Average Rate: 96, Burst Cells: 48
AAL5-VOICE, etype:0xC, Flags: 0x29, VCmode: 0x0
OAM frequency: 0 second(s), OAM retry frequency: 1 second(s), OAM retry frequency:
1 second(s)
OAM up retry count: 3, OAM down retry count: 5
```

continues

Example 9-10 show atm pvc {*vpi/vci*} *Command (Continued)*

```
OAM Loopback status: OAM Disabled
OAM VC state: Not Managed
ILMI VC state: Not Managed
InARP DISABLED
InPkts: 0, OutPkts: 5, InBytes: 0, OutBytes: 75
InPRoc: 0, OutPRoc: 5, Broadcasts: 0
InFast: 0, OutFast: 0, InAS: 0, OutAS: 0
OAM cells received: 0
F5 InEndloop: 0, F5 InSegloop: 0, F5 InAIS: 0, F5 InRDI: 0
F4 InEndloop: 0, F4 InSegloop: 0, F4 InAIS: 0, F4 InRDI: 0
OAM cells sent: 0
F5 OutEndloop: 0, F5 OutSegloop: 0, F5 OutRDI: 0
F4 OutEndloop: 0, F4 OutSegloop: 0, F4 OutRDI: 0
OAM cell drops: 0
Compress: Disabled
Status: UP
Voice FRF11 Information
Total Bandwidth 96 Used Bandwidth 0

cid cid-type          input-pkts    output-pkts   dropped-pkts
4   data                 0             0             0
5   voice call-control 0             5             0
```

show interface [atm0 | atm1/*0*]

The syntax of the **show interface [atm0 | atm1/0]** command is dependent on the platform, so make sure you are using the proper interface syntax with your hardware. This command will show information about the ATM interface; sample output is displayed in Example 9-11.

Example 9-11 show interface atm0 *Command*

```
p1r2#show interface atm0
ATM0 is up, line protocol is up
  Hardware is PQUICC Atom1
  Internet address is 192.168.1.1/24
  MTU 1500 bytes, sub MTU 1500, BW 1536 Kbit, DLY 20000 usec,
     reliability 255/255, txload 1/255, rxload 1/255
  Encapsulation ATM, loopback not set
  Keepalive not supported
  Encapsulation(s):, PVC mode
  1024 maximum active VCs, 2 current VCCs
  VC idle disconnect time: 300 seconds
  Last input 00:33:12, output never, output hang never
  Last clearing of "show interface" counters never
  Input queue: 0/75/0 (size/max/drops); Total output drops: 0
  Queueing strategy: weighted fair
  Output queue: 0/1000/64/0 (size/max total/threshold/drops)
     Conversations  0/0/256 (active/max active/max total)
     Reserved Conversations 0/0 (allocated/max allocated)
  5 minute input rate 0 bits/sec, 0 packets/sec
```

Example 9-11 show interface atm0 *Command (Continued)*

```
5 minute output rate 0 bits/sec, 0 packets/sec
    51 packets input, 4504 bytes, 0 no buffer
    Received 0 broadcasts, 0 runts, 0 giants, 0 throttles
    0 input errors, 0 CRC, 0 frame, 0 overrun, 0 ignored, 0 abort
    58 packets output, 4674 bytes, 0 underruns
    0 output errors, 0 collisions, 2 interface resets
    0 output buffer failures, 0 output buffers swapped out
```

debug Commands

An additional warning on the impact of the **debug** command is always in order—so be careful with this command set. The command yields an amazing amount of information that is helpful in troubleshooting, but it can create a burdensome load on the CPU. If you need to do long-term or intensive cell decodes you should look toward purchasing a separate, stand-alone protocol analyzer. Example 9-12 lists the **debug** commands that you may find helpful in troubleshooting your VoATM configuration:

Example 9-12 *debug atm ? Command (Helpful ATM* **debug** *Commands)*

```
p1r2#debug atm ?
    aal-crc      Display CRC error packets
    errors       ATM errors
    events       ATM or FUNI Events
    packet       ATM or FUNI packets
    pvcd         Show PVCD events
    state        ATM or FUNI VC States
```

For a full description of these commands, see the *Debug Command Reference*.

Summary

This chapter covers the basics of ATM operation and theory, as well as Cisco hardware that supports VoATM and some of the commands that are used in the configuration of VoATM. There are several classes of service—CBR, VBR-rt, VBR-nrt, ABR, and UBR—that can be configured depending on the QoS required by the application. Video and circuit emulation services are best supported by CBR, while VBR-rt is typically used for voice and VBR-nrt or UBR are most frequently used for data. Configuring the Cisco device to support VoATM requires configuring the ATM interface, and the dial peers and may require configuring the controller. There are several commands that can be used to monitor and troubleshoot the operational VoATM network, and some of those commands were illustrated.

Review Questions

The following questions should help you gauge your understanding of this chapter. You can find the answers in Appendix A, "Answers to Review Questions."

1 What are the two general categories of ATM service classes and what are they typically used for?

_____

_____

_____

2 What traffic service class is generally recommended for voice traffic and why?

_____

_____

_____

3 Describe an ATM cell, including header bytes and payload bytes.

_____

_____

_____

4 Describe the formula for calculating the bandwidth required for a particular codec.

_____

_____

_____

5 What are the guidelines for calculating the ATM PVC traffic-shaping parameters for a VC used to carry voice calls?

_____

_____

_____

6 What ATM encapsulation types are used for voice and data?

_____

_____

_____

7 What is the overhead factor used in calculating the total bandwidth necessary for data PVCs?

_____

_____

_____

8 On the Cisco MC3810, what must be configured in order to permit traffic on the ATM0 interface?

_____

_____

_____

9 How many channels are used for ATM on a T1?

_____

_____

_____

10 Which **show** command will display the ATM configuration for each PVC?

_____

_____

_____

After reading this chapter, you should be able to perform the following tasks:

- Identify the basic elements of a VoIP network, including the VoIP protocol stack (H.323 protocols and components, RTP, RTCP, RSVP, MGCP, SIP, SAP, and SDP), header compression, and queuing techniques.

- Explain features of Cisco voice-capable routers that support VoIP.

- Configure VoIP on the Cisco 2600, 3600, and MC3810 using RTP/RTCP, cRTP header compression, QoS, IP RTP Priority, and LLQ.

- Use Cisco IOS commands to configure VoIP over Frame Relay and VoIP over Multilink PPP.

CHAPTER 10

Configuring Cisco Access Routers for VoIP

In this chapter we will examine the challenges associated with transporting Voice over IP networks, with solutions proposed and used in certain installations. We will explore the options used to reorder out-of-sequence packets, options to minimize delay, and protocols used to assist in proper delivery of Voice over IP.

Understanding IP Networks

To understand how voice will be transported over IP networks, it is first necessary to understand the nature of IP networks. The Internet, the grandest IP network of all, promises only "best effort" delivery. This description assumes that the network layer of the TCP/IP stack cannot always guarantee delivery of packets in sequence and without delay. Routers in TCP/IP networks, as in the Internet, route packets based on the "best path." As shown in Figure 10-1, the result is that packets from the same session may take different paths to reach the same destination. When they show up at the receiving end, the packets will need to be placed back in their proper sequence and delivered to the receiver. If they don't show up, Layer 3 (the Internet layer of the TCP/IP protocol stack) can do nothing to help. It will then be necessary to employ upper layers of the protocol stack to assist in packet ordering and reliable delivery without delay.

Figure 10-1 *IP Networks Assume Packet Ordering and Delivery Problems.*

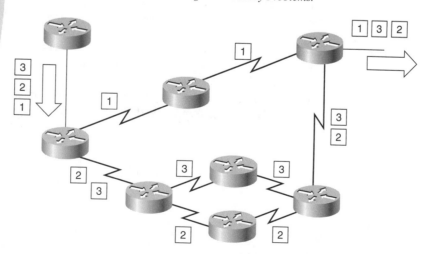

Introduction to VoIP

Earlier chapters described the critical steps necessary to transport delay-sensitive voice-over networks that were originally designed to carry data. Both Frame Relay and ATM need help to ensure that voice arrives without delay and delay variation. Like Frame Relay and ATM networks, IP networks were also originally designed to carry only data and were not designed to allow for delay-sensitive traffic. IP networks assume that there will be delay and that there will be packet-ordering problems due to native routing, which may route packets along different paths only to arrive at the same destination.

In IP networks, Transmission Control Protocol (TCP) can resolve packet ordering problems by using sequence and acknowledgement numbers, but voice will be carried using User Datagram Protocol (UDP). It is necessary to use UDP as the transport protocol for VoIP so the sending device does not need to wait for acknowledgements from the receiving device before sending more packets. VoIP will be sent in a manner similar to that used to send streaming audio or video over the Internet. Loss of some voice packets (less than one percent) is acceptable and can be "recovered" by some coding/decoding mechanisms, as well as by using speech interpolation methods (that is, filling in missing sounds by using DSP technology to intelligently track waveshapes and predict what is missing). We will see that VoIP requires some extra help, and that help will come in similar methods employed by VoFR and VoATM.

Voice Delay Budget

The ITU-T has devoted extensive studies to the delay problem with voice networks. As a result, ITU-T standard G.114 recommends that one-way voice delay should not exceed 150 milliseconds (ms) in order to be acceptable to users of a telephony system. It further recommends that 150 to 400 ms can still be considered acceptable, as long as both the speaker and the listener understand that there is a delay, and that both are able to tolerate the delay. At 400 ms one-way delay, people start to notice. You may have experienced a delay like this if you've ever spoken on a telephone to another continent and the telephone carrier used satellites to transport your voice. The one-way delay of a satellite link is about 170 ms, and that's not counting the delays experienced in the earthbound transmission facilities. The standard also states that over 400 ms delay is unacceptable and cannot be tolerated for voice communications. As we design and deploy our VoIP networks, it will be necessary for us to ensure that we do not exceed the delays set forth in ITU-T G.114. In the following sections, we will look at some of the options we can employ in order to minimize delay. We will use the Open System Interconnection (OSI) seven-layer model as a basis for our exploration.

VoIP Protocol Stack

Because this chapter discusses TCP/IP, you need to understand the protocols used to assist in delivery of voice packets in proper sequence and without delay. If we examine the TCP/IP protocol stack as it relates to voice, a clearer understanding of the tools used at each layer of the stack will emerge. Figure 10-2 shows the protocol stack (traditionally we see seven layers when we use the OSI model to study and understand protocols) with the top layer missing. The figure omits the seventh layer because we assume that it represents the spoken human voice, in any language. Please note that the figure refers to transmission of voice itself, not signaling to connect the call.

Figure 10-2   *The VoIP OSI Reference Model*

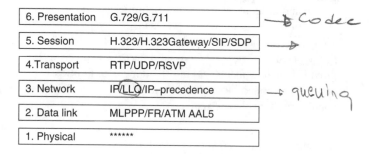

6. Presentation	G.729/G.711	→ Codec
5. Session	H.323/H.323Gateway/SIP/SDP	→
4.Transport	RTP/UDP/RSVP	
3. Network	IP/LLQ/IP–precedence	→ queuing
2. Data link	MLPPP/FR/ATM AAL5	
1. Physical	******	

Beginning at the bottom with the physical layer, it should be noted that IP networks can run on almost any physical layer, and therefore VoIP will be run on any physical medium that can carry pure IP. This includes, but is not limited to, twisted-pair cabling (such as used in traditional Ethernet), serial interfaces (such as V.35 and EIA/TIA 232), leased-line telephone wires (such as those used for Frame Relay and T-1), and ISDN.

The second layer, the data link layer, indicates that IP, and therefore VoIP, can use many different framing formats. These include, as indicated in Figure 10-2, Multilink PPP (MLPPP), Frame Relay (FR), and ATM. We are not limited to these choices, either, as Ethernet and other LAN technologies will carry voice just as well.

At the third layer, things begin to get very interesting. The network layer will naturally use IP as its method for carrying voice, but pure IP will need some help. Because of the delay and delay-variation problems discussed above, IP will need to employ some type of queuing mechanism to ensure voice does not wait to be transmitted when competing with data. Low-Latency Queuing (LLQ) or some other advanced queuing scheme will have to be used in routers to ensure voice gets sent ahead of data. In addition, a marking or coloring scheme, called IP Precedence, will also be used to ensure that voice is considered more important than data traffic. These Layer 3 components will be discussed in more depth in upcoming sections of this chapter.

The next layer is the transport layer. Since we will use UDP to carry voice, we are lacking the packet-ordering mechanism used by TCP to ensure that packets are delivered in their proper sequence regardless of their arrival order at the receiver. The Real-Time Transport Protocol (RTP) will add a sequence number and time-stamp mechanism to make sure it happens. Optionally, we may also choose to use the Resource Reservation Protocol (RSVP) to reserve bandwidth along the IP voice path. This protocol makes sure that enough bandwidth is allocated to voice conversations all along the path, and excludes data packets from using that reserved bandwidth.

The fifth layer, the session layer, also has some interesting ways to help. Many VoIP networks will choose to use ITU-T standard H.323, in which components called gatekeepers and gateways grant permissions for voice connections and make sure that the source and destination agree on capabilities (such as voice and/or video) and that the connection and disconnection events take place properly. Other methods to accomplish the same thing include the Session Initiation Protocol (SIP) and the Session Description Protocol (SDP). These methods are Internet Engineering Task Force (IETF) standards that provide announcements and information about multicast sessions to applications on an IP network. Although incompatible with H.323 methods, the IETF methods are gaining popularity as viable alternatives.

The sixth layer is the presentation layer. As it is defined in the International Organization for Standardization's (ISO's) OSI reference model, the presentation layer understands and interprets data formats. In terms of *data* transmission, Layer 6 concerns itself with data coding, encryption, and file formats. In terms of *voice* transmission, the presentation layer provides for coding and compression methods used for transporting the human voice. Coding and compression methods such as ITU-T G.729 and ITU-T G.711 are two examples of this. (For a complete description of voice coding and compression, see Chapter 3, "Introduction to Digital Voice Technology.")

All of the layers of the protocol stack are used collectively to ensure that delay and packet ordering are no longer a problem. Not all VoIP applications use all the methods at all of the layers, however, and their uses and limitations are described in the following sections. We will first examine the most commonly mentioned signaling and connection method for VoIP, H.323. Next, we'll discuss one alternative to H.323, called MGCP, which serves to coordinate connections outside the VoIP network. Then we'll examine the RTP and RTCP protocols, which are used for sequence numbering and time-stamping to assist the efforts of UDP. After that, we'll look at an optional method to reserve bandwidth along the path of a call using the RSVP protocol. Finally, we'll explore a popular alternative to H.323—SAP, SIP, and SDP—that is gaining popularity in VoIP networks.

H.323

The standard developed by the ITU-T for carrying multimedia applications over IP is called H.323. When it was developed, H.323 was not designed to exclusively carry voice telephony, but was designed to carry any multimedia application, such as video and sound, over IP internetworks. The standard defines many different components or hardware devices that are used to make the system work. Figure 10-3 shows some of the components that can be used in a voice H.323 environment.

Figure 10-3   *Examples of H.323 Components*

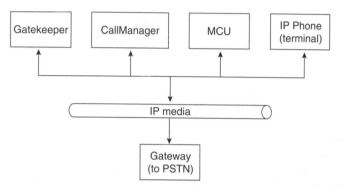

In Figure 10-3, the Cisco router assumes the role of gatekeeper by configuration. Various models of Cisco routers may be used as gatekeepers on the network depending on IOS revision. The H.323 gatekeeper acts as a sentry on a VoIP network, accepting requests for connections from callers and searching for limitations for the requested connections (such as time-of-day restrictions and capabilities limitations), and then ensuring that the called party can be called. It provides for address translation between LAN aliases and IP addresses and also provides for basic telephony services, such as directory services and other PBX functions. It then marries the caller to the called party and steps out of the way. Its exact connection sequence is discussed later in this section. The CallManager's job is to retain knowledge of all the other components in the system, to grant permissions for them to connect to each other, and to update components that fail and then are restored.

The multipoint control unit (MCU) is a device that serves to interconnect multiple users in a conference setting. It can serve to connect not only voice conferences but video teleconferences as well. It is made up of a multipoint controller and an optional multipoint processor. The multipoint controller is the supervisor of all requested conferences. The multipoint controller handles negotiation between the requesting H.323 terminals (such as phones) to ensure common capabilities between endpoints. It also controls conference tasks such as managing resources for multicasting. The multipoint controller does not oversee or manage any of the voice or video streams directly, but relies on the multipoint processor for tasks such as combining voice streams and switching video views.

The IP phone, known as a terminal in H.323 terms, is an endpoint on the H.323 network that provides real-time voice communications. A terminal must be capable of carrying voice but may also have video capabilities. A terminal does not even have to look like a telephone. (Please note that we are talking about generic H.323 endpoints here, not Cisco

IP Phones.) A terminal may also be an application on the user's desktop computer that can provide voice and video services. An example of such an application is Microsoft's NetMeeting. It is an H.323-compliant voice and video terminal that is designed to connect users of IP networks to other users for two-way communications.

Gateways are entrance points to other networks. A gateway is also an endpoint, just as a terminal is, but it can serve to allow connection between an H.323-compliant network and other types of voice and video networks, such as a company's voice PBX or the Public Switched Telephone Network (PSTN). It is responsible for accepting call requests between the H.323 network and the non-H.323 network and performing all necessary conversions to allow the connection to occur.

Signaling on H.323 networks uses a subset of protocols to accomplish a connection. Figure 10-4 shows the sequence of events used for an H.323 connection. The connection sequence consists of four phases. The four phases are:

1   Admission request

2   Connection setup

3   Capabilities exchange

4   Media opened

Each of the four phases is discussed in detail in the following paragraphs.

Figure 10-4   *H.323 Signaling Sequence*

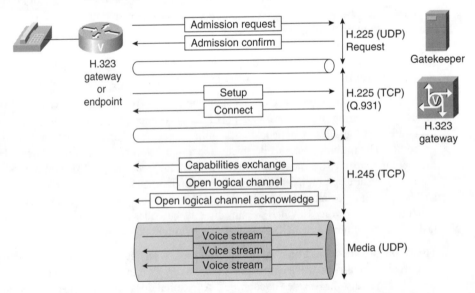

As a subset of H.323, the ITU-T protocol H.225 serves to set up the connection request and exchange capabilities. The H.225 protocol uses UDP first, for an admission request, to ask for a connection to another endpoint. This admission request is sent by the originating endpoint to the gatekeeper. The admission request consists of a desired telephone number with which to connect, along with information about the calling party. In some situations, such as when a dial plan is installed in the originating router, this step may be skipped completely. The gatekeeper examines the request, determines the originating endpoint, and determines whether resources are available to complete the call. The gatekeeper next performs address translation for the calling party, looking up the destination's telephone number and converting it to the destination's IP address. The gatekeeper then forwards its response back to the calling party. The first phase of the H.323 connection is now complete.

The second phase of the H.323 connection is also part of the H.225 protocol. The H.225 protocol under TCP as used in this phase of the connection is derived from the ITU-T Q.931 protocol used in ISDN connections. It must use TCP instead of UDP to make its connection because at this point the originating endpoint is setting up its actual connection to the destination endpoint. A setup message is sent to the destination requesting that a communication channel be opened. The destination verifies the identification of the originating endpoint, then searches its database for existence of the destination telephone number. If the destination telephone number exists, and no security or time-of-day limitation is currently imposed, the destination responds to the originating endpoint with a connect message that grants the originating endpoint permission to proceed with the next phase of the connection sequence. The destination's response of the connect message indicates that the originating endpoint may proceed with the connection, not that the connection is open and ready for use. This completes the second phase of the connection sequence.

In the third phase of the connection sequence, known as H.245, the originating and destination endpoints are tasked with an exchange of capabilities. This may include deciding whether each end is requesting video as well as voice and, for example, one end or the other declining video capability because it is an IP Phone supporting only voice. Also included in the exchange are coding and compression capabilities, and negotiation of requested coding schemes versus capable coding schemes. Once capabilities are agreed upon, the originating endpoint sends an alerting/proceeding message to the destination requesting that the logical media channel be opened. It is at this point that the destination notifies its user of the incoming call, such as sounding a ringing alert that the user can hear. At this point, the communication path is not opened, but each end agrees that the channel is ready to be opened.

In the fourth and final phase, the destination user answers the incoming call by going off-hook. At this point, H.225 is again used to finish up the job. When the destination user answers, connect messages are exchanged and confirmed between the originator and the destination, and the channel is opened and the media is now ready for use.

MGCP

One of the newest methods of implementing VoIP involves an IETF RFC (Request For Comments) called Media Gateway Control Protocol (MGCP). It has been released by the IETF as RFC 2705, and it attempts to coordinate the efforts of multiple gateways on the same network. The primary function of MGCP is to control and supervise connection attempts between different media gateways. As described earlier, a gateway consists of an entrance point to another type of network other than VoIP. These other networks, as described in the RFC, include connections to the PSTN, VoFR networks or VoATM networks, gateways used in residences to provide telephone service from cable or DSL (digital subscriber line) services, or connections to business PBXs. MGCP uses some of the H.323 protocol for connection events and is viewed on the H.323 network as a gatekeeper or as an endpoint. The industry view of the MGCP is that it provides for reliable connection event services when compared to allowing endpoints to connect to each other by themselves. The RFC defines that MGCP actually consists of two components, the MG (media gateway) and the MGC (media gateway controller). The MG serves to perform the translations between the differing media types and maintain proper connection of the media (voice) streams themselves. The MGC is the switching portion of the controller, and its purpose is to oversee the connection requests and track successes and failures. The MGC can track sources and destinations and provide for alternate paths if necessary.

An evolution of MGCP, called Megaco, further defines the MGC by allowing the MGC itself to listen and respond to some of the media stream itself, such as tones input on a telephone keypad, to route calls as an interactive voice messaging system would, or to conference parties together by listening to other tone sequences. Of concern in the industry is whether Megaco will completely replace MGCP or simply augment it. With all of the signaling protocols available, it is difficult to decide which will prevail.

SAP, SIP, and SDP

Yet another set of signaling protocols was born of a need to allow users to contact one another using the structure of the existing Internet. This set of protocols was originally conceived in 1996 as part of the MBONE, or Multicast Backbone, idea. The thought here was to overlay, over the existing Internet, a way to invite participants to attend conferences, talks, seminars, or video presentations using multicast addresses. An important factor in the development of this set of protocols was the ability to locate a user in traditional Internet location terms (via URL) and in real-world physical terms. An invited participant might be in their office with their desktop, at home with the PC in the study, or travelling with a laptop. Addressing the user, then, had to be mobile and in Internet expression. So SIP uses URLs or DNS (Domain Name System) formats of addresses to locate users. Just as an e-mail address might be expressed as jbrown@pickled-pea.com, the SIP address of that user would be sip:jbrown@pickled-pea.com.

SIP is defined in RFC 2543. Once the user is located using SIP, the Session Announcement Protocol (SAP) takes over. SAP is still in draft with the IETF as of this writing and bears no official RFC number. The SAP packet bears an identifier that defines the type of session being initiated. The identifier, described in RFC 2327, is carried via the SDP, which can also add or remove video, perform mute functions, and put sessions on hold. As an example, if a user decides to use an application, such as video teleconferencing, the SIP protocol initiates connection using multicast addresses and specified port numbers. The payload of the resulting packets uses the SAP packet information to announce the type of session via the SDP. SDP provides a means for the source and destination to agree on capabilities and thereby keep the call up and running. In comparison to H.323, SIP uses much less overhead to set up the call—one or two round trips versus seven or eight. In addition, SIP uses existing protocols (such as SAP and SDP), while H.323 relies on Q.931 and must involve other elements as well, such as H.225. But while SIP uses UDP, an unreliable method for connection, H.323 initially uses TCP as was discussed earlier.

RTP and RTCP

Since VoIP is carried using UDP, we need a way to sequence packets. As shown earlier in Figure 10-1, IP networks have packet-ordering problems. One popular method used extensively in VoIP networks is RTP, which is defined in IETF RFC 1889. Running at the transport layer of the VoIP seven-layer model, RTP provides sequencing and time-stamp services to the UDP voice packet. Variations of RTP are used extensively today on the Internet to provide streaming of real-time audio and video channels to users with compatible desktop players. As shown in Figure 10-5, the RTP header adds significant overhead to the voice packet. A total of 12 bytes of RTP header is necessary to carry voice, and more may be necessary to carry other media streams, depending on the use of the CSRC field.

Figure 10-5   *RTP Header: Twelve Bytes and Optional CSRC Field*

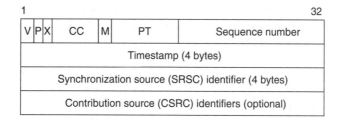

The fields in the RTP header have the following meanings:

- V (Version)—Represents the version of RTP.

- P (Padding)—When this bit is on, the RTP payload contains extra padding bytes necessary to round out the header. The padding bytes are not part of the payload.

- X (Extension)—Indicates that the header contains a header extension.

- CC (CSRC [contributing source] count)—Indicates the number of CSRC identifiers following the main header.
- M (Marker)—Used to delineate frame boundaries.
- PT (Payload type)—Code that represents the current coding and compression scheme carried in the payload.
- Sequence number—Sequential serial number of packet used to track packet ordering and packet loss.
- Time stamp—Clock code that represents the instant of time of the first coded sample in the payload. This stamp is used by the dejitter buffers to overcome delay variation impairments.
- SRSC (Synchronization source identifier)—A randomly chosen number that identifies one RTP session from another to allow for multiplexing.
- CSRC (Contributing source identifier)—Used to provide a list of all sources who are mixed in this RTP session. Allows for conferencing.

RTP also uses a monitoring protocol, also described in RFC 1889, called RTCP (Real-Time Transport Control Protocol). Used to monitor the RTP session, RTCP updates all participants in the session on the progress of the real-time stream. RTCP accompanies RTP by default on Cisco voice routers in order to comply with RTP specification standards.

RSVP

Lauded by some but misunderstood by most, the Resource Reservation Protocol (RSVP) is an older attempt at ensuring bandwidth allocation for voice packets traversing an IP network. RSVP is specified in RFC 2205. At the transport layer of the VoIP reference model, RSVP is a protocol that reserves bandwidth along the path of the voice call. Reservation of bandwidth is a beneficial thing to do for voice, but may be very detrimental to data sharing the same network. RSVP makes its reservation at the expense of data, thereby possibly "starving" data applications that need to get their packets through. Although RSVP can be implemented over private IP networks, expecting reservation of bandwidth through the public Internet is impractical. Routers on the Internet will deliver packets on a best-effort basis, and cannot and will not make a reservation for your particular voice call. Figure 10-6 shows the action of RSVP as it installs a reservation. RSVP sends PATH messages to the destination on initiation of the voice call and returns RESV (Reservation) messages to the source.

When the source initiates a call to a destination, the first router, independent of its signaling protocol to get the call connected, sends PATH messages designed to discover and record the path taken by the voice call. At this point, a reservation is not being made because the call is not set up. When the destination answers the call (assuming it is not already engaged and there is someone there to answer), the destination router returns RESV messages to the source. It is these RESV messages that reserve the bandwidth. If the network must reroute packets due to any network failures, the PATH and RESV messages will naturally follow the reroute because they themselves carry IP headers and therefore IP addresses. In later versions of RSVP (Cisco IOS 12.1.5T and later), the reservation is confirmed before the destination phone is rung. This is done in case the reservation fails, in which case the call can be re-routed, or a reorder tone may be delivered to the originator.

Figure 10-6 *RSVP PATH and RESV Messages*

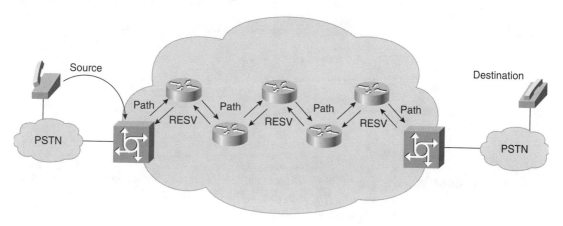

Basic VoIP Signaling

VoIP calls normally start with a user. The user who wishes to place such a call traditionally uses an instrument such as a telephone, or, in more modern terms, a desktop application with voice capabilities. For purposes of the following discussion, we will use the standard telephone for simplicity. If the call is to be successful, the source must signal its locally attached telephony switching equipment that it wishes to place the call. The local equipment is then responsible for propagating the connection request (signaling) all the way to the destination. The destination must then return some indication of status—that is, has the destination been found, and is it successfully being notified of an incoming call. There are three elements involved in basic Cisco VoIP calls:

1 Signaling from PBX to router

2 Signaling between routers

3 Signaling from router to PBX

Each of these steps is discussed in the sections below.

Signaling from PBX to Router

Assuming that the connection consists of two PBXs interconnected with routers, we will assume that this application is a tie-line (see Chapter 5, "Applications for Cisco Voice-over Routers"). We will also assume that the interface between the source PBX and the source router is E&M and that the interface between the destination router and the destination PBX is also E&M (see Chapter 2, "Introduction to Analog Technology"). Finally, we will assume that the routers are interconnected with VoIP via H.323. Figure 10-7 shows the connection.

Figure 10-7   *Signaling from the Source PBX to the Source Router Is Called Trunk Signaling.*

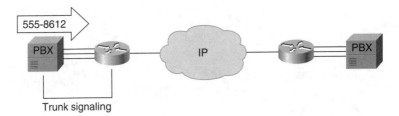

When a user on the source PBX wishes to call a user on the destination PBX, the source PBX signals the router—in this case using E&M signaling. This local signaling must be provided by the router for compatibility with the source PBX and must adhere to the rules and standards set forth for its operation. The source router, therefore, must understand how to perform this signaling method. (For detailed information on this signaling method, see Chapter 2.) The router must understand that a call request has been made, must have the destination telephone number in its local dial plan database, must be able to correlate the telephone number with a destination IP address (and therefore an outbound interface), and must have the ability to propagate the address (telephone number) to the destination. Once the router has been properly notified of a request for a call, it must now prepare to signal the IP network.

Signaling Between Routers

As shown in Figure 10-8, the source router initiates an H.225 call request under H.323 to ask for a connection. The signaling has been converted by the router from E&M to, in our case, H.225. The IP network propagates the destination telephone number through the network via traditional IP routing. This means that the routers within the IP network must have the capability to route IP by traditional means, with each router having a routing table and valid outbound interface on which to send the H.225 request. Eventually (and ideally), the request will reach the destination router.

Figure 10-8 *Signaling Between H.323-Compliant Routers Using H.225*

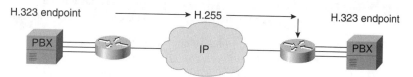

Signaling from Router to PBX

The destination router, realizing that it has a call request to process, must search its dial plan database to determine if it knows where the destination is located. The router must understand that the incoming call is associated with a connection to the destination PBX; once it knows, it can make a connection to the destination PBX, it returns a H.225 acknowledgement to the source router. It then proceeds to signal the destination router using the E&M interface and forwards the dialed telephone number to the destination PBX. Figure 10-9 shows the progression of the call and the acknowledgement returned to the source.

Figure 10-9 *Signaling from the Router to the Destination PBX and Returned H.225 Acknowledgement*

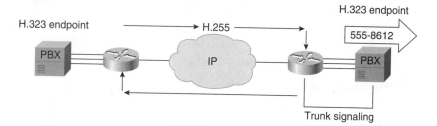

Understanding Cisco VoIP

Cisco's implementation of VoIP is designed to provide solutions to classic challenges facing network administrators using the technology. Previous chapters covered transmission of voice over Layer 2 technologies, specifically Frame Relay and ATM. Although these Layer 2 technologies may be well suited for some voice applications, VoIP adds the following two distinct advantages:

- Automatic any-to-any switched calling (calls just follow IP routing and no tandeming or explicit call routing has to be configured in the network).

- Interoperability with other VoIP applications, such as NetMeeting, soft phones on PCs, and a host of others. VoFR and VoATM are toll-bypass technologies only, whereas VoIP opens a whole world of new voice-enabled applications.

Some of the challenges of carrying voice over Layer 2 technologies have been mentioned in previous chapters with their respective Layer 2 solutions, but VoIP at Layer 3 has its own sets of solutions described in the following sections. Some of the challenges include:

- Delay and delay variation—This issue is addressed through two offered methods: reservation of bandwidth through the RSVP protocol and through prioritization. Configuration of these options is discussed in following sections.

- Packet loss and packet ordering—This issue can be solved by using RTP and RTCP.

- Bandwidth—This issue is solved using various compression options.

- Queuing—This can be solved using various leading edge queuing mechanisms described below.

Remember that IPv4 was not designed to carry real-time traffic such as voice, and configuration options have been provided in Cisco's VoIP implementation that allow network administrators to choose which combination will work best for their network. In the following sections, we will explore ways to solve the issues listed previously. We will first examine the building blocks of basic voice networks in terms of VoIP. Then we'll look at specific compression, queuing, fragmentation, and bandwidth solutions and considerations used to solve the issues. Finally, we'll discuss an implementation of VoIP using Multilink PPP as an example.

VoIP QoS Building Blocks

Various building blocks have been identified by Cisco Systems as keys to successful voice implementations. We will examine each of them in terms of VoIP:

- Backbone speed and scale
- Packet classification
- Bandwidth management and admission control
- Congestion management
- Queue management

These building blocks are the tools you should consider using when planning IP installations that will include voice. Although most network administrators will be adding voice to an existing network, these building blocks should still be considered even though the installation is not new. Their effect on existing IP applications will be far-reaching. Overlooking these tools will be a detriment to any installation, and network administrators should consider carefully the building blocks they will use on their networks. Fortunately, Cisco Systems has provided enough of these tools to allow for smooth and successful implementations. These building blocks are described in the following sections.

Backbone Speed and Scale

If you are adding voice to an existing network or developing and planning a new one, you will need to analyze the network's bandwidth requirements. As with any IP network, a close look at traffic patterns and their bandwidth requirements is necessary to ensure that enough bandwidth is available when the applications need it. Voice is no exception. In addition to bandwidth requirements, performance of the hardware involved is critical. Cisco's voice-capable routers are high-performance devices that, when implemented with properly planned bandwidth backbones, perform with little strain on the router and the network with minimal effect on voice quality. Cisco has achieved this feat by using high-performance processing techniques, including specialized digital signal processors (DSPs), in voice-capable routers. The DSPs serve to compress and decompress the voice audio with no strain on the router's main processor.

Packet Classification

Since voice traffic travels in real time, the network needs to know that the voice traffic is more important than the data traffic. Implementing an identification scheme to allow the network to tell the difference between voice and data is critical. One way to do this is to use the ToS (type of service) bits in the IP header to identify that the voice traffic is more important than the data traffic (packet marking). This technique ensures that the voice traffic will get priority through the network (packet classification) and provides for Differentiated Service, or DiffServ, through the network. Configuration of this voice precedence technique is described later in this chapter.

Bandwidth Management and Admission Control

The amount of bandwidth consumed by voice across any type of network is an important issue for network administrators. Voice cannot consume so much bandwidth as to make the cost of carrying it on the IP network prohibitive. The building block that can address this need includes the various compression techniques seen in Chapter 3 as well as compression of the RTP header contained in the transport layer. Compression of the voice signal can lead to 8:1 compression ratios, and more bandwidth can be achieved if silence suppression is employed. Compression of the RTP header (discussed below) can trim the cost of transmitting the IP and RTP header from 40 bytes to about 2 or 4. As you can imagine, these compression methods will have a great impact on the bandwidth required for VoIP. Also, admission control as supervised by the gatekeeper in H.323 networks will control who can connect to whom and when. Using the gatekeeper in this way can significantly reduce the cost of bandwidth on the IP network.

Congestion Management

One way that a router controls congestion at the network layer is management of the Window Size parameter in the IP header. Cisco routers use a method called Random Early Detection (RED) and Weighted Random Early Detection (WRED), which can monitor individual applications to ensure that each session does not consume too much buffer space in the outbound queue and overwhelm the router. When various sessions must share bandwidth on an outbound interface (as is the case in most IP networks), management of buffer space can get tricky. RED and WRED are two optional tools in the hands of network administrators who need better control over router buffer performance. RED and WRED are used as congestion management tools for data traffic, and should not be considered for voice. They are mentioned here as an example of congestion management for purposes of understanding this building block.

Queue Management

Ensuring that voice gets special treatment and priority on the outbound interface is critical to successful VoIP implementations. Cisco has developed creative methods to ensure that queuing techniques meet the needs of the application demands on the network and that critical traffic gets priority in the outbound queue. Techniques such as LLQ have been developed to satisfy that need. A section below will describe this queuing method in detail along with other queuing options offered on the router.

RTP Header Compression

As noted earlier in this chapter, VoIP uses UDP as its transport layer mechanism. It is implemented in this way because voice will not need receipt for each and every packet sent and does not need all of the extra "baggage" associated with using TCP at Layer 4. But we learned earlier in this chapter that an RTP header will be necessary to solve the packet-ordering problem encountered as a result of using UDP. Unfortunately, the RTP header adds 12 bytes to the overall IP header for a total of 40 bytes of overhead. Figure 10-10 shows the 40-byte header and the resulting 2- or 4-byte header achieved with RTP header compression. RTP header compression must be specifically configured on the router, and its command syntax is shown later in this chapter.

Figure 10-10 *RTP Header Compression*

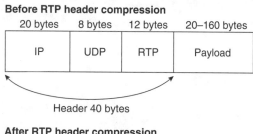

The RTP header compression option is obviously beneficial because VoIP packets will, by default, use relatively small (when compared to IP data traffic) payloads. Typically, 20 to 30 ms of voice is sent in each packet by design. This is done so that small amounts of voice in a given payload, which might be lost in the IP network, are not noticed and will be compensated for in the DSP circuits with Digital Speech Interpolation (a means by which the DSPs compensate for and "fill in" lost instances of voice playout by using prediction methods). Left alone, a VoIP packet would have more header than payload. RTP header compression helps to save the necessary bandwidth over the network and ensure that data gets more time on the wire.

To configure IP RTP header compression, use the following command in EXEC mode:

```
Router(config-if)#ip rtp header-compression [passive]
```

Table 10-1 explains the elements of the **ip rtp header-compression** command.

Table 10-1   **ip rtp header-compression** *Command Elements*

Element	Description
passive	(Optional) Compresses RTP header only if received packets are also compressed.

The Evolution of WAN Edge Queuing Techniques

Cisco has devoted considerable time and effort to developing queuing methods that ensure that proper bandwidth is available to applications when they need it. Different approaches have been developed over the years for both voice and data, as shown in Figure 10-11. The following sections describe some of the methods developed for data transmission:

- First in, first out (FIFO)
- Priority queuing (PQ)

- Custom queuing (CQ)
- Weighted fair queuing (WFQ)
- Class-based weighted fair queuing (CBWFQ)

Figure 10-11 *Evolution of Cisco Systems Queuing Techniques*

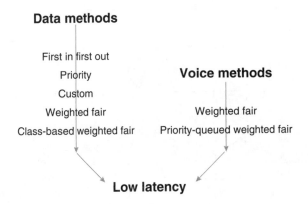

As you will see, the queuing methods developed for data transmission are not suitable for voice because they don't include a priority mechanism to ensure that voice is transmitted first, without starving other applications or causing jitter problems.

FIFO

FIFO has been around for as long as computer systems have existed. It runs on the premise that the first packet or frame to be deposited in the outbound queue will be transmitted, regardless of the size or priority of the packet or frame. This method is obviously detrimental to voice because voice must be recognized as a priority flow over data and must get to the front of the queue at all times.

PQ

With PQ, the network administrator identifies priority traffic by using a priority list configured on the router. The priority list can configure four queues, named High, Medium, Normal, and Low. The priority list can be enhanced by using access control lists (ACLs) to further identify traffic not provided for in the priority list command syntax. Each time the router dispatches a packet, it returns to the High queue to see if a priority packet needs to be sent. PQ is not an acceptable solution for VoIP because it tends to starve other applications that are not queued as high as voice needs to be queued. The goal in setting up a queuing mechanism is to be fair to all flows, but to give voice some prioritization, not exclusive prioritization.

CQ

Developed as a method to queue traffic by bandwidth demand, CQ allows for up to 16 queues to be set up on an outbound interface with a mechanism to empty each queue based on byte count. The custom queuing method dispatches packets in the current queue until a "service order threshold" (SOT) is reached. The SOT is configurable for each of the 16 queues and represents bytes transferred for each queue. For example, consider that three queues are set up, the first with a SOT value of 1500 (the default), the second with a SOT value of 3000, and the third with a value of 4500. Total SOV value for all three queues, by adding all three SOT values together, is 9000 bytes. The first queue will be allowed to transfer 1500 bytes of traffic (regardless of packet size), and therefore is allowed 1500/9000 or 1/6 bandwidth in the outbound queue. The second queue has a SOT value of 3000, and is therefore allowed 3000/9000 or 1/3 the bandwidth. The last queue's SOT value is 4500 and is allowed 4500/9000 or half the bandwidth on the outbound interface. The queues are serviced in a round-robin sequence, one after the other. Though the CQ method might seem a good choice for VoIP, it can cause delay variation problems (jitter) at the receiving end because voice may have to wait in its outbound until other queues are emptied.

WFQ

The WFQ method lends itself well to voice because it can examine flows of traffic and differentiate them using key identifiers. The identifiers may include source and destination port numbers and source and destination IP addresses. For voice, WFQ can calculate the bandwidth required by identifying the voice flows using the TOS or precedence bits. Keep in mind that the calculation made by WFQ does not reserve or guarantee bandwidth, but only serves to figure out how much bandwidth each flow needs. How much bandwidth each flow gets is a function of the total number of flows and their associated TOS values. In WFQ, bandwidth of a flow of traffic can be described as follows:

BW of a flow = Circuit BW × (FlowA [1 plus IP Precedence])/Sum of All Flows

For example, consider a 56-kilobits-per-second (kbps) outbound bandwidth that must service two VoIP flows at 24 kbps each and two FTP flows at 56 kbps each. The FTP flows are considered to be at the full bandwidth rate of the interface because they will use all the bandwidth that they can all the time. The voice flows will only use about 24 kbps because of compression and packet header overhead. If IP precedence is not configured, then the resulting bandwidth available for one VoIP flow will be as follows:

BW = 56 kbps × 1/4 = 14 kbps

This will not be acceptable because voice needs the full 24 kbps to perform properly. If the precedence is set higher for voice, at a value of, say, 5, then the resulting bandwidth available to the voice flow is as follows:

BW = 56 kbps × 6/14 = 24 kbps

As a result of setting the precedence to 5, the calculation now more closely matches the bandwidth required by the voice flow. Now that WFQ realizes that the voice flow needs 24 kbps of bandwidth, it will receive some measure of priority over other flows with lesser precedence values, but its full bandwidth requirement still may not be met, depending on the demands of other flows and their respective precedence. So Cisco set about to do more development on queuing methods that ensure voice performs at optimum levels through the network.

CBWFQ

CBWFQ is designed to classify flows based on identifying factors that can be customized by using ACLs. Each of the flows is identified in a map-class, and then a policy-map is configured for an interface. Example 10-1 is an example of CBWFQ configuration.

Example 10-1 *CBWFQ Configuration*

```
class-map data
  match input-interface ethernet 0/0
class-map voice
  match access-group 101
class-map class-default
  match any

policy-map wan
  class voice
    bandwidth 80
  class data
    bandwidth 48

interface serial 0/0
  ip address 10.10.10.1 255.0.0.0
  bandwidth 128
  service-policy output wan

access-list 101 permit ip any any precedence critical

dial-peer voice 1 voip
  destination-pattern 5551212
  session target ipv4: 10.10.10.2
  ip precedence 5
```

The class-map called data identifies traffic that originated on interface ethernet 0/0. The class-map called voice identifies traffic that is defined in access-list 1—that is, traffic that carries a critical precedence (or TOS value) of 5. The precedence value is applied to dial-peers. The class-map called class-default is a safety net that identifies all other traffic not specifically identified in the other class-maps. Next, a policy-map is created with the freely chosen name wan. It provides for 80 kbps of bandwidth for the voice class and 48 kbps of bandwidth for the data class. Then the service-policy is applied to the outbound interface—

in this example, interface serial 0/0. CBWFQ is the first time we get BW guarantees. The fundamental differences between WFQ and CBWFQ (they use the same internal algorithm) are:

- For WFQ the weight of the flow is given (derived from IP Precedence) and the BW calculated
- For CBWFQ the BW is given, and the resulting weight is calculated

As in other queuing methods, CBWFQ may seem to be a viable solution for voice, but voice still does not get an explicit priority over data.

PQ-WFQ

A specialized queuing method was developed specifically for Voice over Frame Relay purely at Layer 2. This method, called priority queuing-weighted fair queuing (PQ-WFQ), automatically places the VoFR traffic in the priority queue, and also allows you to assign the entire outbound queue a percentage of bandwidth (75 percent by default) of the outbound WAN network for use by all traffic. In Frame Relay terms, we set the bandwidth percentage for the priority queue to the mincir value. The reason we configure to the mincir for Frame Relay is because mincir is guaranteed bandwidth on the WAN after BECN (backward explicit congestion notification) events. However, to implement PQ-WFQ at Layer 3 for VoIP (independent of the Layer 2 protocol), configuration of PQ-WFQ involves a variation of the **ip rtp priority** command. It is necessary to use the **ip rtp priority** command because Layer 3 needs a mechanism to distinguish the voice packets from the data packets. The operation of PQ-WFQ is shown in Figure 10-12.

Figure 10-12 *Operation of PQ-WFQ*

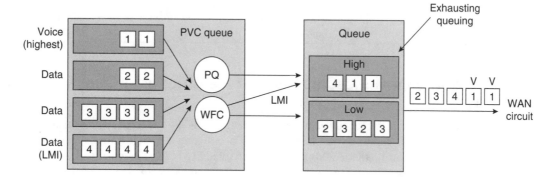

As is shown in Figure 10-12, the voice traffic gets an exhaustive priority queue, serviced until it is empty, while the data traffic is independently handled via WFQ. Note that Frame Relay Local Management Interface (LMI) traffic, although managed by WFQ, also uses the high queue due to the fact that Frame Relay requires that LMI run every 10 seconds and cannot be delayed. PQ-WFQ configuration is configured with the following EXEC mode command:

```
Router(config-if)#ip rtp priority starting-rtp-port-number port-number-range
bandwidth
```

Table 10-2 explains the elements of the **ip rtp priority** command.

Table 10-2 ip rtp priority *Command Elements*

Element	Description
starting-rtp-port-number	Defines the starting port number that RTP will prioritize. Cisco's starting UDP port number for VoIP is 16384.
port-number-range	Defines the total number of ports, starting with the starting-rtp-port-number, that will be prioritized. Cisco's range of UDP ports for VoIP is 16383.
bandwidth	The amount of bandwidth to be allocated to each prioritized session. Total voice bandwidth is dependent on compression technique used plus overhead and should include five to ten percent more.

For both PQ-WFQ and LLQ, by default, 75 percent of the WAN bandwidth can be reserved for the outbound queue. Cisco has provided an optional command that you can configure to change the default value. The syntax of that command is:

```
Router(config-if)#max-reserved-bandwidth percent
```

Table 10-3 explains the elements of the **max-reserved-bandwidth** command.

Table 10-3 **max-reserved-bandwidth** *Command Elements*

Element	Description
percent	Maximum amount of bandwidth that can be reserved for the outbound queue.

LLQ and Priority Queuing-Class-Based Weighted Fair Queuing (PQ-CBWFQ)

Specialized queuing techniques were developed specifically for voice to provide for balanced access between voice and data while prioritizing the voice traffic. Cisco's preferred queuing method for VoIP networks is called LLQ). LLQ is actually a combination of two previously described queuing methods—PQ and CBWFQ. Some Cisco documentation even refers to LLQ as PQ-CBWFQ. As shown in Figure 10-13, LLQ provides a fast lane for the voice traffic in a similar way that some interstate highways in the U.S. provide a fast HOV (high-occupancy vehicle) lane. The HOV lane is provided for vehicles that meet a specified criteria (carrying more than two occupants, for example). In LLQ, the fast lane is provided for packets that carry voice.

Figure 10-13 *LLQ's Fast Lane for Voice Traffic; Balanced Access for Non-Voice Traffic*

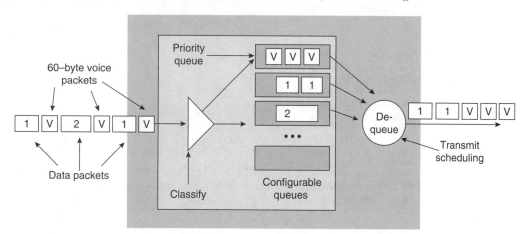

Operation of LLQ is similar to that of CBWFQ. As traffic arrives in the outbound queue, it is first sorted into conversations. The identifiers used for the sorting may include specific access lists, source and destination ports, source and destination IP address, or specific IP protocol. Voice will be identified using an access list as was shown in the earlier section "CBWFQ." Once the traffic is sorted, voice gets a priority queue, class-based, which gets serviced before data traffic is serviced (the "fast lane"). The data traffic is independently queued using an independent WFQ process. Logically separate from the queuing process, the data traffic is fragmented and interleaved as necessary to provide for balanced access to the WAN. Configuration of LLQ is almost the same as configuration for CBWFQ, except that the keyword bandwidth is replaced by the keyword priority in the policy-map. The priority keyword triggers the router to set aside the fast lane, which will be used for the voice traffic. Example 10-2 shows an example of LLQ configuration.

Example 10-2 *Configuration of LLQ*

```
class-map data
  match input-interface ethernet 0/0
class-map voice
  match access-group 101
class-map class-default
  match any

policy-map wan
  class voice
    priority 80
  class data
    bandwidth 48

interface serial 0/0
  ip address 10.10.10.1 255.0.0.0
```

continues

Example 10-2 *Configuration of LLQ (Continued)*

```
     bandwidth 128
     service-policy output wan

access-list 101 permit ip any any precedence critical

dial-peer voice 1 voip
   destination-pattern 5551212
   session target ipv4: 10.10.10.2
   ip precedence 5
```

In this configuration, the policy-map defines that the class called voice should be given priority over the class called data. The class called data is configured with the keyword **bandwidth**, which identifies it as subject to WFQ. The class called voice is configured in the policy-map with the keyword priority, which allows it to use the fast lane.

Serialization Delay, Fragmentation Size, and Bandwidth Consumption

Although configuring a queuing method for prioritization and balanced access is a great idea and mostly necessary, VoIP still does not address delay problems due to serialization delay. Serialization delay is defined as the time it takes for the router to deposit the frame onto the outbound WAN. Chapter 8 shows that, assuming a 56-kbps WAN link speed, it will take 144 ms to apply a 1024-byte frame onto the WAN. Although voice uses smaller frame sizes, data traffic can and will take up too much time to transmit longer frames. In Chapter 8, "Configuring Cisco Routers for VoFR," you learned that fragmentation methods are employed in the FRF standards to circumvent the serialization delay problem. VoIP also needs to employ some type of fragmentation because of longer frames, which take too much time to transmit.

Your task now is to determine how much to fragment the frames in order to provide acceptable levels of service for voice. You don't want to fragment too small, as that might overwhelm the data application. You will want to fragment based on acceptable and tolerable delay budgets as outlined in the ITU-T standard G.114. Chapter 8 has a guide to deciding on fragmentation sizes, with acceptable delay budgets versus links speeds. The chart is a guide in calculating the fragment size, based on the maximum delay the low-speed link can add to the delay budget. For example, if the maximum acceptable delay the low-speed link can add to the budget is 20 ms, and the link speed is 64K, then the maximum fragment size may be set to 160 bytes. In any case, you should never fragment below 70 bytes as it becomes counterproductive. Also, you should never fragment so small that the voice packets begin to be fragmented.

To configure fragmentation for VoIP, you use the following from EXEC mode:

```
Router(config-if)#mtu size
```

Table 10-4 explains the element of the **mtu** command.

Table 10-4   **mtu** *Command Element*

Element	Description
size	Maximum fragmentation size to be used for this outbound interface.

NOTE   Note that configuring the MTU size on an interface will affect all traffic that uses the interface, not just voice. Adjusting the MTU on an outbound interface should only be done as a last resort when other fragmentation methods such as FRF.12 are not available.

A third and final important factor you need to consider when configuring queuing options for VoIP, especially for the commands that need a bandwidth statement for voice, is the total amount of bandwidth that will be consumed per voice call over IP. The basic formula used to calculate bandwidth needed for a voice call is as follows:

BW = (payload + Layer 2 Overhead + Layer 3 Overhead) × Compression Rate Payload

As an example, given a VoIP call that uses G.729 compression over Ethernet framing, the total bandwidth required is:

BW = (20 + 14 + 40)/20 × 8000 = 29,600

Table 10-5 shows various bandwidth requirements based on different compression techniques and different Layer 2 encapsulations.

Table 10-5   *VoIP Bandwidth Consumption per Channel*

Compression	Voice BW (kbps)	DSP Output (bytes)	DSP Outputs in Each Packet	L3 Header Size	L2 Technology Used	L2 Header Size	Total BW
G.711	64	80	2	40	Ethernet	14	85.6
G.711	64	80	2	40	MLPPP	6	82.4
G.711	64	80	2	2 (CRTP)	MLPPP	6	67.2
G.729	8	10	2	40	Ethernet	14	29.6
G.729	8	10	2	40	MLPPP	6	26.4
G.729	8	10	2	2 (CRTP)	MLPPP	6	11.2

The first column of Table 10-5 represents the chosen compression scheme. Both G.711 and G.729 are shown in the chart as examples. Other compression schemes are available. The second column represents the number of bits generated per second for one voice call for the listed compression scheme. The third column represents the number of bytes that the DSP will output to generate 10 ms of audio. For example, for G.711, the DSP outputs 80 bytes representing 10 ms of audio. It takes 100 10 ms samples to get one second of audio. Thus, 80 bytes times 100 is 8000, times 8 bits per byte is 64000 bits per second. The third column states the default number of DSP outputs (samples) that are included in one IP packet. Please note that this parameter is configurable and will affect total bandwidth. The fifth column is the Layer 3 (IP) header size including RTP (two of the examples show compressed RTP). The sixth column simply shows the Layer 2 technology used, with the seventh column showing the Layer 2 overhead required for the respective technology. Finally, the last column shows the total bandwidth necessary for VoIP.

Using the formula, you can calculate your individual bandwidth requirements depending on your chosen compression technique and Layer 2 technology.

To adjust the payload size for VoIP, use the following EXEC mode command:

```
Router(config-dial-peer)#codec codec-type [bytes payload-size]
```

Table 10-6 explains the elements of the codec command.

Table 10-6   **codec** *Command Elements*

Element	Description
codec-type	Specifies the coder rate of speech. Examples of valid values include **g711alaw** and **g729r8**.
payload-size	(Optional) The number of bytes in the voice payload of each frame or packet. Enter a ? character after the keyword **bytes** to get a list of valid payload values for your specific dial peer.

Remember that adjusting the payload value will alter the bandwidth requirements for each voice call. Also, adjustment of the parameter may cause delay and delay variation problems. Be sure to consider carefully the impact of changing this particular parameter in your VoIP network.

Multilink PPP Considerations

In some VoIP networks, it may be necessary and even beneficial to use PPP (Point-to-Point Protocol) as the Layer 2 framing method. Use of PPP may include the use of Multilink PPP (MLPPP) depending on how the network is planned. When MLPPP is used, we will need to configure the router for Multiclass Multilink PPP (MCML PPP). By using MCML PPP, fragmentation can be specified and interleaving needs to be specifically configured. Unfortunately, these two parameters cannot be configured directly on a physical interface. We will need to use a virtual template or Multilink Groups to accomplish such a configuration. The three key commands to configure MCML PPP are shown in the next paragraphs.

To configure Multilink PPP, use the following EXEC mode command from interface configuration mode:

```
Router(config-if)#ppp multilink
```

To configure the PPP Multilink fragmentation, use the following command from interface configuration mode:

```
Router(config-if)#ppp multilink fragment-delay delay-time
```

Table 10-7 explains the element of the **ppp multilink fragment-delay** command.

Table 10-7   **ppp multilink fragment-delay** *Command Element*

Element	Description
delay-time	Maximum delay in milliseconds allowed for transmission of a packet fragment on a multilink PPP bundle.

To configure Multilink PPP for interleaving, use the following EXEC mode command from interface configuration mode:

```
Router(config-if)#ppp multilink interleave
```

Configuring **ppp multilink interleave** commands the router to provide for real-time packet interleaving amongst the elements of the multilink bundle.

To illustrate how the multilink group is used, a look at a configuration example is appropriate here (see Example 10-3).

Example 10-3 *Use of Multilink Group*

```
NYC-3640#

interface Serial0
 bandwidth 56
 no ip address
 encapsulation ppp
 no ip route-cache
 no ip mroute-cache
 no fair-queue
 ppp multilink
 multilink-group 1


interface multilink 1
 ip address 10.1.1.1 255.255.255.252
 no ip directed-broadcast
 ip rtp priority 16384 16383 26
 no ip mroute-cache
 fair-queue 64 256 1000
 ppp multilink
 ppp multilink fragment-delay 10
 ppp multilink interleave
 multilink-group 1
```

Configuring Cisco VoIP

Configuring VoIP in a network depends on a number of factors, as you have learned. You need to consider existing traffic patterns, type(s) of traffic in your network, tolerable delays for both voice and data, voice compression schemes, and many other factors. As you can see from these many factors, varying VoIP scenarios may emerge. A simple example of VoIP over Frame Relay is shown in Figure 10-14 and Example 10-4. Remember that your configuration will probably vary, and that this is a simple example only.

Figure 10-14 *Example: VoIP over Frame Relay*

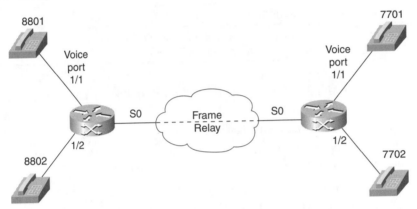

The accompanying configuration for Figure 10-14 is shown in Example 10-4.

Example 10-4 *Configuration of VoIP over FR*

```
voip1(config)#interface serial0/0
voip1(config-if)#no ip address
voip1(config-if)#encapsulation frame-relay
voip1(config-if)#frame-relay ip rtp header-compression
voip1(config-if)#fair-queue
```

Monitoring and Troubleshooting Commands

Several monitoring and troubleshooting commands are available to help troubleshoot VoIP networks. These commands, coupled with your personal knowledge of your network, will help you understand performance of your network and help you to troubleshoot your network when anomalies occur. Use **show** commands to view momentary snapshots of your network's performance when attempting to gather facts. Use **debug** commands when you wish to view events and packets in real time. **debug** commands will aid in tracking what is happening with your router as it happens. Following are some recommended monitoring and troubleshooting commands that may be helpful for your VoIP network.

show Commands

Table 10-8 lists some common **show** commands for VoIP along with a synopsis of expected output and use of the commands. Other **show** commands that relate to IP routing may also be helpful.

Table 10-8   *VoIP* **show** *Commands*

Command	Description
show ip route	Shows the IP routing table. This command may be helpful in determining IP routes taken by the VoIP packets.
show voice dsp	Displays the use of DSPs and their assigned codec compression scheme.
show voice call summary	Displays current voice call statistics.
show voice port	Displays extensive statistics about the current status of voice ports. Interesting parameters include on-hook and off-hook conditions and analog information.
show dial-peer voice	Displays configuration of all dial-peers. With the added **summary** argument, displays the configured dial-peers in a table format.
show ip protocol	Displays the IP routing protocols currently configured on the router that are configured to support IP.
show call active voice	Displays active voice calls with their associated dialed destination.
show call history voice	Displays a log of attempted and connected calls.

debug Commands

Although the **show** commands can help to monitor performance of a VoIP network, sometimes you need to look at real-time packets and events to see what is happening and when it is happening. The **debug** commands designed for use in VoIP networks are powerful tools that allow you to see real-time events as they happen in your network. Some of the most useful **debug** commands are described in Table 10-9. Note that not all **debug** commands work on all platforms with all versions of IOS. For further information on VoIP debugging, see www.cisco.com/warp/public/788/voip/voip_debugcalls.html.

Table 10-9   *VoIP* **debug** *Commands*

Command	Description
debug voip ccapi inout	Outputs strings of event information that represents the end-to-end connection sequence for VoIP calls.
debug vpm [*option*]	Displays signaling, connection, and DSP events that occur on the Voice Processor Module. Options may include **signal**, **spi**, **dsp**, **port**, or **all**.
debug voice cp	Outputs events related to the call processing engine of the voice router.

continues

Table 10-9 *VoIP* **debug** *Commands (Continued)*

Command	Description
debug voice eecm	Displays events related to the end-to-end CallManager.
debug voip aaa	Enables debugging messages for gateway AAA to be output to the system console.

Summary

In today's ever-changing networking landscape, more and more enterprises are searching for creative ways to cut expensive IT costs. Convergence of voice and data networks is one such way to save costs, but presents its own set of problems when attempting to carry real-time traffic over existing IP networks. Solutions to delay problems, variable delay problems, frame-size problems, queuing choices, and packet-ordering solutions—as well as other problem-solvers—are found on Cisco routers to ensure that network administrators have the tools necessary to converge voice and data. As more enterprises install this new technology, creative configuration will be necessary to ensure that the network performs within specified tolerances. VoIP networks may be new, but they're here to stay. Network administrators must prepare themselves by understanding the issues surrounding the technology, understanding the building blocks for the technology, and understanding the configuration commands necessary to deploy it.

Review Questions

The following questions should help you gauge your understanding of this chapter. You can find the answers in Appendix A, "Answers to Review Questions."

1 How many bytes does the RTP header add to the UDP header?

_____

_____

_____

2 What is the ITU-T standard that specifies recommended one-way voice delay?

_____

_____

_____

3 In configuring VoIP networks, what command is used to enable fragmentation, but fragments all frames leaving the interface?

_____

4 What ISDN signaling protocol does the ITU-T H.323 protocol use for connection requests?

5 What IETF RFC does RSVP follow?

6 Which **show** command displays the dial-peer information in a summarized table?

7 RTP header compression can reduce the IP/UDP/RTP header from _____ bytes to about 2 or 4.

8 Which queuing method uses four queues, called High, Medium, Normal, and Low?

9 By default, a VoIP packet contains how many bytes of payload?

10 At Layer 2, how many bytes of overhead in VoIP applications does Ethernet use?

PART IV

Changing Voice Technologies

Upon completion of this chapter, you will be able to perform the following tasks:

- Explain the features of the hardware and software components of the Cisco IP Telephony (CIPT) solution, a component of Cisco's Architecture for Voice, Video, and Integrated Data (AVVID).

- Describe functionality offered by AVVID components.

- Describe how the AVVID components interoperate over an IP network.

- Explain the features offered by Cisco Conference Bridge, Cisco WebAttendant, and uOne.

CHAPTER 11

Cisco IP Telephony Solutions: New-World Technology

This chapter provides an introduction to the theory and operation of Cisco IP Telephony (CIPT) and the new-world technology model into which it fits. We will characterize the traditional public switched telephone network (PSTN) system as old-world technology and compare that to the exciting possibilities offered by converged networking and new-world technology. We will discuss the components of the AVVID IP Telephony solution, the protocols they use to accomplish their tasks, and how the AVVID components work together to provide a scalable solution, and then will provide detail on several of the components.

Reviewing Old-World Telephony

In old-world office communications, there were two separate networks: the proprietary-based telephone network and the data network. This configuration typically compartmentalized job responsibilities into separate divisions or workgroups. It also required duplicate circuits between locations: one for voice and another for data. If no calls were made on the voice circuits, the voice circuits were not available for data transport, hence money was wasted on unused capacity. PBXs, voice-mail systems, and other telephony infrastructure were proprietary, leading to increased acquisition costs, increased maintenance costs, and limited flexibility. Training and other personnel costs were also higher in the old-world technology model, since voice network skills could not be readily transferred to the data network.

Cisco Architecture for Voice, Video, and Integrated Data (AVVID) eliminates the need for proprietary PBXs and, in fact, for a dedicated telephony infrastructure based on TDM and circuit-switched technology. The fixed costs for a voice-only network are eliminated, and the knowledge necessary to understand the voice transport is the same as required to understand data transport, since they are the same transport, network, and data link layer protocols. The telephone becomes just another network device, as are the PBX and the voice-mail system.

New-World Applications

Cisco AVVID is a standards-based flexible architecture that is comprised of four distinct building blocks, as illustrated in Figure 11-1:

- IP infrastructure systems such as multilayer intelligent switches, routers, and gateways with vital network services like QoS, security, and management
- Call processing platforms and solutions
- New-world applications such as unified messaging, new collaborative capabilities, IP contact centers, and so forth
- Intelligent IP-enabled clients including IP telephones and software-based phones, and video clients based on standards such as TAPI and JTAPI

Figure 11-1 *New-World Applications*

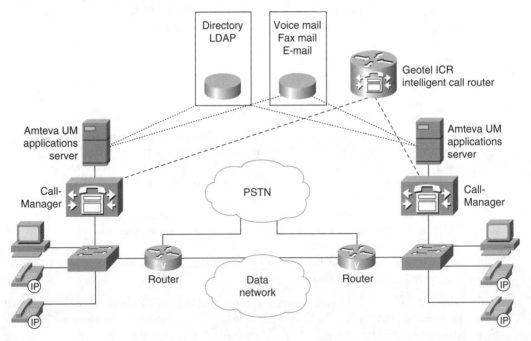

Cisco AVVID is a significant milestone in the evolution of IP networking in the enterprise. It enables customers to move from maintaining a separate data network and a closed, proprietary voice PBX system to maintaining one open, standards-based converged network for data, voice, and video. Cisco AVVID eliminates the need for additional, traditional, old-world proprietary PBXs.

Understanding AVVID

The CIPT network is part of Cisco's AVVID for the corporate enterprise. AVVID signals the availability of new-world voice, video, and data solutions, and is the culmination of Cisco's five-phase multiservice strategy.

The CIPT network provides a cost-effective voice communication by replacing PBXs and bypassing long-distance voice carriers or leased-line costs. The AVVID network may be connected to a global system telephone network (GSTN) by using Cisco gateways. This allows voice communication between old-world networks and new-world networks. (GSTN is the term used in the International Telecommunication Union Telecommunication [ITU-T] Standardization Sector recommendations when referring to what in North America is typically called the PSTN.)

The new network uses a variety of components, protocols, and hardware to accomplish its task. We will examine these in the next sections of this chapter.

AVVID Components

The core components of AVVID include:

- CallManager server—For providing call processing and H.323 gateway services for the IP Phones.

- uOne—A messaging application containing the application logic to interact with uTel, directory, and unified messaging servers

- IP Phones—Phones with an Ethernet connection to connect to the network that use the same variety of standards as any other Internet device, as well as standard protocols for voice compression and signaling.

- Access gateway devices—Devices used to connect to the GSTN for access to the public network and—in case the service provider's network fails—for transparent fail-over.

Each of these components is explained in greater detail in subsequent sections of this chapter.

CallManager

CallManager is an application that runs on Windows 2000 and is accessed using either Microsoft Internet Explorer or Netscape Navigator. CallManager is the call processing engine of the AVVID solution. The primary functions of the Cisco CallManager are the following:

- Call processing (digit analysis and resolution)
- Signaling and device control

- Features, capabilities, and dial plan
- Operation, Administration, Maintenance, and Provisioning (OAM&P)
- Programming interface to external voice processing applications

Cisco CallManager extends enterprise telephony features and functions to packet telephony network devices such as IP Phones, software phones, and VoIP gateways.

Supplementary and enhanced services such as hold, transfer, forward, conference, multiple-line appearances, automatic route selection, speed dial, last-number redial, and other features are extended by Cisco CallManager to IP Phones and gateways. Because CallManager is a software application, enhancing Cisco CallManager involves upgrading software, thereby avoiding expensive hardware upgrade costs. Further, Cisco CallManager configuration allows all phones, gateways, and applications to be distributed across a routable IP network, providing a single, distributed, virtual telephony network.

uOne

uOne is a corporate unified messaging system that includes many features, which are discussed in the following paragraphs.

Call answer services allows callers to leave messages for users, and when a user's phone is busy or ringing with no answer, calls are transferred by Cisco's CallManager to uOne. uOne then plays the user's activated greeting and takes the message. This feature replaces the standard voice-mail storage service of many PBXs.

uOne also permits user messaging services—where users can call the access number for uOne, sometimes referred to as the *pilot number*. After logging in with a PIN, the user's new messages are inventoried and the user can either listen to voice messages or administer mailbox preferences. The feature permits retrieval of voice mail from any phone, anywhere.

Another feature of uOne are the calling services that permit callers to zero out to a predefined attendant and transfer to another number. Users can return a call while listening to their messages or they can make a call to someone else. When the call is completed, the user is returned back to uOne, right where they left off.

The goal of uOne is to provide:

- A single-box solution running all uOne components on a single Windows server.
- An IP-based messaging system to supply voice and unified messaging services for IP-based telephony services.
- Support for standards and off-the-shelf components so customers can choose the products and vendors they want to use.

- An application that operates as simply another client/server application, running on the same computers, networks, and e-mail messaging infrastructure that customers already purchase, deploy, and maintain.
- Simplified Web-based administration.
- Interoperability with other Cisco messaging solutions, so you can mix and match messaging solutions and still share the same set of back-end servers and data.

IP Phones Overview

Cisco IP Phones bring state-of-the-art technology to voice communication solutions. Cisco now offers new opportunities for rapid deployment of old- and new-world voice applications by providing high-quality voice instruments that leverage IP as the transport technology. This allows the consolidation of voice and data into a single network infrastructure, such as a single cable plant, a single switched Ethernet fabric for campus or branch offices, or unified operational systems for Operation, Administration, and Management (OA&M) for voice and data.

IP Phones are covered in greater detail later in this chapter, in the section "IP Phones."

Overview of Access Gateways

The Cisco AVVID telephony solution offers multiple methods of connecting an IP telephony network to the PSTN or legacy PBX and key systems.

Every gateway selection is made by combining common or core requirements with site- and implementation-specific features. The three core requirements for an AVVID CIPT gateway are dual tone multifrequency (DTMF) relay capabilities, support for supplementary services, and the ability to handle clustered CallManagers. Any gateway selected for a large campus deployment should have the ability to support these features. Additionally, every AVVID CIPT implementation will have its own site-specific feature requirements.

Analog access gateway devices have *not* been homologated outside North America and Canada. This may cause restrictions on both use and importation, depending on local law. A gateway can reside behind a homologated device, such as a PBX, in most jurisdictions. The DT-24+ and the DE-30+ have been homologated in many countries outside North America. The DT-24+ and DE-30+ are covered in more detail below, in the section "Digital Trunk Gateways."

Gateways are covered in greater detail later in this chapter, in the section titled "Gateways."

IP Protocols Used

A number of IP protocols are used throughout the life cycle of the AVVID system, including the following:

- Initialization and CallManager registration can use DHCP, DNS, and Cisco's TFTP Server.
- Call and media control use Skinny Station Protocol over TCP/IP from the device to CallManager.
- Content transport uses RTP, UDP, and IP between devices.
- Calls between CallManagers use H.323 for setup
- Access Gateways use Media Gateway Control Protocol (MGCP), H.323, or Skinny Station Protocol, depending on the Access Gateway.

Each of these protocols and its use in IP Telephony are covered in greater detail in the following sections.

Dynamic Host Configuration Protocol (DHCP)

DHCP automatically assigns IP addresses and tells devices where to find the TFTP server. DHCP is a client/server system that is available with Windows NT Server. DHCP automatically assigns an IP address to a Cisco IP Phone whenever you plug one in. For example, this allows you to connect multiple phones anywhere on the IP network and DHCP will automatically assign IP addresses to them.

NOTE For more information on DHCP and DHCP options, see the following RFCs:

- 1534, "Interoperation Between DHCP and BOOTP"
- 2131, "Dynamic Host Configuration Protocol"
- 2132, "DHCP Options and BOOTP Vendor Extensions"

TFTP

TFTP is a simplified version of FTP that transfers files. TFTP is used to obtain a configuration file that tells each device the location of its CallManager.

CallManager runs a TFTP server that stores the configuration files for individual IP Phones. The IP Phones receive their configuration from a TFTP server. The Cisco TFTP Server program that comes with CallManager is a special implementation of a TFTP server that builds configuration files for IP Phones and must be used to serve configuration files to the IP Phones.

NOTE   For more information on TFTP and TFTP extensions and options, see the following RFCs:

- 1350, "The TFTP Protocol"

- 1785, "TFTP Option Negotiation Analysis"

- 2347, "TFTP Option Extension"

- 2348, "TFTP Blocksize Option"

- 2349, "TFTP Timeout Interval and Transfer Size Options"

H.323

H.323 is an ITU-T standard for multimedia service and equipment connected to IP-based networks. H.323 defines the following components, as illustrated in Figure 11-2:

- A gatekeeper, which is optional in an H.323 system, provides call control services to the H.323 endpoints. More than one gatekeeper may be present, and gatekeepers may communicate with each other in an unspecified fashion.

- A gatekeeper, when present, must provide address translation, admissions control, bandwidth control, and zone management.

- The gatekeeper may also perform additional, optional functions.

Figure 11-2 *AVVID H.323 Devices*

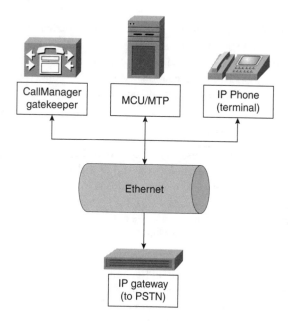

Media Gateway Control Protocol (MGCP)

MGCP is a protocol developed specifically for controlling VoIP gateways. MGCP describes an architecture, a set of application programming interfaces, and a protocol for controlling VoIP gateways. MGCP is a master/slave protocol that assumes that the call-control intelligence is outside of the gateway device itself. This permits the gateways to be relatively simple devices while the intelligence is contained in the call agent. Call agents can be synchronized among themselves in order to send coherent messages to the gateways.

A gateway provides the appropriate translation between transmission formats and between communications procedures. The gateway also performs call setup and clearing on both the LAN side and the switched-circuit network (SCN) side. Translation among video, audio, and data formats may also be performed in the gateway. In general, the purpose of the gateway (when not operating as a multipoint control unit [MCU]) is to reflect the characteristics of a LAN endpoint to an SCN endpoint, and the reverse, in a transparent fashion.

An MCU is an endpoint that provides support for multipoint conferences. It consists of one (required) multipoint controller (MC) and zero or more multipoint processors (MPs).

A terminal (that is, a client) includes one or more of the following functions and protocols:

- A video coder/decoder (codec; H.261 and others) encodes the video from the video source (for example, a camera) for transmission, and decodes the received video code, which is output to a video display.

- An audio codec (G.711 and so on) encodes the audio signal from the microphone for transmission, and decodes the received audio code that is output to the loudspeaker.

- A data channel supports telematic applications such as electronic whiteboards, still-image transfer, file exchange, database access, audiographics conferencing, and other applications. The standardized data application for real-time audiographics conferencing is Recommendation T.120. Other applications and protocols may also be used via H.245 negotiation as specified in the standard.

The default basis of data interoperability between an H.323 terminal and other H.323, H.320, or H.310 terminals is the T.120 protocol.

The H.245 control function uses the H.245 control channel to carry end-to-end control messages governing the operation of the H.323 entity, including capabilities exchange, opening and closing of logical channels, mode preference requests, flow control messages, and general commands or indications. H.245 signaling is set up between two endpoints: an endpoint and an MC or an endpoint and a gatekeeper. The endpoint sets up only one H.245 control channel for each call in which the endpoint is participating.

The H.225.0 layer is used after logical channels of video, audio, data, or control information are established based on Recommendation H.245. Logical channels are unidirectional and are independent in each direction of transmission. Some logical channels, such as for data, may be bidirectional. Any number of logical channels of each media type may be transmitted, except for the H.245 control channel, of which there is only one per call. In addition to the logical channels, H.323 endpoints use two signaling channels for call control and gatekeeper-related functions. The formatting used for these channels is defined in the H.225.0 recommendation.

The RAS channel carries messages during the gatekeeper discovery or endpoint registration processes, which associates an endpoint's alias address with its call-signaling channel transport address.

H.323 media formats supported by a Cisco IP Telephony network include:

- G.711—Pulse code modulation (PCM) at 64 kilobits per second (kbps)

- G.723.1—Multipulse maximum likelihood quantization (MP-MLQ) and algebraic code excited linear prediction (ACELP) at 6.3 and 5.3 kbps

- G.729—Conjugate structure-algebraic code excited linear prediction (CS-ACELP) at 8 kbps

---

NOTE For more information on H.323 and its supporting protocols, see the following ITU-T standards:

- Recommendation H.323, "Packet-based multimedia communications systems"

- Recommendation H.245, "Control protocol for multimedia communications"

- Recommendation H.225.0, "Call signaling protocols and media stream packetization for packet-based multimedia communications systems"

- Recommendation T.120, "Data protocols for multimedia conferencing used with associated collaborative applications"

Skinny Station Control Protocol (SSCP)

SSCP is a proprietary protocol developed by Cisco Systems, Inc., that provides for call setup, teardown, and signaling from an IP Phone. The protocol has less overhead than H.323 and hence was named "skinny." The messaging set for the control of "skinny" stations includes three sections. Section one describes the messages used during the registration process. Section two describes the messages used to control a call. Section three describes the messages used to control the media stream (audio stream) of the conversation. Message types are as follows:

- Registration and management messages—More than 20 registration and management messages are defined for the skinny station system. When an instrument (that is, a client) first powers up, it must register with the CallManager (that is, its controller).

Messages are passed between the client and controller to verify a range of parameters, including IP port, time, date, capabilities, station ID, line status, and configuration status.

- Call control messages—Call control messages fall into two groups: those from the client and those to the client. These messages provide the status of calls. Messages from the client result from actions originated by the client, and include keypad buttons pressed, speed dial usage, station hook status, and use of hook flash. Messages to the client are messages originating from the CallManager. These messages command the client to carry out an activity such as start or stop tone, set ringer, set lamp, set a microphone to a specified state, and clear the call display.

- Media control messages—Most of the media control messages originate from the CallManager. These messages control the audio stream. Messages include items such as start media transmission, stop media transmission, stop session transmission, and multicast media reception.

Signaling and Control

Devices communicate to CallManager using a lightweight stimulus/response protocol for registration, call control, and media management. To implement a full H.323-compliant terminal requires a high expenditure for computer power and memory size.

When communicating from an IP Phone to an H.323 device, the IP phone signals to the CallManager using SSCP and the CallManager signals to the H.323 terminal using H.323. The data stream is set up directly between the IP phone and the H.323 endpoint is an IP/UDP/RTP data flow.

The CallManager-based system is H.323 compliant in the following ways:

- CallManager is an H.323 endpoint to other H.323 devices.

- The IP Phone appears to other H.323 devices as an H.323 device through the use of CallManager.

Cisco CallManager (CCM)

CallManager is a real-time service that runs on the server. CallManager Administration is an application used to control the system environment. CallManager Administration maintains the CallManager database, which in turn is referenced by the CallManager service.

If you make database changes for a device that registers with CallManager, then only that device needs to be reset. For devices that don't register (for example, Cisco IOS gateways), a CallManager reset is necessary.

Admission Control

Using Cisco CallManager admission control, you can specify the amount of bandwidth allocated for voice communications over each WAN link (such as a PRI or T1/E1 line) in your network. Each type of call consumes a known amount of bandwidth. Thus, by specifying the bandwidth allocated on each WAN link, you also specify the total number of calls that can be active on that link at one time.

The following are some suggestions for using CCM for call admission control:

- Once bandwidth for calls across a WAN is configured in the database, the CallManager assumes that the bandwidth is guaranteed.

- You *must* have a hub-and-spoke network topology to use CallManager admission control.

- You *must* have a CallManager at each spoke to have fail-over to the PSTN.

Figure 11-3 shows a typical scenario for the use of admission control. A main call processing location, or hub, is linked to several remote locations by means of a WAN. The primary Cisco CallManager is at the hub, and it controls the hub as well as all the remote locations. There may also be backup Cisco CallManagers at the remote locations.

Figure 11-3 *Typical Use of Admission Control*

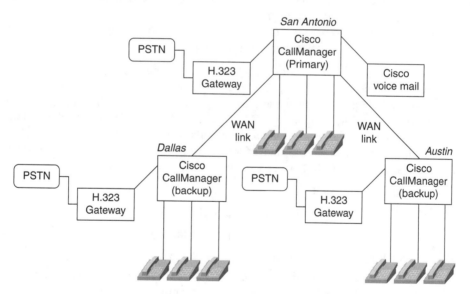

The available bandwidth between devices in the same location is assumed to be unlimited, so there is no need to use admission control to limit bandwidth consumption on calls within the same location.

Without admission control, additional calls above the capacity of the link will cause the quality of existing calls to degrade. With admission control, additional calls will not be permitted once the configured voice capacity of the link is exceeded.

For example, assume that the locations in the network shown in Figure 11-3 are defined as follows:

Location	Bandwidth to Hub (kbps)
Hub	Unlimited
Dallas	200
Austin	100

Calls that use G.723 (and G.729) compression consume 24 kbps of bandwidth, and calls that use G.711 compression consume 80 kbps. Cisco CallManager continues to admit new calls to a link until the available bandwidth for that link drops below zero. Thus, the link to Austin could support two G.711 calls at 80 kbps each, five G.723 (or G.729) calls at 24 kbps each, or one G.711 call and one G.723 (or G.729) call. Any additional calls that try to exceed the bandwidth limit are rejected, and the calling party receives a reorder tone.

The maximum bandwidth that you can allocate for a link is limited to the committed information rate (CIR) for that link. For example, if the Austin link uses a T1 line with a CIR of 1.5 megabits per second (Mbps), then you could allocate a bandwidth of 1.5 Mbps or less for voice communications over that link.

CallManager continues to allow calls to complete over a link until the available bandwidth for that link becomes negative (that is, drops below zero). For example, if a particular link has only 15 kbps of bandwidth left, one more call (either a G.723 or a G.711 call) can be made over that link. If another G.711 call is made, the available bandwidth for that link would then be –65 kbps. At that point, any additional calls would fail, and the calling party would receive a reorder tone. In performing bandwidth calculations for purposes of admission control, CallManager assumes the following:

- All calls are full duplex.
- Calls using G.723 consume 24 kbps of bandwidth.
- Calls using G.729 consume 24 kbps of bandwidth.
- Calls using G.711 consume 80 kbps of bandwidth.

There are some situations that are exceptions to the bandwidth assumptions. It is important to be aware of these situations because they can cause an oversubscribed situation. In these cases, additional calls can complete even though the available bandwidth has gone negative. The situations are discussed in the following paragraphs.

Calls made through a media termination point (MTP) can complete even if they exceed the available bandwidth limit.

If a feature—such as call transfer or hold—temporarily stops call streaming, the bandwidth from that call is temporarily available for use on another call. For example, when a user places a call on hold, the bandwidth from that call becomes available again. If new calls consume all the available bandwidth while the original call is on hold, the user can still retrieve the held call even though it exceeds the available bandwidth limit. This causes voice quality to degrade, because we are now in an oversubscribed condition.

Simultaneous calls on the same link can exceed the available bandwidth limit.

AVVID Hardware

Figure 11-4 demonstrates a call between an IP telephone and a NetMeeting H.323 user device. The system consists of the CallManager Windows 2000 service resident on the server and the installed devices, which terminate voice streams. Examples of devices include IP Phones, gateways, and H.225 devices.

Figure 11-4 *Call Between IP Phone and MS NetMeeting*

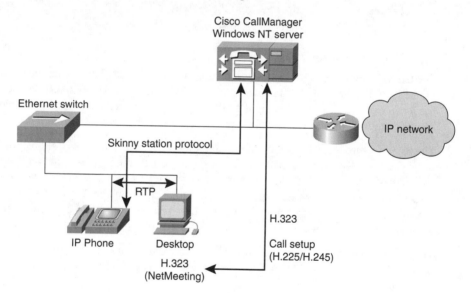

Control is accomplished from the phone to CallManager using SSCP and, independently, desktop to CallManager as an H.323 gatekeeper using H.225/H.245 over TCP. The content of the conversation between the NetMeeting user and the IP Phone user is transported by RTP over UDP/IP between the phone and the desktop.

Recognize that the CallManager server is serving as a call-processing engine and the CallManager is fulfilling the duties of a PBX.

IP Phones

There are several IP Phone models available today and Cisco Systems, Inc., will continue to develop and release new models of IP Phones as market needs dictate. In this section we will cover the most common of the IP Phones, providing a general description and a list of the features specific to each model. At the time of this writing, all IP Phones currently on the market provide the following features:

- Hearing-aid-compatible (HAC) handset with ADA-compliant volume
- G.711 and G.729a audio compression
- H.323 and Microsoft NetMeeting compatibility
- DHCP and BootP are supported
- DHCP automatically assigns IP addresses to devices when you plug in the phone
- Comfort noise generation and voice activity detection (VAD) programming on a system basis
- Designed to grow with system capabilities; features will be able to keep pace with new changes via software updates from the system

The Cisco IP Phone 7910

The Cisco IP Phone 7910 has a single-line appearance. The display area on the Cisco IP Phone 7910 is 2 lines by 24 characters per line (2×24 display). The Cisco IP Phone 7910 has a message-waiting indicator light on the handset and comes with other basic features. An illustration of the 7910 IP Phone is provided in Figure 11-5.

The basic-feature member of the second-generation Cisco IP phone portfolio is the 7910, primarily designed for common-use areas such as lobbies, break rooms, and hallways that require basic features. This single-line phone also provides four dedicated feature buttons, located prominently under the display for Line, Hold, and Transfer. A system administrator can program an additional group of feature-access keys. The standard configuration for these keys includes speed dial, redial, messages, and conference.

The 7910 also provides a large character-based 2×24 character LCD display. The display provides features such as date and time, calling-party name, calling-party number, and digits dialed.

Additional buttons for call monitor speaker (used for on-hook dialing) and handset volume control, and a ringer and mute button for the handset microphone, are arranged at the bottom of the set.

The Cisco IP Phone 7910 plugs into a standard RJ-45 Ethernet with one 10BaseT interface. The 7910+SW model also supports 10/100 BaseT and has 2 RJ-45 connections.

Figure 11-5 *Cisco IP Phone Model 7910*

The foot stand of the 7910 is adjustable from flat to 60 degrees to provide optimum viewing of the display and comfortable use of all buttons and keys.

The Cisco IP Phone 7960

The Cisco IP Phone 7960 includes an information button, six programmable line or feature buttons, and four soft-key buttons providing access to features such as additional call detail or access to Web-based information, such as stock quotes. The Cisco IP Phone 7960 includes an LCD display, which is used to display call detail and soft-key functions. An illustration of a Cisco IP Phone 7960 is provided in Figure 11-6.

The Cisco IP Phone 7960 includes two RJ-45 connectors:

- One connector can be used to connect the phone to a switch that provides 10/100 Mbps connectivity and receive power from that switch.

- Another connector can be used for network connectivity to a desktop device, such as a computer.

Figure 11-6 *Cisco IP Phone Model 7960*

Gateways

There are a variety of voice-access gateways available to meet customer needs. These gateways support a number of different protocols and offer various line and trunk connections using diverse signaling methods. In this section we will discuss some of the current popular voice-access gateways. For information on the latest products available, see the Cisco Web site.

Digital Trunk Gateways

The DT-24+ and DE-30+ digital gateways are a second-generation digital access solution for the Cisco IP Telephony solution products. These gateways provide a standard T1 ISDN PRI or E1 ISDN PRI interface to external voice networks, while providing 10BaseT access into the VoIP telephone network. Each gateway consists of a single circuit board, which can be plugged into a standard PCI bus in a PC, where it only draws power. After a gateway has been connected to the Cisco IP Telephony network, it automatically obtains an IP address via DHCP and locates and registers with the Cisco CallManager. From this point on, the gateway is fully configured and controlled by the Cisco CallManager. The DT-24+ and DE-30+ are completely self-contained, with all necessary control logic, voice DSPs, and interface circuitry on one board.

The following are features of digital trunk gateways:

- Self-contained operation means gateways only draw power from the host PC. No special software or drivers are needed. The gateway card runs independently from the host PC and does not affect PC performance.

- Both gateways support standard ISDN PRI Layer 3 protocols, including National ISDN-2, 5ESS, and DMS variants.

- Both network-side and user-side ISDN signaling is supported (via software selection from Cisco CallManager) for universal connectivity.

- Native support for Cisco IP Telephony solution type supplementary services is available via the Skinny Gateway Protocol. This allows Cisco IP Phones to place outside calls on hold, to forward calls, and so on.

- The gateways are already widely homologated. The DT-24+ is available for sale in the U.S., Canada, and Japan. The DE-30+ is homologated throughout most of Europe.

- The DT-24+ and DE-30+ are fully managed and controlled using an Internet browser connected to the Cisco CallManager.

VG 200 Voice-Access Gateway

The Cisco VG200 VoIP gateway is a next-generation voice-conversion device that provides powerful interoperability and advanced features in an affordable package. It is used to connect a Cisco IP Telephony Solution Network to traditional telephone trunks, legacy voice-mail systems, and other analog devices. These telephone trunks may be connected to the PSTN or existing PBX. Analog devices include legacy telephones, fax machines, and voice conference units.

On the data network side, the Cisco VG200 provides an autosensing 10/100 Ethernet port. Internally, the Cisco VG200 is equipped with digital signal processors (DSPs) that convert analog and digital voice into IP packets for transport through the IP network using standard codecs, including G.711, G.723.1, G.729(A), and others.

Cisco VG200 VoIP gateways provide a cost-effective solution to meet new generation telephony needs for:

- PBX and PSTN connectivity
- Analog and digital dial-access services
- Voice-mail connectivity to legacy voice-mail systems

The Cisco VG200 VoIP gateway modular architecture allows interfaces to be upgraded to accommodate expansion or changes in technology as new services and applications are deployed. Cisco VG200 gateway is managed, controlled, and administered using the Cisco CallManager, but can also be accessed directly using the same command-line interface (CLI) as other Cisco IOS Software-based products.

The Cisco VG200 hardened chassis contains open slots for installation of various combinations of voice interface cards. Cisco VG200 allows businesses to extend cost-effective, seamless network infrastructures to branch offices or small and midsize offices.

The Cisco VG200 voice gateways provide the following benefits:

- Investment protection—The Cisco VG200 is an integrated Cisco AVVID component and offers a modular design with support for the same range of interfaces in voice-enabled Cisco 2600 and 3600 routers and gateways. Modules are field upgradable and can easily be replaced by the customers.

- Lower cost of ownership—The Cisco VG200 VoIP gateway communicates with the Cisco CallManager using MGCP or H.323 Version 2. It allows connectivity to legacy telephony equipment, PSTN, PBX, and voice-mail systems. Cisco VG200 provides a space-saving solution that can be managed remotely using network management applications such as CiscoWorks.

- Reliability—The Cisco VG200 uses proven hardware and Cisco IOS Software and can be managed with Simple Network Management Protocol (SNMP) and Telnet access.

Catalyst 6000 Family Voice T1 Module

The Catalyst 6000 family brings data, voice, and video integration onto the campus for fully integrated communications on every desktop. Campus multiservice networking, or convergence, provides voice support using the IP network infrastructure rather than the traditional PBX. This increases the leverage of telephony spending into overall infrastructure spending, reduces capital and operational costs, and opens the environment to innovation by supporting new applications.

The Voice T1 and Services Module, also available in an E1 variant, introduces the following new features for campus multiservice networking:

- Digital T1 or E1 PSTN and PBX gateways
- Transcoding (G.711, G.729a, G.723)
- Conference bridging
- Campus Multiservice Networking

The Voice T1 and Services Module allows larger enterprises to connect the PSTN and legacy PBXs directly into the campus multiservice network. Telephony signaling types supported include:

- Common channel signaling (CCS)—In this mode there are 23 DS0 channels for T1 and 29 for E1 for voice traffic; the 24th T1 DS0 or 16th E1 channel is for signaling. Any channel can be configured for common channel signaling.

- Integrated Services Digital Network (ISDN) Primary Rate Interface (PRI) signaling—Each interface supports 23 channels for T1 and 30 channels for E1. The default mode is for the 24th T1 channel or 16th E1 channel to be reserved for signaling. Both network-side and user-side operation modes are supported.

The Voice T1 and Services Module provides transcoding and conferencing services for the multiservice network. Transcoding enables a full voice-compression solution by offering transcoding services to endpoints not capable of supporting compressed voice or a different encoding type to the remote end.

Media Convergence Servers

There are two Media Convergence Servers that are offered for use in production networks at the time of this writing. These are the MCS-7822 and the MCS-7835. In this section we will provide a short introduction to each.

The MCS-7822

The Cisco Media Convergence Server 7822 (MCS-7822) is a low-cost server platform for Cisco AVVID. The MCS-7822 is a member of the MCS-7800 series server family, which also includes the high-availability MCS-7835. The MCS-7822 is intended to be a pilot or test platform, when it is used as a stand-alone server or a production server when part of a Cisco CallManager 3.0 server cluster. The MCS-7822 comes with Cisco CallManager 3.0 software on an automated installation CD to make the deployment of IP telephony simple and cost effective.

The MCS-7822 is a high-performance server platform designed for today's IP telephony applications. The processor is an Intel Pentium III with a clock speed of 550 megahertz (MHz). Memory is a robust 512 megabytes (MB) of ECC RAM.

High availability can be achieved with the new clustering capability of Cisco CallManager Version 3.0. Cisco CallManager 3.0 significantly enhances the availability of the enterprise IP telephony solution by allowing you to cluster up to eight servers and manage them as a single entity. Cisco CallManagers can back up each other even if they are physically in different locations. The capability of clustering multiple call-processing servers on an IP network is unique in the industry and highlights the industry-leading architecture of Cisco AVVID.

The Cisco AVVID architecture allows for a great deal of scalability to meet the requirements of almost any organization. Cisco CallManager clustering technology allows you to start with ten IP Phones and grow your network to tens of thousands of users at your own pace. This is very different from legacy PBX technology that forces you to decide the ultimate scalability of your PBX on the day of purchase. We think it makes more economic sense for your voice infrastructure to grow as your business does.

The MCS-7835

The MCS-7835 features a 733-MHz Intel Pentium III processor and is expandable up to 4 gigabytes (GB) of 133-MHz registered SDRAM, extending the high performance you will require to roll out current and future Cisco AVVID applications. There is hardware RAID support for dual 18.2-GB Ultra2 small computer serial interface hot-plug hard drives to improve overall system performance. All of this power is delivered in a space-saving rack-mountable form factor (3U), smaller than the MCS-7830, designed to save precious rack space in your data center.

Availability, or the percentage of time that a system is available to provide service, was assumed in old-world networks. Availability is a key requirement in the new-world networks Cisco is building today. The high-availability design of the MCS-7835 will deliver a robust platform for your mission-critical Cisco AVVID applications. The MCS-7835 comes standard with a redundant hot-plug power supply and two redundant 18.2-GB SCSI hot-plug hard drives running RAID-1 disk mirroring to ensure maximum availability. If a hard drive or power supply fails, it can be replaced without powering down the server, and the failure will not affect service. In the case of the SCSI drive, as soon as the replacement drive is inserted, the integrated RAID controller will restore the image to the new drive, without any user intervention.

Whether you start your Cisco IP telephony network with five telephones or 5000, the MCS-7835 server allows the customer's network to grow at a manageable pace. A MCS-7835 server can serve as a Cisco CallManager server or a Cisco uOne voice-messaging server. Additional applications are planned for the platform in the future. As a Cisco CallManager 3.0 server, each MCS-7835 can handle up to 2500 IP telephones (total number of IP Phones dependent on N+1 redundancy configuration). Remote sites can also be interconnected through an H.323 interface, using an H.323 gatekeeper.

The MCS-7835 is configurable to run either Cisco CallManager software or Cisco uOne voice-messaging software. The MCS-7835 was also designed to run future Cisco application packages that will become part of the Cisco AVVID solution. The MCS-7835 has an optional internal 12/24-GB DAT tape drive to back up critical data, and also offers the flexibility of saving important user data to a separate server located elsewhere on the IP network.

Cisco Conference Bridges

There are two types of conference devices: unicast and multicast. For conferencing, you must determine the total number of concurrent users, or audio streams, required at any time. Then you create and configure a device to support the calculated number of streams. These audio streams can be used for one large conference or several small conferences.

For example, a conference device that was created with 20 streams would provide for 1 conference of 20 participants, or 5 conferences with 4 participants each (or any other combination up to 20 total participants at any time). The total number of conferences supported by each conference device is calculated by taking the total number of streams (for example, 20) and dividing by 3. Therefore, in the example, you can have 20 divided by 3, or 6 conferences supported by the conference device.

Unicast conference devices can be installed on the same PC as the CallManager, or on a different PC. The available system resources will determine the location of the Conference Bridge application. For unicast, a conference bridge must be running for the conference feature to work. For multicast, CallManager must be running for the conference feature to work.

Each conference device, whether unicast or multicast, can be configured as an Ad-Hoc conference bridge (where the conference controller adds conference participants from their own phone) or a Meet-Me conference bridge (where all users dial in to a specific conference directory number).

There are four significant differences between unicast and multicast conference devices:

- Unicast devices physically reside on a PC and register with CallManager when started.

- Multicast devices are virtual devices and don't actually register with CallManager.

- Multicast conferences cannot include calls that join the conference through a voice gateway, because the analog and digital gateways do not support multicast.

- Conference devices only support G.711 calls.

Meet-Me conferences require that a range of directory numbers be allocated for their exclusive use. When a Meet-Me conference is set up, the conference controller selects a directory number from the range of directory numbers and advertises it to the group that will join the conference. The users may then call that directory number to join the conference. Anyone who calls that directory number while the conference is active will be added to the conference (provided that the maximum number of participants specified for that conference type has not been exceeded, and sufficient streams are available on the conference device).

Ad-Hoc conferences are controlled by the conference controller, and the conference controller individually calls each participant and conferences them in. Any number of parties can be added to the conference up to the maximum number of participants specified for Ad-Hoc conferences, provided that sufficient streams are available on the conference device. Ad-Hoc conferences will sum the last four voices that were heard.

Cisco WebAttendant

Cisco WebAttendant is a client/server application. The client components are the Cisco WebAttendant and the Attendant Console Administration. The server components are the line-state server and the telephony call dispatcher.

- The line-state server provides real-time statistical data about line activity for all four- or five-digit directory numbers in the system.

- The telephony call dispatcher (TCD) provides statistical information about the number of redirected calls and the number of Cisco WebAttendants that are online. The TCD dispatches calls that have arrived at pilot point numbers to the appropriate destination based on hunt groups (hunt groups are configured in Attendant Console Administration). TCD must be installed on the same PC as the Cisco CallManager. The line-state server is a prerequisite for TCD.

Cisco WebAttendant allows a company to post one or more live attendants to answer and handle inbound and outbound calls that are not serviced by direct inward dialing (DID), direct outward dialing (DOD), or automated attendant functions. The Cisco WebAttendant operates using a Microsoft Internet Explorer 4.0 or later Web browser with an associated Cisco IP Phone, which provides voice-stream termination. Functions are signaled to the Cisco CallManager with a Cisco Telephony Application Programming Interface (TAPI) call. The Cisco WebAttendant is dependent on an associated Cisco IP Phone's keypad template for operation. Use the default Cisco WebAttendant template that is provided for you, or be sure that any custom keypad template you create has a maximum number of lines and buttons for hold, transfer, and answer/release.

Cisco WebAttendant uses information in the Cisco CallManager database to direct calls. For complete call dispatch capabilities, be certain all users are entered in the User area of Cisco CallManager Administration, and that resources, such as conference rooms with phones, are also entered in the User area of Cisco CallManager Administration. Cisco WebAttendant can be used by a receptionist or secretary to handle call traffic for multiple Cisco IP Phones, or by an individual to handle call traffic for a single Cisco IP Phone.

The Cisco WebAttendant client can be used to manage calls for Cisco IP Phones belonging to a single user, or for company-wide receptionist duties.

Summary

The Cisco IP Telephony system is an integral part of AVVID. AVVID provides enterprises with a converged network that eliminates old-world technology such as PBXs, analog or digital telephones, and legacy voice-mail systems. Enterprises that implement the converged AVVID network, including CIPT, have the advantage of a system that is based on the IP protocol suite and other standards-based protocols such as H.323. The Cisco IP Telephony system includes Cisco IP Phones, H.323 endpoints such as CallManager, voice-access gateways such as the DT-30+, and the uOne unified messaging system. Cisco IP Phones use a variety of protocols such as DHCP, TFTP, DNS, and the IP suite of protocols to download their configuration and perform their Voice over IP functions. Admission control is used to ensure that you will place only phone calls that you have the bandwidth to support. The H.323 protocol stack and components are covered here in a shortened version to provide you with a basic understanding of the multimedia protocols components of the AVVID solution. Several AVVID products are discussed to give you a good background in the product line and capabilities.

Review Questions

The following questions should help you gauge your understanding of this chapter. You can find the answers in Appendix A, "Answers to Review Questions."

1 What two Cisco IP Telephony components are used to connect to the PSTN?

_____

_____

_____

2 What are three skinny station protocol message types?

_____

_____

_____

3 What are two functions of a Cisco IP Phone?

_____

_____

_____

4 Which standard protocol does the Cisco IP Phone use to communicate with the Cisco CallManager?

_____

_____

_____

5 What protocols are used to carry Voice over IP traffic?

_____

_____

_____

6 What are the two Media Convergence Servers covered in this book?

_____

_____

_____

7 List three voice-access gateways.

_____

_____

_____

8 Which IP Phone(s) discussed in the text has an Ethernet switch integrated in it?

_____

_____

_____

9 How much bandwidth does a G.729 call consume?

_____

_____

_____

10 What two types of conferences are supported in the CallManager architecture?

_____

_____

_____

Upon completing this chapter, you will be able to perform the following tasks:

- Describe PBX functionality and explain how it fits in an enterprise communications network.

- Describe basic PBX technology found in popular PBX systems on the market today.

- Identify internal processing functions of a PBX including switching, traffic handling, and signaling.

- List and describe common user and system-wide application features offered by a PBX.

CHAPTER 12

Old-World Technology: Introduction to PBXs

Telephony has come a long way since its widespread introduction using analog technology over 60 to 80 years ago. As businesses realized they could gain better cost control and increased productivity through the use of their own internal switchboards, a new industry was born. This industry filled the need for switchboards at first, eventually making telephone switches similar to the switches at the central office (CO). As technology proliferated and miniaturization increased, once-costly switches shrank, lost some features, and because widely deployed in even small businesses with fewer than a dozen employees. Such is the history of the private branch exchange (PBX) and key system businesses.

It is important to understand a bit of the history, especially the technology of the PBX and key system world so that you can understand what functions the new technology is replacing, when it is appropriate to leave the old technology in place, and how the old and the new can coexist until the economic life of the old technology is completed.

This chapter covers basic technology and functionality of PBX telephone systems, and the hardware and applications typically associated with PBX and PBX-like systems. The chapter discusses legacy telephone PBX systems, their technology, design, operation, and features. It also covers key systems, which originally were analog systems but have evolved into a hybrid of both analog and digital lines to the PSTN and the use of digital features internally.

Private Phone Systems

The two types of private phone systems deployed in most of the world today are PBXs and key systems. Some organizations actually install the same switches that the CO uses, but these are very large organizations.

Key systems were developed to provide an economical means of adapting the PBX technology to smaller organizations. Just as computers were first deployed in government, educational, and large corporate environments that had the budgets to afford them, so PBXs were first installed in the same market. Advances in technology increased the functionality and decreased the price of these switches. Businesses saw the market potential for selling a product with similar features to smaller customers. By focusing on the needs of these smaller customers, these businesses making key systems created a new market. Even though PBXs were developed first, we'll address the details of key systems first in this chapter.

Key systems were originally much simpler than PBXs, although, as pointed out in the chapter introduction, the line between key systems and PBXs is rapidly blurring. PBXs typically service a much larger number of users than a key system does, provide for more connections to the CO, and offer a greater array of features than key systems do. Table 12-1 gives a brief overview of the major differences between key systems and PBXs.

Table 12-1 *PBXs and Key Systems Compared*

Feature	Key System	PBX
Technology	Digital or analog	Mostly digital (some analog)
Switch	Not a switch	Similar to CO switch
Typical site	Small company or branch office (typically 50 or fewer users)	Large company or site
Method for accessing outside trunks	Press a button to access an outside line	Dial 9 or other access number to get trunk

Key Systems

A key telephone system is often used in smaller organizations and small branch offices where a PBX offers functionality and complexity that may not be required. The term *small* is relative, but typically a PBX would not be installed for fewer than 100 users, and an organization would probably need 500 or more internal lines to find a PBX cost effective. A key system offers small businesses distributed answering from any telephone. The major difference between the two is that the PBX appears to the CO as just another phone switch, while the key system appears to be multiple lines. Figure 12-1 illustrates the major components of a key system.

Today's key systems are microprocessor-based, and are either analog or digital. Key systems are typically used in offices with 30 to 40 users, and some scale to support well over 100 users. Keep in mind that—more than the number of users—what really determines the most cost-effective solution is the amount of time spent on the phone. A sales organization with 50 full-time telemarketers may well need a PBX, while an engineering firm of over 100 with limited need for outside lines may find a key system meets their needs perfectly.

Figure 12-1   *A Sample Key System*

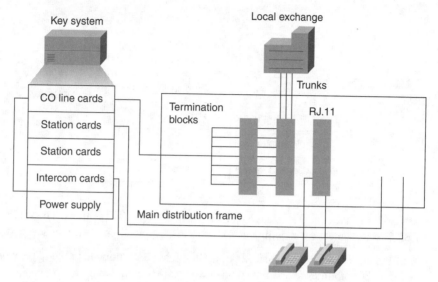

A key system contains three major components:

- Key service unit—The key service unit is the centrally located box that stores the logic for the system.

- System software—System software provides the operating system and calling feature software.

- Telephones (instruments or handsets)—These devices are required to support each user. With a key system, users press a button on their phone to access an available outside line.

With a key telephone system, each telephone has multiple line appearances so users can access outside lines to their CO. As calls come into the company, a line, or key, lights up on the phone to indicate that a particular line is in use. An intercom function (overhead paging or speaker phone) is used to call another extension or to let another person know where to pick up a call.

PBXs

A PBX is a small version of the phone company's large central switching office. A PBX is considered private because a company owns it. Figure 12-2 illustrates a PBX system and its components.

Every office has a telephone system, which may be a PBX, a key telephone system, or a Centrex service. Most large offices with more than about 50 telephones (or handsets) choose a PBX as their telephone system to connect people internally and to the outside world.

Figure 12-2   *An Example of a PBX System*

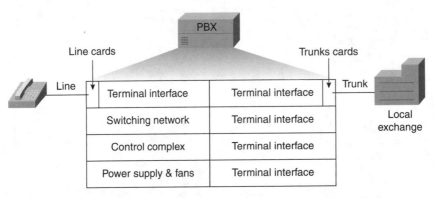

PBXs come in a variety of sizes, typically from 20 to 20,000 stations. The selection of a PBX is important to most companies, since a PBX typically has a life of seven to ten years.

All PBXs offer a standard, basic set of calling features. Optional software provides additional capabilities. A PBX contains three major components:

- Terminal interface—A terminal interface provides the connection between terminals and the PBX features. Terminals may include a telephone handset, trunks, lines, and so on. Common PBX features include dial tone and ringing.

- Switching network—A switching network provides the transmission path between two or more terminals in a conversation. For example, two telephones within an office communicate over the switching network.

- Control complex—A control complex provides the logic, memory, and processing for call setup, call supervision, and call disconnection.

Switching and PBX technology has evolved through three generations: *First-generation systems* used wired logic and analog step-by-step or crossbar switching fabric. First-generation telephones were non-proprietary rotary dial or dual tone multifrequency (DTMF) analog systems. These were phone company switches, not PBXs. During this period most enterprises had a central switchboard with operators literally connecting wires into holes to complete calls. *Second-generation systems* used either standard analog or proprietary phones (usually analog) to control a limited number of features. Second-generation PBXs use pulse amplitude modulation (PAM) switching networks. Some second-generation systems are still in operation but they cannot support the North American numbering plan, so they have either been replaced or will be replaced soon. *Third-generation systems* support end-to-end digital transmission, employs pulse code modulation (PCM) switching technology, and supports both analog and digital proprietary telephones.

PBXs typically connect to three types of external networks:

- Local—A local network provides local telephone service, usually within a limited local area. Local service providers offer local telephone service.

- Interexchange—An interexchange network provides long-distance calling service, which may cover virtually any area from statewide to nationwide to international.

- Private—A private network is leased to a company, usually to provide long-distance calling service to a single location at a fixed cost. The same service providers that offer interexchange networks also lease provide networks.

Many systems support special services such as T1 lines and foreign exchange trunks to provide local calling in distant cities.

Computer telephony interfaces (CTIs) connect PBXs to mainframes, minicomputers, or servers so that the PBX can supply call information to the computer and the computer can send call routing instructions to the PBX. Some of these implementations are illustrated in Figure 12-3.

Figure 12-3 *PBX Implementations Within an Enterprise*

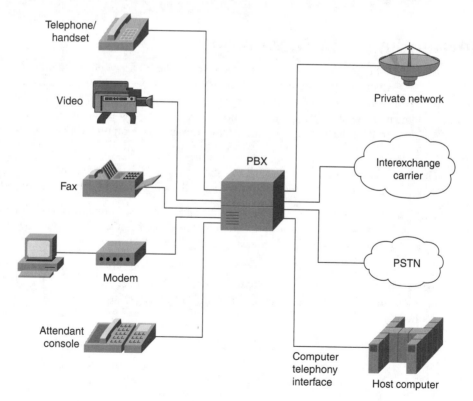

Virtual Private Networks

Virtual private network (VPN) is a general term that describes specific services that different service providers offer. In essence, the service provider offers the equivalent of a private leased-line network over the PSTN.

Examples of VPN services available from service providers include:

- Software Defined Network (SDN), from AT&T, which provides access to both voice and data services
- V-Net from MCI
- VPN Service from US Sprint
- Featurenet from British Telecom

To create a VPN, the service provider stores an image of the customer's voice network in a database. This database is accessible at the customer's premises so the customer can administer their own data. PBXs are connected to the provider's switches, generally using ISDN PRI links. When a call is set up, the PBX sends the call to the network switch, which then connects to the customer's database.

Understanding PBX Technology

A number of components compose a PBX system. The PBX performs the same functions as a CO switch and frequently provides more services as well. The typical PBX does not have nearly the capacity of a CO switch, but is still an incredibly complex piece of equipment. This section describes the hardware and software components of PBX systems. An illustration of the components and interactions among components of a PBX is shown in Figure 12-4.

The PBX is typically housed in a PBX equipment room. All trunks from the local exchange terminate in the PBX equipment room, as does all wiring to the telephone equipment.

The *main distribution frame* (MDF) connects telephone lines coming from outside (from the local exchange) on one side and the internal lines on the other. It may include other protective devices such as a central testing point for trunks from the local exchange.

The *intermediate distribution frame* (IDF) is a metal rack that connects cables. It is usually located in an equipment room or closet. It provides the connection between the MDF and individual telephone wiring. Changes in wiring are generally done at the IDF to avoid confusion.

Figure 12-4   *An Overview of a PBX System*

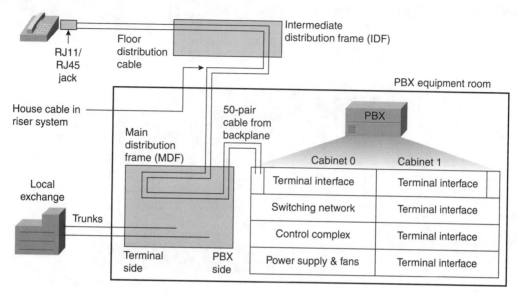

Switching Network Design

PBXs come in various engineering design styles. The design style determines the placement of the components of the PBX, as well as the cost and expandability. PBX vendors typically implement one of four basic switching network designs. These switching network designs are characterized by the type of control—centralized, distributed, dispersed, or adjunct—they exhibit. The following sections describe each of these types of control.

Centralized Control PBXs

The centralized control PBX switching model is characterized by the following features:

- A main cabinet controls all PBX functionality including the switching matrix, TDM bus, and time-space-time.

- There are no separate port cabinets or shelves to support telephone lines and CO trunks. Lines and trunks terminate at the main cabinet.

- All processors are housed in the main cabinet.

- All port circuit card microprocessors are housed in the main cabinet.

Figure 12-5 illustrates a centralized control PBX.

Figure 12-5 *A Centralized Control PBX*

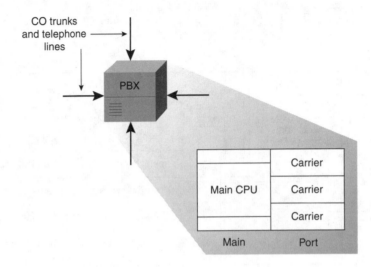

Distributed Control PBXs

The distributed control PBX switching model is characterized by the following features:

- Two or more main processor element controllers act in a peer-to-peer architecture. These processors may be located in a single cabinet or in separate cabinets.

- Each processor controls its own PBX functionality, including the switching matrix and TDM bus.

- An Inter-Switch Link (ISL) between PBXs links the independent TDM bus segments in each cabinet to each other.

- All port circuit card microprocessors are housed in each cabinet included in the distributed design.

Figure 12-6 illustrates a distributed control PBX.

Figure 12-6   *A Distributed Control PBX*

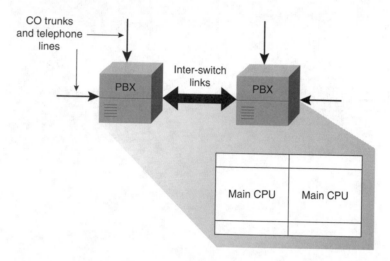

Dispersed Control PBXs

The dispersed control PBX switching model is characterized by the following features:

- A main processor controls all PBX functionality within each cabinet.
- A centralized, center-stage switch controls distributed TDM buses.
- An ISL between PBXs and the center-stage switch links the independent TDM bus segments in each cabinet to each other.
- Each cabinet contains cabinet and shelf controllers.
- All port circuit card microprocessors are housed in each cabinet included in the distributed design.

Figure 12-7 illustrates a dispersed control PBX.

Figure 12-7   *A Dispersed Control PBX*

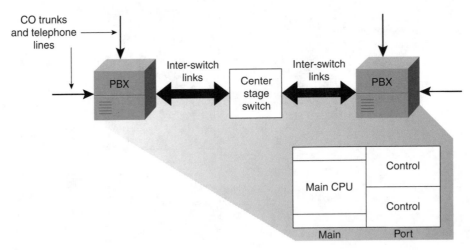

Adjunct Control PBXs

The adjunct control PBX switching model is characterized by the following features:

- The main processor controls all PBX functionality.
- An adjunct control cabinet, which connects to the main processor, is under the control of the main CPU. It offers additional PBX software or optional software. Optional software may include systems management, messaging, MIS reporting, wireless support, Internet telephony, and so on.

Figure 12-8 illustrates an adjunct control PBX.

Figure 12-8   *An Adjunct Control PBX*

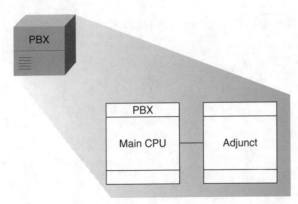

Processing Network Design

The processing network design PBXs use servers such as Windows NT, Windows 2000, UNIX, Linux, or mainframe computers as processors coupled to the traditional PBX hardware. The processor handles call processing and/or specialized applications, while the traditional PBX hardware handles the trunk connections, line connections, and switching fabric. We look at two such designs in this section.

Windows NT, Windows 2000, or UNIX-based Server PBX Design

A Windows NT, Windows 2000, or UNIX-based server may be used in a PBX network design to offload some of the call-processing functions from the PBX. The Windows NT server, in this case, is responsible for call processing, and the PBX connects to telephones (handsets).

An illustration of a server-based PBX is seen in Figure 12-9.

Figure 12-9   *Server-based PBX Design*

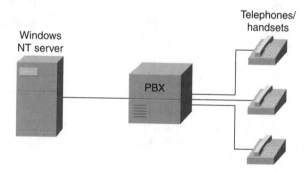

Client/Server PBX Design

In a client/server design, the local-area data network and the voice PBX network are integrated using application servers, which can provide any of the following functions:

- Call processing (offloading this process from the PBX)
- Conversion of voice to data packets for transmission over the data network
- Voice-mail services
- Interactive (database) voice response

Clients on the network may work from a traditional PC or other workstation that offers both voice and data functionality. PC applications may actually act as a telephone (handset) on the network, with access to both PBX functionality and data applications. New applications that may be available using a client/server design include multimedia, video conferencing, rich e-mail, and others.

Meanwhile, IP Phones, which convert analog voice to digital data packets within the phone, can use the data network to transport voice signals. In this case, application servers or the PBX may take on the call-processing functions.

Figure 12-10 illustrates a client/server design.

Figure 12-10   *Client/Server Design*

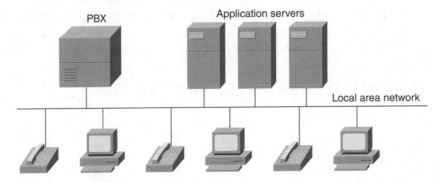

A server farm may provide more distributed processing, with its application servers, network interface servers, integrated servers, and third-party servers each taking on part of the call processing functions. For example, the network interface server can provide access to the Internet, PSTN, LAN, and WAN. At the same time, the application server can provide call-processing functionality.

Basic PBX Architecture and Components

As mentioned earlier in the chapter, PBXs are highly complex devices. They include a central processing unit, memory, line connections, and trunk connections, and provide both intra- and interconnections between the lines and trunks. We briefly introduced the PBX world in the section "PBXs" earlier in this chapter. This section describes the major processing elements within a PBX in much greater detail. The elements of a basic PBX are illustrated in Figure 12-11.

Figure 12-11   *Basic PBX Processing Elements*

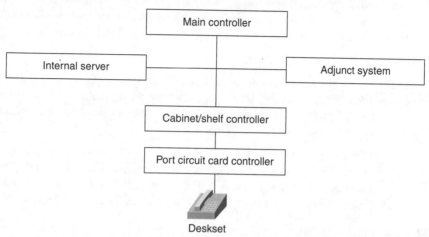

The main controller performs all centralized PBX control functions. It sometimes generates dial tone but usually commands the line cards to supply dial tone, and it directs the setup of switch connections, system features, and functions. The main controller includes database and memory elements, which include a generic program and user database parameters.

The main CPU resides within the main controller. Early PBXs had 8-bit processors, but today's PBXs have 32-bit processors using a high-level complex instruction set computing (CISC) or reduced instruction set computing (RISC) design. Many CPUs use off-the-shelf third-party technology from vendors like Intel, Motorola, and Silicon Graphics and run operating systems including UNIX, Windows NT, and other proprietary operating systems.

Large PBXs have redundancy built in, but smaller systems may not have redundancy and backup systems. Redundancy in intermediate to large systems may include hold/cold standby, shadowing, and crossover.

The cabinet/shelf controller provides the link between the PBX main control and port functions. It includes localized functions, such as switch network access, and maintenance and diagnostic functions. The cabinet/shelf controller performs mundane and repetitive tasks to offload the processing burden from the main CPU. To provide redundancy, the cabinet/shelf is available in many different versions and with different options. The goal is that any processor failure will not affect local ports and functions.

The system processor bus, usually an Ethernet bus, provides communications between the main controller and local control elements within the PBX. The goal of vendors over the last few years has been to provide higher speeds on the system processor bus.

The port circuit card controller uses microprocessors to provide a link between the PBX system and peripheral equipment, including telephones and computers. Features include diagnostics, recognition of touch-tones, and telephone display features and functions. The port circuit card controller has no redundancy built in and offers a fixed number of ports.

The internal server provides PBX software processing.

The PBX main system controller controls an adjunct system. An adjunct system may be optional PBX system packages sold separately from the main PBX or may include third-party systems. Typical adjunct systems include automated call distributors (ACD) and interactive voice response (IVR) systems.

These processing elements compose the PBX. Since all PBXs must perform the same tasks and connect to the same kinds of trunks and lines, they are all of a similar design. Figure 12-11 shows a digital PBX with a switching fabric that connects to line and trunk interface circuits. The architecture of a basic PBX is illustrated in Figure 12-12.

Figure 12-12   *Basic PBX Architecture*

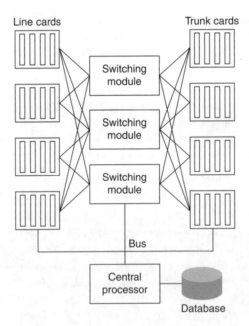

Although the basic structure of PBXs is similar, different manufacturers' products offer significantly different features. However, most systems have universal card slots into which either line or trunk cards can plug in. The basic PBX architecture includes the following components:

- Central processor—Operates the PBX software, which is stored in memory.

- Line and trunk interface circuitry—Circuitry contained on cards that slide into slots that mount in modules or cabinets.

- PBX backplane—Ties the lines, trunks, and central control circuits to the switching fabric and buses over which the circuits communicate. Cards plug into the PBX backplane.

Because of the different service needs of PBXs, they come in two basic types of cabinets:

- Single carriers—Carriers used for smaller systems. Single carriers are stackable. Each cabinet contains a common backplane and supports one function, such as the control, switch, or port.

- Multi-carriers—Carriers used for intermediate to large-scale systems. Multi-carrier cabinets usually offer multiple functions, with each cabinet performing a specialized function.

PBX cabinet carriers are available in many different types, including control, port, switching network, power, and application server. Early PBXs used large cabinets to house cards. Today, most PBXs provide one cabinet per shelf. This modularity provides easier handling, redundancy of critical components (such as power supplies), and easier installation.

Trends in PBX cabinet design include:

- Port cards continue to be reduced in size or they allow more ports per card than in earlier systems.

- Additional ports or stations are provided for each card, up to 32 stations or lines per card.

- T1/E1 trunk cards typically support two trunks (rather than one on older cards).

- Modular wiring is common providing a self-documenting system. Patch panels in the wiring closet are used to improve maintenance and reconfiguration later.

Customers may have a series of cabinets to create a single PBX system. Typically, four to six carriers make up a total stack or cabinet, offering 16 to 24 slots for line and trunk cards. Some slots are reserved for specific cards, while others are dedicated to specific functions. Customers typically segment slots for power, networking, and processing.

You can expand the capacity on a PBX system by adding cabinets or carriers to provide additional ports, more processing power, the ability to handle additional traffic, redundancy and duplication, and other processing functions.

Cabinets are typically connected using high-speed transmission or processing links, typically using fiber-optic cabling. These links may or may not be duplicated to provide redundant links for fail-over.

The circuit cards that provide processing, line connections, and trunk connections are installed in the cabinet. These PBX circuit cards are typically available in three types:

- System—Includes processor controllers, memory, switch network, and maintenance cards.
- Service—Includes touch-tone receivers, conference, input/output, cabinet links, and host and client/server computer telephony integration (CTI) links.
- Port—Includes line and trunk cards.

Port circuit cards to provide the line and trunk services typically range from 1 to 26 circuit terminations for each card, and come in those two types.

The *line circuit card* is typically available in an analog, digital, or IP model. The digital circuit card may be either a TDM model using T1/E1 service or an ISDN BRI model.

The *trunk service card* is typically available in analog or digital models. The analog models are used to provide FXO, FXS, and E&M signaling in traditional analog telephony environments. Other services, such as direct inward dialing (DID) and Centrex services, may also be provided via analog trunks. The digital models can provide a plethora of transport and signaling options, including T1/E1, CAS, CCS, ISDN PRI, QSIG, R2, and several others.

Both of these port circuit cards are covered in more detail in the sections below, "PBX Line Interfaces" and "PBX Trunk Interfaces."

PBX Line Interfaces

PBX systems typically offer three different types of line interface cards:

- Analog line interfaces—Analog line interfaces support telephone sets that are independent of the PBX manufacturer. The standard for analog phones, developed by AT&T, is the 2500 set for DTMF service.
- Digital line interfaces—Digital line interfaces support proprietary telephones that work only with a particular manufacturer's PBX.
- ISDN basic-rate interfaces—ISDN line interfaces support telephone sets that are independent of the PBX manufacturer. However, ISDN standards do not define all of the features that the PBX may offer, so ISDN phones are usually tied to a particular manufacturer's PBX.

Line-card density varies between PBX manufacturers. Typical line densities range from 8 to 32 ports.

All PBX systems need one cable pair for analog telephones, but some systems require two cable pairs for digital telephones. The number of pairs required is usually not an issue for installations.

PBX Trunk Interfaces

PBX trunk interfaces connect the PBX with the outside world, including the public network and other private networks. PBX systems typically offer three different types of trunk interface cards:

- Analog trunks—Analog trunk cards typically support from 4 to 16 trunks per card. Analog trunks support two-way CO trunks, foreign exchange lines, WATS lines, and 800 lines.

- Digital (T1/E1) trunks—T1 trunks typically support 24 timeslots; E1 trunks support 30 timeslots.

- Digital (PRI) trunks—Some PBXs use a single card for T1/PRI, while others offer separate cards.

- Tie trunks—Many organizations that operate multiple PBXs link them through tie trunks, which are intermachine trunks that terminate on the trunk side of the PBX. Tie trunks can be privately owned or obtained from a service provider. Tie trunks connect PBX to PBX, even though the actual line may be procured via a service provider.

A separate trunk card is typically required for DID trunks, while some manufacturers provide universal trunk cards.

PBX Wiring Issues

Every PBX must conform to the standard EIA-464 interface, which specifies technical and performance criteria for the interface between the local exchange and the PBX.

In addition to the EIA-464 interface, the following interfaces should be provided in a PBX:

- PRI and BRI interfaces

- Computer telephony integration interface

- Interface to local area networks

- 1.544-megabits-per-second (Mbps) interface to external trunk groups or to internal devices, such as remote access servers, in North America only. In the remainder of the world, one would find 2.048-Mbps interfaces.

All PBXs have a two-wire station interface to a standard analog DTMF telephone.

When designing a system installation, building code regulations, floor distribution, and conduit size are important considerations. Typical floor distribution can be through floor ducts, in the ceiling below the floor served, in the ceiling of the floor served, in a raised computer room–type floor, or flat under a carpet.

When planning a PBX installation, environmental issues are just as important as they are in computer installations. A PBX usually resides in a dedicated PBX equipment room. Additional requirements are:

- Floor space requirements—PBXs can be very large, depending on the number of cabinets included. The cabinet requires sufficient space for the basic cabinet as well as work space to add and remove circuit cards and perform other maintenance tasks.

- Weight support requirements—The PBX equipment room must support the floor-loading requirements of the PBX. Each equipment manufacturer will define the floor-loading requirements.

- Power requirements—Typical power requirements include AC/DC power available within the country. Additionally, each PBX should have a battery backup and an uninterrupted power supply (UPS). These redundancy features are critical, just as the PBX is mission-critical within companies today.

- Heating and humidity—As with computers, heating and humidity regulation are important in the PBX environment.

- Heat dissipation, air conditioning, and ventilation—Heat dissipation and ventilation are important in the PBX environment. Each equipment manufacturer will define these environmental requirements.

Processing Inside the PBX

The PBX needs to handle inbound and outbound calls and switch some calls to voice mail, and do it all in a fast manner. To do these functions the PBX must transparently process several bits of information every second. Here we describe the processing that takes place inside the PBX. In the paragraphs that follow, there are discussions of switching types (circuit, packet, and fast packet), traffic handling and measurements, and signaling.

Circuit and Packet Switching

Circuit switching is the process of setting up a circuit for a call and keeping it open so that two or more users have the exclusive use of the circuit until the circuit is released. Circuit switching provides a temporary physical path for the duration of a telephone call. The entire line or trunk is in use for the entire term of the conversation.

Circuit switching provides a method to connect users by assigning a fixed amount of bandwidth on an exclusive use basis for the duration of the connection. Thus, one path is used for one conversation.

Circuit switching is like having a railroad track that is available for a single train to travel on as long as the track is open. Once that train is gone, the next train gets to use that track.

Normal analog trunks use circuit switching. Thus, when the trunk is seized (usually when a telephone goes off-hook), only one conversation can take place.

Packet switching is traditionally a data-switching technique, but can now be used for other types of traffic. In packet switching, a conversation (which may be voice, video, images, data, etc.) is converted into small packets of information. Each packet is assigned a unique source address and destination address. Each packet may go through a different route to arrive at its destination and packets may arrive in a different order than the order in which they were sent.

Packet switching connects users by routing data from a source to a destination in groups of bits. The packets are transmitted over a shared communications facility, or channel. Thus, one path on a packet-switched network can support multiple conversations.

In packet switching, switches read the addresses assigned to packets and send the packet to the next switch on the network. A switch reads each packet to send it where it needs to go.

When using packet switching for voice traffic, each packet can be sent along a different track that is open. Then packets are assembled at the other end, typically in the last switch, before the packets reach the end user.

Fast packet switching is a wide-area network (WAN) technology that switches data at a very high rate of speed. The packet switch digitizes and places analog data into packets before presenting the data to the network for transport and switching. Common fast packet-switching services include:

- Frame Relay
- Switched Multimegabit Data Service (SMDS)
- Asynchronous transfer mode (ATM)

With SMDS and ATM, the packets are in the form of fixed-length cells (53 octets). With Frame Relay, packets are in the form of variable-length frames (0 – 4096 octets).

The underlying switching technology of a fast packet-switching network is based on the statistical multiplexing of the data contained within the cells or frames. The goal of fast packet switching is to minimize the processing load of the packet switches in an effort to maximize throughput and reduce overall network transit delay.

Fast packet switching is a very effective way to use the available bandwidth on the wide-area network.

Busy-Hour Call Handling

A network that contains fewer paths than terminations is called a *blocking network*. In a blocking network, not all users can make or receive calls at one time. Thus, with high call volume, some users will be denied service.

A nonblocking network makes a connection between any two ports, independent of the amount of traffic. With digital networks where the switching medium is entirely solid state, nonblocking networks are economically feasible; vendors tout nonblocking networks.

A nonblocking switch network does not guarantee that users will not encounter blockage. If the PBX is configured without enough common equipment, users will encounter delays that are viewed as blockage. PBXs can also experience processor overloads, which can result in call-processing delays.

Traffic Handling Measurements (Callout)

PBXs are rated based on the number of *Busy Hour Call Attempts* (BHCA). Since voice networks must handle peak loads, you need to know the heaviest load period, called the busy hour. The busy hour is actually a composite network design point that represents the design peak.

A common factor used to determine busy-hour call handling is the *10-high-day* (10HD) *busy period*, which is determined by averaging the amount of traffic in the busiest hour of the 10 days during the year when the highest traffic load is experienced. The *average bounding busy hour* (ABBH) is often used to size trunk groups. The ABBH is the average of the daily busy hour over the study period, which is usually one week.

The busy hour for the network as a whole may not be the busy hour for all circuit groups or all equipment. Each circuit group will have its own busy hour because the traffic flow varies between nodes and time zones.

The busy hour is always used to calculate the required number of trunks required for an installation. Using a number lower than the busy-hour traffic will result in poor service to subscribers.

The busy-hour traffic gives us a feel of past traffic, but does not reveal how the traffic is offered or what happens when there are no trunks available.

Determining the amount of voice traffic and the number of CO trunks required for a PBX can be a huge task. Several measurements are available to help traffic engineers determine how to provision a PBX and the connections to the service provider.

Erlang is one of the most common measurements of telephone traffic. One erlang equals one full hour of use or $60 \times 60 = 3600$ seconds of phone conversation. Numerically, traffic on a trunk group when measured in erlangs equals the average number of trunks in use during the hour under study. Therefore, if a group of trunks carries 12.35 erlangs during an hour, a little more than 12 trunks were busy, on average.

A *centum call second* represents 1/36 of an erlang. To calculate a centum call second, multiply the number of calls per hour by their average duration in seconds and divide the result by 100. A system port that can handle continuous one-hour calls has a traffic rating of 36 centum call seconds (or about 3600 seconds). Station traffic varies greatly among users, but the typical range is about 6 centum call seconds to 12 centum call seconds per port. When provisioning a PBX, most traffic engineers assume that the mix will be 1/3 internal calls, 1/3 inbound calls, and 1/3 outbound calls, though these amounts can vary significantly. In general, trunk traffic is assumed to be 30 centum call seconds to 36 centum call seconds per port.

Time and talk slots are another important factor in determining capacity. The holding time per call is another very important element. It includes the time required for dialing and ringing (call establishment), conversation time, and the time required to end the call.

Users vary widely in the number of calls they attempt per hour and the length of time they hold a circuit. The attempts and holding time of any user are independent of the activities of other users. To determine traffic capacity, a common method is to plot the number of simultaneous calls on the network to determine the probability that exactly x simultaneous calls will occur. PBX systems may be provisioned to allow the maximum number of simultaneous conversations expected at the busiest time of day.

Signaling Dynamics

In any voice system, some form of signaling mechanism is required to set up and tear down calls. When a caller from an office desk calls someone across the country at another office desk, many forms of signaling are used, including the following:

- Between the telephone and PBX
- Between the PBX and CO
- Between two COs
- Between two PBXs

Each instance of signaling may be—and usually is—different than the next one in the link. For example, the signaling from a PBX to the phone may be FXS, from the PBX to the CO may be what we call "robbed-bit" signaling or channel-associated signaling (CAS), and the signaling from one CO to another will probably be SS7.

Five basic categories of signals are commonly used in a telecommunications network:

- Supervisory—Used to indicate the various operating states of circuit combinations. Also used to initiate and terminate charging on a call.
- Information—Inform the customer or operator about the progress of a call. They are generally in the form of universally understood audible tones (for example, dial tone, busy, ringing) or recorded announcement (for example, intercept, all circuits busy).
- Address—Provides information concerning the desired destination of the call. This is usually the dialed digits of the called telephone number or access codes. Typical types of address signals are DP (dial pulse), DTMF, and MF.

- Control—Interface signals that are used to announce, start, stop, or modify a call. Control signals are used in interoffice trunk signaling.

- Alert—Ringing signal put on subscriber access lines to indicate an incoming call. Signals such as ringing and receiver off-hook are transmitted over the loop to notify the customer of some activity on the line.

Some PBXs support more than 80 kinds of signaling. It is not possible in a short chapter like this to cover all kinds of signaling. The most common kinds of signaling are probably loop start, ground start, wink start, common channel signaling (CCS), QSIG, and Q.931. For a fuller treatment of this subject, see Chapter 2, "Introduction to Analog Technology."

Loop Start or Ground Start

A trunk in a voice network is started, or seized, when it receives a supervisory signal from the phone system. The signal to seize the trunk typically occurs when a user takes the phone off-hook. There are two ways to do that, as shown in Figures 2-3 and 2-4 in Chapter 2:

- Loop start—With loop start, you seize a line by bridging through a resistance of both wires in the telephone line. This typically occurs when a user takes a phone off-hook that connects to the local telephone company CO.

- Ground start—With ground start, one side of the 2-wire trunk is momentarily grounded to seize the trunk. Most PBX systems require ground start. In this case, the CO switch looks for a signal from the PBX before the circuit is seized and the dial tone is delivered.

DID Wink Start

A *wink signal* is sent between two telecommunications devices as part of a handshaking protocol. It is a momentary interruption in the single-frequency tone indicating that one device is ready to receive the digits that have just been dialed.

With a DID trunk, a wink signal from the CO indicates that additional digits will be sent. After the PBX acknowledges the wink, the DID digits are sent by the CO. Wink-start signaling is illustrated in Figure 2-6 in Chapter 2.

CCS

CCS is a signaling method used between computer-controlled switching machines that make up the PSTN. Examples of CCS include Q.931 PRI and BRI, QSIG (based on Q.931), and SS7.

CCS allows carriers on the PSTN to provide intelligent switching services for the future, allowing trunk-terminating equipment to be simpler and less expensive than previous alternatives. Each 64-kilobits-per-second (kbps) time slot will support the signaling needs for up to 1500 trunks. CCS also significantly speeds up call setup times from 8–15 seconds to about 1–3 seconds.

The DTMF tones that a user generates on their handset are converted to PCM samples, which are then propagated in the time slot that becomes the talk path.

QSIG

QSIG provides a powerful way to connect private integrated services network exchange (PINX) equipment in a corporate network. It provides:

- A method for interconnecting PBX equipment from different vendors.
- Synergy with the public ISDN network and business applications developed for the public ISDN network.
- A flexible, low-cost way to link PBX equipment.

The basic QSIG call is based on Q.931, which is defined at Layer 3 of the OSI reference model and in QSIG is further subdivided into three sublayers. The first sublayer is *QSIG Basic Call* (BC), which extends the public ISDN access protocol to use in providing ISDNs. With QSIG BC, both the user side and the network side of the interface are identical to support peer-to-peer operation. The second sublayer in the generic functional protocol is referred to as *QSIG GF*. QSIG GF provides a standard method to exchange signaling information to control supplementary services and additional network features over a corporate network. QSIG GF does not provide supplementary services, but provides generic layer services to specific supplementary service-control entities. The third sublayer is currently in development and will provide procedures for supplementary services, such as hold, call transfer, and conferencing.

Q.931

Q.931 is a message-oriented signaling protocol used in the ISND PRI D-channel. It is also referred to as ITU-T Recommendation I.451. With Q.931, companies can integrate PBX features with the features available in the public network.

Q.931 describes the contents of the signaling packet and defines the message type and content. Specifically, Q.931 specifies:

- Call setup and takedown
- Called party number information, including type of number indication (private or public), as well as privacy and authenticity indicators
- Bearer capability, to distinguish, for example, voice versus data for compatibility checks between telephones
- Status checking for recovery from abnormal events, such as protocol failures or the manual busying of trunks

Services provided using Q.931 include:

- Access to the public network such as wide-area telecommunications service (WATS), direct distance dialing (DDD), dial-800, and other special number services and operator-assisted calls

- Access to and from private networks, tandem PBX networks, and extension-dialing networks

- Integration of voice and circuit-switched data traffic

PBX User and System-wide Features

PBX manufacturers continually add new features to their systems. PBXs can offer over 200 features controlled by the user or by the administrator.

The most popular and most used features to support users include:

- Number and name display (Caller ID)—Displays the phone number and name of the incoming caller on the phone.

- Ring again/callback—When a station is busy, the PBX will monitor the status of a busy extension and the call will be repeated when the recipient is available.

- Call forwarding—Redirects a call to another extension or location when a station is busy or on a ring/no answer.

- Last number redial—Allows a single button to redial the last number called.

- Distinctive ringing—Allows the phone to give distinctive ringing sounds based on the origin of the call. For example, internal calls may ring differently than external calls.

- Third-party conference—Allows a user to conference two or more people into a conference.

Having looked at PBX user features, let us turn now to some of the system-wide features found in PBXs. PBXs from different manufacturers offer different features as a means of product differentiation. Some typical features found on PBXs include:

- DID—A method of enabling callers from outside a PBX to reach an extension by dialing the access code plus the extension number.

- Voice and data integration—For organizations that are large enough to justify the cost of a T1 to a distant PBX, voice and data can be integrated at the facility level by dedicating some channels on a T1 to voice and some to data.

- Automatic call distribution—This feature allows PBXs to route incoming calls to a group of extensions.

- Least-cost routing (LCR)—A feature that lets the system determine the least expensive route for a call and dials the digits to place the call. Also referred to as automatic route selection (ARS).

- Call detail recording (CDR)—Provides the equivalent of a detailed toll statement for PBX users. Sometimes also referred to as station message detail recording (SMDR).

- Power failure transfer—Feature that connects central office trunks to standard DTMF phones and provides an effective way to obtain minimum service when a power failure occurs.

PBX Applications

There are several different application categories that PBX vendors can package with their PBXs or add on later. These packages provide services that add value by making the PBX a tool that increases productivity. Some of the applications in this arena are an automated attendant, automatic call distributor, interactive voice response, voice mail, and computer-telephony integration.

Automated Attendant

Traditionally, an attendant answers incoming calls. An *automated attendant* offers the same functionality, but answers calls automatically using a specialized attendant function. The automated attendant handles the call by transferring it to a process in the PBX that uses voice prompts (see the section "IVR" later in this chapter) to transfer the call to voice mail, a fax server, another automated process, or—typically as a last resort—an attendant (human operator).

An automated attendant answers callers with a digital recording and allows callers to route themselves to an extension or other destination through touch-tone input in response to recorded prompts. An automated attendant can replace a person with an attendant console, thereby reducing overhead costs.

Many automated attendants allow for single-digit dialing. Callers press a single digit, usually introduced by an audio menu, to reach a pre-selected phone, group of phones, or voice mailbox.

ACD

An *ACD* is typically featured on a PBX, although in at least one case it is a specialized type of PBX, and this feature is typically used in what is known as a "call center."

Three characteristics distinguish an ACD from a regular PBX:

- Calls are answered by representatives in the order in which the PBX receives the calls.

- Statistics are available through the ACD that provide information on the level of customer service and the productivity of the people answering the calls.

- Smaller groups of people handle a greater number of trunks, assuming that callers will wait for some period of time for the "next available representative."

ACD systems generally offer inbound-call routing. Inbound calls can be routed to a group, and then routed to a specific agent within a group.

Inbound call routing uses artificial intelligence to recognize and accommodate call center traffic and agent performance.

IVR

IVR refers to the use of a touch-tone telephone to request information from a computer database. DTMF sounds are converted to digital signals understood by the computer. In areas where DTMF penetration is low or where firms want to provide easy access for those who may have physical disabilities, voice recognition IVR systems are also common. It is becoming quite common to hear a prompt that says "Press or say 'one.'" This is an IVR system that permits both DTMF IVR and voice recognition IVR. Digital signals coming back from the computer are in turn converted into a voice that speaks the requested information.

Voice Mail

A *voice-mail* system allows callers to leave a voice message for an extension when the user is on another phone call or away from their phone. A voice-mail system may accompany a PBX or it may be a separate physical system.

Users can retrieve, edit, and forward messages to one or more voice mailboxes in a company. With voice mail, employees typically have their own private mailboxes.

A voice-mail system helps companies and individuals because it can improve communications, eliminate time-zone and business-hour issues, reduce callbacks, improve message content, and offer 24-hour-a-day service.

CTI

CTI allows an external computer to communicate with a PBX, and provide voice and data integration for the end user. The basic components of a CTI system include a PBX and a database/CTI server. IVR systems are often added to enhance functionality.

A typical CTI installation includes a voice system (the PBX/ACD) and a data system (the CTI server).

CTI applications are used today for office productivity, switching-system enhancements, and vertical market applications.

Detailed uses for CTI applications are as follows:

- Office productivity applications include voice messaging and corporate directory dial-in, and integrate voice and electronic e-mail.

- Switching system enhancements include enhanced automatic call distribution, calling feature activation and deactivation, and network performance testing.

- Vertical market applications include customer-assistance call centers, catalog sales, and telemarketing.

Summary

In this chapter, you have learned about the basic functions of a PBX and how it fits into an enterprise, and how it integrates with the PSTN, private leased-line networks, tandem PBX networks, and virtual private networks.

This chapter describes a typical PBX equipment room and the types of switching network design, including centralized control, distributed control, dispersed control, and adjunct control. You learned about the basic architecture and cabinet design of a PBX.

This chapter also describes typical PBX functionality, including switching, traffic handling, signaling, and user and system-wide application features.

Review Questions

The following questions should help you gauge your understanding of this chapter. You can find the answers in Appendix A, "Answers to Review Questions."

1 What are the four primary differences between a PBX and a key system?

_____

_____

_____

2 What are the major components of the PSTN in the North American market and who provides them?

_____

_____

_____

3 What are the four types of switching-network designs for a PBX?

_____

_____

_____

4 What types of circuit cards are typically installed in a PBX and what are their functions?

_____

_____

_____

5 What is busy-hour call handling?

_____

_____

_____

6 What are the main distribution frame (MDF) and intermediate distribution frame (IDF) used for?

_____

_____

_____

7 Where is call processing handled in the client/server PBX design?

_____

_____

_____

8 What functions do line cards provide in a PBX?

_____

_____

_____

9 What are the five categories of signaling in a telecommunications network?

_____

_____

_____

10 What is the purpose of the PBX backplane?

_____

_____

_____

Upon completion of this chapter, you will be able to perform the following tasks:

- Design a voice-over data network using the six steps of the network design process.

- Evaluate voice traffic and select the best voice-over technology (that is, VoFR, VoATM, or VoIP) to support different voice traffic profiles.

- Determine voice quality and voice delay requirements for a voice-over network.

- Given customer case studies, redesign the network using Cisco voice-over solutions and describe the benefits of your solution.

CHAPTER 13

Network Design Guide

This chapter presents a six-step network design process to help you perform an audit of existing voice equipment, set objectives and goals for a network, evaluate available technologies, and formulate a technical network design using Cisco's voice-enabled devices, such as 2600 and 3600 series routers and MC3810 access routers.

This chapter covers all the essential elements needed to connect branch and regional offices with voice-over Frame Relay (VoFR), ATM (VoATM), and IP (VoIP). It also covers a six-step process for designing integrated voice and data networks. This process shows how to choose products based on needs and cost savings. The chapter concludes with three case studies that use the six-step process to show how to integrate networks and the cost benefits of performing the integration.

Using Voice Versus Integrated Voice/Data Networks

Many characteristics are common to voice networks and integrated voice/data networks, such as signaling, addressing, routing, and cost-effectiveness. Voice and data networks share many functions and similarities. One network characteristic exists in both networks but is very significant in voice networks: delay. Delay factors into data networks also, but much less significantly: In a data network, users tend to complain a little about delay, but they learn to live with periodic delays. In a voice network, delay is intolerable; if the delay gets bad, conversations become impossible due to breakup, echo, or delay in the spoken words. The trick to integrating voice and data is a balance of cost-effectiveness and QoS, which means it is important to control delay.

Cost efficiencies in an integrated network are of little value if they come at the expense of satisfactory voice quality. The shared resources must be properly engineered to accommodate the critical requirements of both traffic types.

Not all voice traffic is necessarily as delay-sensitive; for example, fax, voice mail, and modem traffic need not be handled in real time. Adding fax and modem traffic to a data network may be justification alone for creating a voice-over data network.

In a traditional voice network, delay is not a major issue, but it is recognized as a factor affecting voice quality. The largest contributor to delay is propagation delay, which is the period of time required for the electronic signal carrying voice to travel the distance across the physical network medium. When distances are short, propagation delay is negligible. For example, a phone call from Munich to Frankfurt has a 1-millisecond (ms) delay, but a call from Munich to Los Angeles has a much higher delay, 32 ms. However, in an integrated voice and data network, transmission priorities can cause delay. The addition of equipment for packetizing voice, and the delays inherent in the packet network, make managing delay a critical factor in integrating voice and data networks.

Current voice and data networks utilize signaling, addressing, and routing to establish communications between end users: Voice networks have been designed to limit delay and delay variation; data traffic, because it is not as delay sensitive as voice traffic, has been designed with throughput efficiency in mind. These two different goals—reducing delay and reducing delay variation (jitter)—can be accommodated in an integrated voice/data network by reducing delay and delay variation as delay-sensitive and delay-insensitive traffic is mixed onto the same network.

The Network Design Process

This network design process consists of six steps:

Step 1   Audit the current network.

Step 2   Set network objectives.

Step 3   Review technologies and services.

Step 4   Establish technologies guidelines.

Step 5   Plan the network capacity.

Step 6   Conduct financial analysis.

The following sections describe each of these steps in detail.

Step 1: Audit the Current Network

The first step in designing a network is to take stock of what currently exists. It is important to know what the network currently uses for voice and data transport. This will help you determine how to plan the integration and give you some ideas about what your requirements will be.

Some of the things you should do as a part of auditing the current network are as follows:

- Look at the existing equipment and evaluate its capabilities and operating costs.
- Determine the existing facility costs and whether the network will meet the planned voice and data needs.

- Identify upcoming projects that will be required in the network and determine their impact on the network.

- Determine the service quality of both voice and data and whether it needs to be improved.

- Conduct a traffic study to look at current traffic patterns and determine whether links can be removed or need to be increased to integrate voice and data. The study should also determine which types of traffic are used and how much bandwidth they use.

After you have gathered this information, you have a guideline for what the voice and data network is doing and what equipment is being used. You can also identify any upcoming needs for the data or voice network. After you conduct the audit, you should make an outline of the current network and identify the needs and requirements.

Step 2: Set Network Objectives

After you understand your current networking situation, the next step is to set the integrated network objectives, which involves doing the following:

- Determine the dominant traffic type you expect to carry on the integrated network. Also consider how closely you wish to tie voice and data functionality. Estimating these points will help you select the appropriate technologies.

- Set voice quality objectives so that you can design a network with acceptable amounts of delay and compression.

- Determine the level of performance expected of the integrated network. Is it permissible to allow the addition of voice to an existing data network to slightly degrade the performance of the data traffic? If not, the data network will require some degree of upgrading.

- Determine what return on investment or payback period you are trying to meet. This allows you to determine if integrating your data and voice networks will cut cost enough to pay for the new equipment and reduce the overall cost and improve productivity and revenues.

Your project objective, along with your outline of your current network, becomes the basis of the integration project. After you document these needs, it is useful to compile a document that spells out the current state of the network and the plans for integration. This can be done in many formats, including a project proposal. The format is not of major importance; the picture of what you are trying to accomplish and what the current foundation for building the network is like are the most important factors.

Step 3: Review Technologies and Services

After you have set network objectives, the next step is to evaluate available technologies and services, and pick the model and technology that best meets your objectives. Integrating voice and data networks should first include an evaluation of VoFR, VoATM, and VoIP, as described in earlier chapters.

The technology alternatives are as follows:

- Frame Relay—FRF.11/FRF.12 has provisions for supporting voice over a public Frame Relay network. It is also relatively inexpensive and quite common in many parts of the world.

- ATM—ATM is connection-oriented. It was designed to handle time-sensitive traffic such as voice. Its signaling, addressing, and routing allow you to build a network that follows the translation model. The routing function in particular is quite robust, allowing users to build connections based on meeting a certain delay and delay variation.

- IP—IP is connectionless. Development in areas of prioritization, resource reservation, and packet fragmentation are all relatively new. IP, like ATM, has robust signaling, addressing, and routing functionality, which makes the translation model a possibility. The most compelling argument for IP is its integration with current data and multimedia applications. VoIP over Frame Relay and VoIP over ATM are also viable alternatives that provide the desired applications interworking as well as leverage current WAN technologies

The technologies you choose depend on several factors, including cost, availability, and the objectives determined in step 2. You should also keep any future growth in mind when selecting technologies.

Step 4: Establish Technologies Guidelines

During step 4 of the network design process, you review the factors that could affect voice quality and determine guidelines to follow. Elements to consider when designing an integrated voice-data network include the following:

- Hardware and software requirements
- The method for calculating delay budget
- Technical guidelines such as the following:
    - Balance voice quality, delay, and bandwidth
    - Determine acceptable delay and delay variation thresholds
    - Calculate delay for the chosen model
    - Avoid tandem (or multiple) conversions

In examining the requirements, you need to establish the acceptable voice quality. Two common network characteristics that affect quality are delay and delay variation, also known as jitter.

Delay can cause two potential impairments to speech:

- Long delays in conversation cause the receiver to start to talk before the sender is finished. Delay exacerbates the problem of echo, which is the reflection of the original signal to the sender. Echo is indiscernible under low-delay conditions. It is noticeable to the point of distraction when the delay becomes too great.

- Jitter causes gaps in the speech pattern that cause the quality of voice to be jerky. Line quality also affects voice quality.

In the following sections we'll discuss how to establish acceptable voice quality, as well as look at how much delay you can expect as you plan your network. Coding and voice compression methods are the first factors that could affect voice quality. Then we need to get a MOS (Mean Opinion Score) and determine what delay components are involved.

Coding and Compression

We use the term *coding* to refer to the entire process of converting between an analog voice signal and its digital counterpart. Pulse code modulation (PCM) is the standard coding method for representing voice as a 64-kilobits-per-second (kbps) bit-stream.

Compression is the method of reducing the amount of digital information below the traditional 64 kbps. Advances in technology have greatly improved the quality of compressed voice and resulted in a spectrum of compression algorithms. As you work with coding and compression, it is important to understand that multiple conversions from analog to digital or changes in compression schemes affect the quality of the original voice signal.

When dealing with voice compression, a trade-off occurs between the level of voice quality delivered and the bandwidth savings achieved. Using voice compression and therefore optimizing bandwidth can lead to significant monthly cost savings.

PCM is the most common form of digital voice coding available. PCM runs at 64 kbps, provides no compression, and therefore provides no bandwidth savings over a traditional voice channel. It is also the sound-quality benchmark against which we compare voice-compression schemes.

Adaptive differential pulse code modulation (ADPCM) provides various levels of compression. The quality difference between 32-kbps ADPCM and 64-kbps PCM is virtually imperceptible. Some fidelity is lost as the compression increases. Depending on traffic mix, cost savings are generally 25 percent for 32-kbps ADPCM, 30 percent for 24-kbps ADPCM, and 35 percent for 16-kbps ADPCM.

Low delay code-excited linear-predictive (LD-CELP) algorithms model the human voice. Depending on traffic mix, cost savings are generally 35 percent for 16-kbps LD-CELP.

Conjugate structure-algebraic code excited linear predictive (CS-ACELP) provides eight times the bandwidth savings of PCM and four times that of 32-kbps ADPCM. CS-ACELP is a more recently developed algorithm modeled after the human voice. It delivers quality comparable to LD-CELP and 24-kbps ADPCM. Dependent upon traffic mix, cost savings generally run 40 percent for 8-kbps CS-ACELP.

Mean Opinion Score

MOS is a widely used subjective measure of voice quality. As shown in Table 13-1, scores of 4 to 5 are deemed toll quality, 3 to 4 communication quality, and less than 3 synthetic quality.

Table 13-1   *MOS Ratings*

Score	Quality	Description of Quality Impairment
5	Excellent	Imperceptible
4	Good	Just perceptible, not annoying
3	Fair	Perceptible and slightly annoying
2	Poor	Annoying but not objectionable
1	Bad	Very annoying and objectionable

Table 13-2 shows MOSs for varying compression algorithms such as ADPCM and CS-ACELP. These high MOS scores are a result of improvements in the algorithms, along with dramatic increases in the power of the digital signal processors (DSPs).

Table 13-2   *MOSs for Compression Algorithms*

Compression Method	MOS	Delay (ms)
PCM (G.711)	4.1	0.75
32-kbps ADPCM (G.726)	3.85	1
16-kbps ADPCM CELP (G.728)	3.61	3–5
8-kbps CS-ACELP (G.729)	3.9	10
8-kbps CS-ACELP (G.729a)	3.72	10

It is possible to integrate voice and data networks while maintaining high-quality voice. What you need to be concerned with is the trade-off between bandwidth compression and higher delay. When you design networks, bandwidth compression and higher delay must be balanced to ensure overall voice quality.

Figure 13-1 shows another way of looking at voice quality guidelines, as of 1995. Since that time, the MOSs for hybrid coders have risen 0.2 point because of improvements in the technology.

Figure 13-1 *Voice Quality Graph*

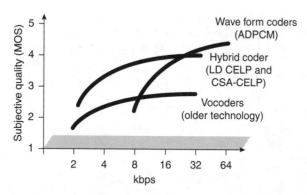

Table 13-3 summarizes the ITU's recommendations for voice delay guidelines. You can see that below 150 ms is considered acceptable for most applications. Delays ranging from 150 to 400 ms are also acceptable, subject to current voice quality.

Table 13-3 *ITU Recommendations for Voice Delay*

One-Way Delay (ms)	Description
0–150	Acceptable for most user applications.
150–400	Acceptable, provided that the administrators are aware of the transmission time impact on the transmissions quality of user applications.
400+	Unacceptable for general network planning purposes; however, it is recognized that in some exceptional cases this limit will be exceeded.

For example, a 200-ms delay from Chicago to New York City is unacceptable, given users' experiences with public networks. However, a 200-ms delay from Chicago to Singapore will likely be acceptable given current conditions. Furthermore, higher delays may be acceptable if they are accompanied by high enough cost savings.

Delay Components

There are many delay components in voice networks. When designing a network, you must take into consideration each of the delay components. While any one may not be a considerable problem, the combination of components can cause considerable delay in the network. These components include fixed-delay and variable-delay components. Once you know the components, a delay budget must be calculated to determine the effect on the voice traffic.

Fixed-Delay Components

As shown in Figure 13-2, three delay components—propagation delay, serialization delay, and processing delay—are fixed in nature and add very little to delay variation.

Figure 13-2 *Fixed-Delay Components*

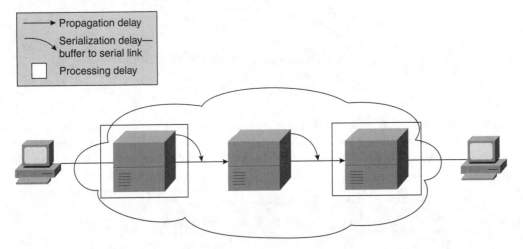

Propagation delay is the delay incurred as a packet travels between two points. Propagation delay is based on the total distance between source and destination. As a planning number, you can use 6 microseconds per kilometer for propagation delay. For example, if a node were 1000 kilometers away, the propagation delay would be 0.006 seconds.

Serialization involves the process of placing bits on the circuit. The higher the circuit speed, the less time it takes to place the bits on the circuit; therefore, the higher the speed, the less serialization delay. For example, it takes 125 microseconds to place 1 byte on a 64-kbps circuit. The same byte placed on an OC-3 circuit will take half a microsecond.

Processing delay can be broken down as follows:

- Coding, compression, decompression, and decoding delay depend on the algorithm used. It is important to note that these functions can be performed in either hardware or software. Using specialized hardware such as DSPs will dramatically improve the quality and reduce the delay associated with different voice-compression schemes. Current voice-over Internet products using software-based compression should not be confused with VoIP.

- Packetization delay is the process of holding the digital voice samples for placement into the payload until enough samples are collected to fill the packet or cell payload.

Variable-Delay Components

The delay components depicted on Figure 13-3—queuing delay and de-jitter buffer—are variable in nature and result in higher delay variation than fixed delays; they are also more controllable than fixed delays.

Figure 13-3 *Variable-Delay Components*

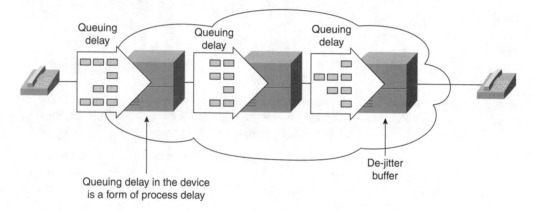

Queuing delay occurs when a packet is waiting for others to be serviced first on the trunk. This waiting time is statistically based on the arrival of traffic; hence, the more inputs, the more likely we will encounter contention for the trunk. It is also based on the size and priority of the packet currently being serviced.

De-jitter buffers are used at the receiving end to smooth out delay variability and allow time for decoding and decompression. They also help on the first talk spurt to provide smooth playback of voice traffic. Setting these buffers too low will cause overflows and loss of data. Setting them too high will cause excessive delay. In effect, a de-jitter buffer reduces or eliminates delay variation by converting it to fixed delay.

Calculating Delay

When the fixed and variable delay components have been identified, you can calculate your delay budget. The *delay budget* is the maximum amount of delay you can have in your planned network while still holding to your voice quality objectives.

Figure 13-4 shows a sample network for calculating a delay budget. For this example, assume that the delay budget is 200 ms. This example is of a private Frame Relay network running over leased lines. For a public service such as public Frame Relay, you should contact the service provider for its guaranteed delay figures and use those figures in your delay budget.

Figure 13-4 *Delay Budget Example*

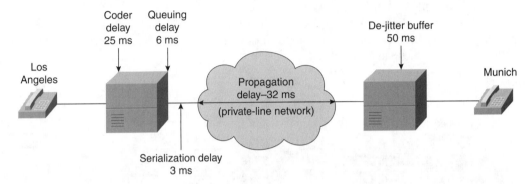

In this example, we are filling a Frame Relay frame with two 10-byte samples. The coder delay for G.729 voice compression is an initial 5 ms for a look-ahead, plus 20 ms for the two 10-byte frames. Each of the delays involved is listed below:

- Packetization delay is the rate at which we fill a packet, which is typically governed by the speed at which voice samples are created. Standard voice is transmitted at 64 kbps, or 1 byte every 125 microseconds. The delay for this transmission is included in the coder delay.

- Queuing delay is variable and occurs during periods of congestion, and with a 64-kbps trunk will equal 3 ms per 20-byte packet already in queue. We used two voice packets in queue for a total queuing delay of 6 ms. However, this assumption will be variable and is cumulative based on the number of devices in the network.

- Serialization delay is the time it takes to place a packet onto the 64-kbps trunk.

As mentioned earlier in the chapter, propagation delay on a 64-kbps private line from Los Angeles to Munich is 32 ms.

The de-jitter buffer is set to twice the coder delay as a general rule to account for the decoding and decompression delay as well as any variable delays in the network. By increasing the de-jitter delay, there is less chance for variation in the delays. Our variable delay is only 6 ms, but a public packet network such as Frame Relay or the Internet can have much higher variable delays. It pays to be able to vary the size of the de-jitter buffer.

Totaling up these figures, as shown in Table 13-4, we get 110 ms of fixed delay, which is well within the delay guidelines set forth by the ITU and our planning number of 200 ms.

Table 13-4   *Budget Delay 1 for Frame Relay Example*

Component	Fixed Delay	Variable Delay
Coder delay G.729 (5 ms look-ahead)	5 ms	
Coder delay G.729 (10 ms per sample)	20 ms	
Queuing delay 64-kbps trunk		6 ms
Serialization delay 64-kbps trunk	3 ms	
Propagation delay (private lines)	32 ms	
De-jitter buffer	. 50 ms	
Total	**110 ms fixed delay**	

Figure 13-5 shows the delay effects of switching voice calls through a central-site tandem PBX. This example is of a private Frame Relay network running over leased lines. The speed of the Site A to Site B link is 64 kbps. The link from Site B to Site C is E1. The first hop is a duplicate of the example in Figure 13-4. So the total for Delay 1 is 110 ms.

Figure 13-5   *Tandem PBX Delay*

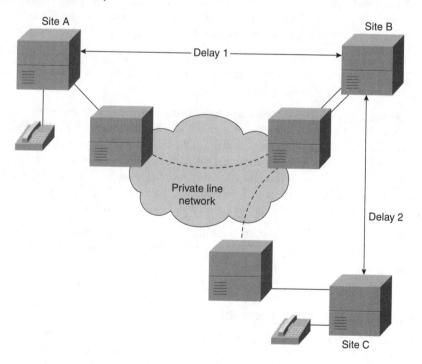

Table 13-5 shows Delay 2 from Site B to Site C. This delay would then be added to Delay 1 to give us the overall delay for the call.

Table 13-5 *Budget Delay 2 for Frame Relay Example*

Component	Fixed Delay	Variable Delay
Coder delay (G.729)	25 ms	
Packetization delay (included in coder delay)		
Queuing delay 2-Mbps trunk		0.2 ms
Serialization delay 2-Mbps trunk	0.1 ms	
Propagation delay (private line)	5 ms	
De-jitter buffer	50 ms	
Delay 2 total	**80 ms**	
Delay 1 total	**110 ms**	
Combined total	**190 ms**	

Notice that Delay 2 is calculated in the same manner as Delay 1, and Delay 2 totals 80 ms. Delay 2 is less than Delay 1 because the distance is shorter from Site B to Site C and because the speed of the private line is 2 Mbps versus 64 kbps.

The total for both hops equals 190 ms, which is in the category "acceptable provided that the administrators are aware of the transmission time impact," as shown in Table 13-3, and it meets the planning delay budget of 200 ms.

Figures 13-4 and 13-5 show the impact of using a tandem PBX at Site B. With the tandem PBX, the voice has to be broken out or decoded at the tandem PBX. If you were to switch the voice without needing to break it out at the tandem PBX, you could lower the aggregate delay. Without the breakout, the aggregate delay could be reduced by 75 ms, lowering the total delay to 115 ms. (Note that the delay associated with the tandem PBX itself has not been included in this estimate. This delay will vary depending on vendor and would have been a part of the overall delay in a traditional voice network.)

As in all designs, a balance must be struck between quality and cost. Given the large MOS quality improvements in today's compression techniques, achieving this balance is easier than ever.

Step 5: Plan the Network Capacity

Step 5 of the network design process involves capacity planning for the integrated voice/data network. First, you need to determine the links that will be needed to interconnect the sites. *Line provisioning* means establishing the number of trunks from the PBX to the integrated voice/data network. This is done based on the requirements of the network determined in earlier steps. Once you establish the number of trunks, the next step is to translate that to the required network bandwidth.

The correct number of PBX or key system trunks is determined by:

- Traffic volume and flow
- The selected grade of service (or blocking factor)
- Your objectives

Some organizations will simply move certain trunks from their current network to the integrated network. Others will take this opportunity to update their traffic engineering information and choose to conduct a traffic study. Either approach can work and is very familiar territory for voice-engineering professionals.

The use of the transport or translate model can have an impact on the number of trunks simulated by the network. The transport network model matches a virtual connection for a tie line on a one-for-one basis. From a voice engineer's standpoint, nothing in the overall network structure has changed. However, the translate model uses the network to simulate a tandem PBX, thereby potentially reducing the number of trunks required.

The voice engineer routes calls over specific trunk groups in which one group could be the integrated voice-data network. One engineer may choose to use the network as the first option, and another may use it as the last choice.

You can calculate the required bandwidth based on the proposed network design and the required number of trunks between locations. Bandwidth calculations should take into account compression, overhead, and utilization. Each of these elements will vary, depending on which voice-over data network technology is chosen.

You need to calculate the delay matrix between locations and confirm that the delay calculations meet the delay budget requirements. If they do not, adjust the bandwidth, or select a different voice-over data network technology. For example, if you are using a technology such as IP to carry the voice traffic but the delays caused by intermediate devices are too high, you may want to choose a technology such as ATM or Frame that provides PVCs for transport and requires less traffic manipulation along the path.

Step 6: Conduct Financial Analysis

After you have completed the first five steps, you must determine whether the cost is justified in relation to the benefits received. Once you've worked through some different aspects of financial analysis, you can calculate the return on the network investment.

To determine the cost benefits, you look at the current costs of data and voice lines, the cost of the equipment, and the cost of maintenance on the equipment. Then you compare these costs to that of the single links that will be used for voice and data as well as the cost of any new equipment and maintenance savings. Typically, the new equipment cost will be heavy but the savings of recurring cost over the life of the project should outweigh any investment in the product. This is the most crucial step in determining whether to integrate data and voice and it should be carefully considered. The case studies in the following sections show how this analysis is done.

Network Design Case Studies

This section describes three different case studies to show the concepts covered so far in this chapter. By using three different technologies (Frame Relay, ATM, and IP), standard voice-engineering principles, actual PSTN and VPN toll charges, and private-line or Frame Relay service costs, you will see the financial benefits of the various solutions.

For all three case studies, the traffic volume, patterns, and assumptions have been held constant. Only the regional PSTN, Frame Relay, or private-line prices vary from one case to another.

Each case represents a main office with several branch offices. The cases explore how to integrate data and voice to provide cost savings to the corporation. The six steps discussed earlier in this chapter are used in the case studies to show how to integrate a voice and data network.

The networks presented in the case studies are overengineered and have more bandwidth than required. Overengineering the network was done to show that the added voice traffic would not negatively affect data traffic. In fact, given that voice and data traffic often peak at different times of the day, the data network will likely perform better with the additional bandwidth that is added for the voice integration.

Case 1: A Global Firm

The global firm described in this case study has a corporate headquarters and five smaller offices that use E1s, T1s, and fractional private lines. For historical reasons the firm has two separate private TDM-based networks: one transporting data traffic and the other transporting voice traffic. Additionally, a relatively new requirement, videoconferencing, has been implemented using the public ISDN network at 384-kbps rates.

Case 1: Auditing the Original Network Design

There are approximately 30 to 50 people working at the corporate headquarters site. In each of the five regional offices there are 8 to 15 people. Figure 13-6 shows the original network design.

Figure 13-6 *Global Firm Original Network Design*

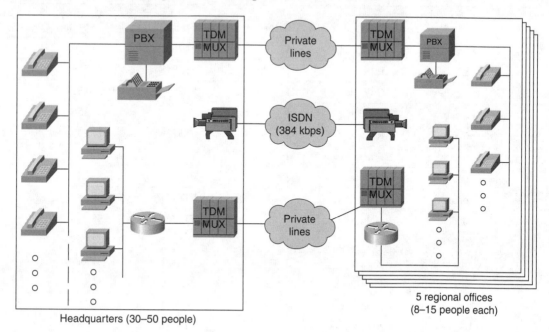

Headquarters (30–50 people)

5 regional offices
(8–15 people each)

Case 1: Setting Network Objectives

The objective of this firm is to reduce costs of the network by merging the three networks as efficiently and cost-effectively as possible while maintaining voice quality. The company would also like to reduce reliance on tandem PBXs for switching its voice calls. Reducing reliance on PBXs would avoid the decompression/recompression cycle associated with the tandem PBX, allowing higher voice-compression ratios and resulting in further reductions in bandwidth expenses.

Case 1: Reviewing Technologies and Services

After reviewing the different voice-over data alternatives, an ATM network was selected because of the large capacity requirements, the videoconferencing needs, and the intent for the network to add voice switching. ATM reduces the company's reliance on tandem PBXs.

Maintaining separate networks is very costly, particularly when the networks are very large. Focusing only on bandwidth costs, Table 13-6 shows the voice network expenses.

Table 13-6 *Case 1 Voice Backbone Configuration and Expense*

Source	Destination	Quantity	Speed	Monthly Expense
London	Manchester	1	FT768K	$4,460
Paris	Lyon	1	FT384K	$6,516
Frankfurt	Cologne	1	F-1920	$10,314
Frankfurt	Munich	1	E1	$9,156
Milan	Rome	1	FT512K	$13,338
London	Frankfurt	1	T1	$40,452
London	Philadelphia	1	E1	$47,950
London	Toronto	1	FT512K	$21,822
London	Dublin	1	FT256K	$8,276
Amsterdam	Paris	1	FT512K	$21,672
Amsterdam	Zurich	1	FT512K	$17,179
Amsterdam	Philadelphia	1	F-1984	$48,581
Amsterdam	Rome	1	FT768K	$32,837
Amsterdam	Manchester	1	FT768K	$22,374
Amsterdam	Cologne	1	FT768K	$19,228
Amsterdam	Budapest	1	FT512K	$29,001
Amsterdam	Oslo	1	FT384K	$12,573
Frankfurt	Milan	1	FT768K	$30,967
Frankfurt	Copenhagen	1	FT384K	$17,377
Frankfurt	Zurich	1	FT384K	$17,362
Frankfurt	Lyon	1	FT512K	$20,259
Copenhagen	Oslo	1	FT256K	$5,275
Philadelphia	Toronto	1	FT512K	$8,829
Manchester	Dublin	1	E1	$22,093
Budapest	Munich	1	FT768K	$19,228
TOTAL		**25**		**$507,119**

Table 13-7 displays the data backbone configuration and monthly expense.

Table 13-7   *Case 1 Data Backbone Configuration and Expense*

Source	Destination	Quantity	Speed	Monthly Expense
Cologne	Frankfurt	1	F-1920	$10,314
Cologne	Munich	1	F-1920	$15,158
Paris	Lyon	1	FT256K	$4,117
Frankfurt	Munich	1	E1	$9,156
Milan	Rome	1	FT256K	$8,927
Amsterdam	Philadelphia	1	F-1984	$38,581
Amsterdam	London	1	FT512K	$17,281
Amsterdam	Copenhagen	1	FT256K	$10,130
Amsterdam	Zurich	1	FT512K	$17,179
Amsterdam	Philadelphia	1	F-1984	$48,581
Amsterdam	Rome	1	FT256K	$15,785
Amsterdam	Manchester	1	FT512K	$17,281
Amsterdam	Budapest	1	FT512K	$29,001
Amsterdam	Oslo	1	FT256K	$8,867
Philadelphia	Toronto	1	T1	$12,277
Cologne	Paris	1	FT256K	$12,269
Cologne	Zurich	1	F-1920	$41,382
Cologne	Lyon	1	FT256K	$12,269
London	Dublin	1	FT256K	$8,276
Milan	Munich	1	FT256K	$18,017
Copenhagen	Oslo	1	FT256K	$5,275
Philadelphia	Toronto	1	T1	$12,277
Manchester	Dublin	1	E1	$22,093
Budapest	Munich	1	FT768K	$19,228
TOTAL		**24**		**$413,721**

As you can see from Tables 13-6 and 13-7, the combined monthly expense for voice and data networks is $920,840.

Case 1: Establishing Technologies Guidelines

By redesigning the network, the different types of traffic can be consolidated across a single redundant infrastructure based on ATM cell switching. Figure 13-7 shows the new network design.

Figure 13-7 *Global Firm Network Redesign*

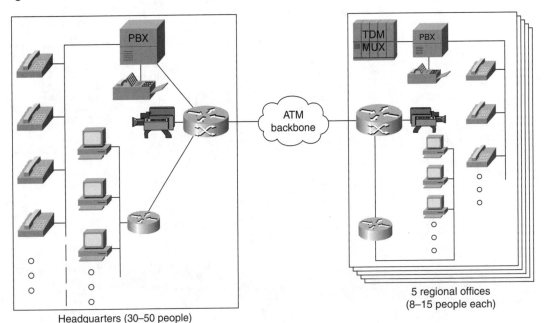

During the redesign process, the firm migrated from TDM networks to take advantage of the efficiencies an ATM network affords. Both voice compression and closed-loop congestion traffic management are implemented to maximize bandwidth utilization. The firm also employed techniques such as VAD to further increase utilization. Given the efficiency gained by deploying this network, voice traffic essentially rides for free.

Numerous card/port types and speeds are supported across the network. The company in Case 1 now has the flexibility to deploy new applications to users quickly.

Case 1: Planning Network Capacity

When planning the network capacity, you need to be aware of how the bandwidth will be utilized. Voice conversations are half-duplex by nature. For telephone conversations, that translates to 60 percent of a 64-kbps voice channel containing silence, because users listen and pause between sentences.

As shown in Figure 13-8, no cells are sent across the link during periods of voice inactivity. VAD allows traffic from other voice channels or data circuits to make use of the silence on expensive WAN bandwidth and provides for additional savings beyond that achieved by voice compression.

Figure 13-8   *VAD for Case 1*

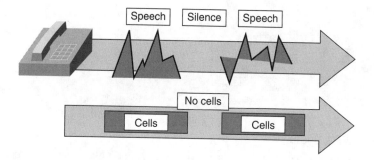

The effects of VAD are highest with the addition of more channels, because the statistical probability of silence increases with the number of voice conversations being combined.

The company in Case 1 has implemented VAD to detect the presence of speech. If a cell does not contain speech, it is not transmitted through the network. So, by implementing VAD together with voice compression, the firm realized even more savings.

Because the link will carry many types of traffic, as shown in Figure 13-9, it is important to manage the traffic in a way that provides optimal bandwidth utilization. Rate-based traffic management helps prevent network congestion from occurring and serves to minimize cell discards and retransmissions.

The switch monitors the network for congestion every 40 ms and adjusts the data admission rate to a speed the network can support. This rate-based traffic management ensures optimal bandwidth utilization.

Figure 13-9 *Optimal Bandwidth Utilization*

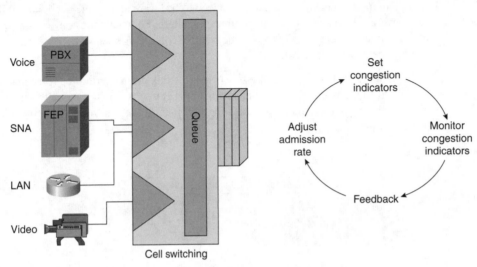

Case 1: Financial Analysis

By redesigning the network to carry voice, data, and video, and implementing special functions in the network, the company reduced its costs by more than 30 percent while preparing itself for growth.

Figure 13-10 summarizes the financial analysis. The capital cost for redesigning the existing network to an ATM network is about $1,205,000 including equipment installation, and the payback period is four months.

The old network cost $920,840 per month to operate. The new combined voice and data ATM network costs $620,280 per month, which represents a savings of $300,560 per month. By combining its voice network with its data network, the company reduced its original yearly network costs from $11 million to about $7.5 million. Table 13-8 shows the redesigned ATM network's monthly bandwidth cost.

Figure 13-10 *Financial Analysis for the Global Firm*

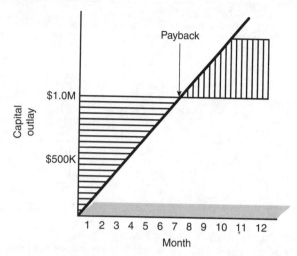

Table 13-8 *Case 1 Redesigned ATM Network Expenses*

Source	Destination	Quantity	Speed	Monthly Expense
Paris	Lyon	1	FT768K	$11,835
Frankfurt	Munich	1	F-1920	$12,776
Milan	Rome	1	FT384K	$19,124
Amsterdam	Philadelphia	1	F-1984	$48,581
Amsterdam	Zurich	1	F-1984	$44,952
Amsterdam	Rome	1	FT512K	$35,993
Amsterdam	Manchester	1	FT768K	$22,374
Amsterdam	Budapest	1	F-1024	$23,456
Amsterdam	Oslo	1	F-1984	$31,357
Philadelphia	Toronto	1	T1	$12,277
Cologne	London	1	F-1024	$38,073
Cologne	Paris	1	F-1024	$32,831
Cologne	Milan	1	FT512K	$32,082
Cologne	Copenhagen	1	F-1920	$41,578

continues

Table 13-8   *Case 1 Redesigned ATM Network Expenses (Continued)*

Source	Destination	Quantity	Speed	Monthly Expense
Cologne	Zurich	1	F-1920	$41,382
London	Toronto	1	T1	$44,090
London	Dublin	1	FT768K	$16,412
Frankfurt	Lyon	1	FT768K	$27,054
Copenhagen	Oslo	1	T1	$17,280
Manchester	Dublin	1	E1	$22,093
Budapest	Munich	1	F-1920	$44,680
TOTAL		**21**		**$620,280**

To determine the payback period, you compute the net monthly savings; in this case, it is simply (old monthly expenses) – (new monthly expenses) = net monthly savings. Once you have determined the monthly savings to get the payback period, divide the total capital cost by the monthly savings. Table 13-9 shows the payback period calculation for Case 1.

Table 13-9   *Case 1 Savings and Payback*

Savings/Costs/Payback	Amount
Monthly data and voice expenses	$920,840
Redesign expenses	$620,280
Net total monthly savings	**$300,560**
Net annual savings	**$3,606,720**
Total capital costs	**$1,205,000**
Payback period (months)	**4.0**

Case 1: Summary

To summarize, Case 1 has shown that after redesigning the network, a 30 percent annual expense reduction was achieved, with no reduction in voice quality. Because of the reduction in equipment and improved robust network design, network availability and manageability have increased.

Case 2: Pan-European Firm

A medium-sized firm has nine locations throughout Europe. The firm's headquarters is in Brussels, and it has offices in eight different countries. Each office is connected to a public

Frame Relay network for data and a VPN for voice networking. To be conservative, we have assumed that the firm generates enough call volume to obtain a VPN contract from a carrier at about a 15 percent discount off standard PSTN rates. Figure 13-11 shows the connections for the pan-European firm.

Figure 13-11   *Pan-European Firm Original Connection Configuration*

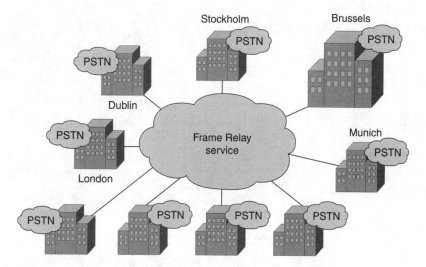

The majority of calling in the firm occurs between branch employees and outside local customers. Most of its international long-distance traffic is from branch offices to Brussels. This is the most expensive traffic on a per-minute basis.

The company's current Frame Relay network is adequate to handle today's data traffic only.

Case 2: Auditing the Original Network Design

There are 40 to 70 people in each regional office. On average, each person uses the phone 1.5 hours per day. The majority of branch calls are between branch employees and local outside customers. Figure 13-12 shows the firm's original network design.

Only 20 percent of the calls are between the offices and headquarters. The interoffice calls amount to about 15 minutes per day between the headquarters and each branch per month per person. The calls also include fax traffic. Although the branch-to-headquarters traffic volume is lower, it is also the most expensive on a per-minute basis because it is billed at international rates. As a result of these high rates, this firm pays more than $50,000 per month or about $600,000 per year for international long-distance services.

Figure 13-12 *Pan-European Firm Original Network Design*

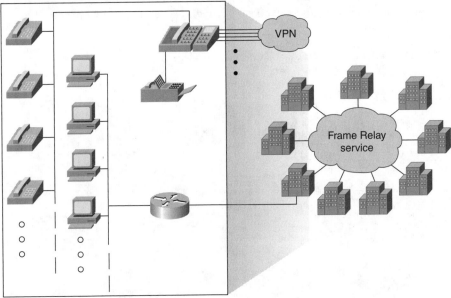

Regional offices: 40–70 people each

The firm is using a public Frame Relay service. Voice services are provided by small PBXs connected over the PSTN. To ensure the financial analysis is conservative, it is assumed that the firm's call volume is large enough to obtain a VPN contract from a carrier at about a 15 percent discount from standard PSTN rates.

Case 2: Setting Network Objectives

The pan-European firm's goal is to reduce long-distance toll charges between the offices and headquarters. The firm is also interested in migrating to an ATM service when it becomes available.

Case 2: Reviewing Technologies and Services

Any of the packet voice technologies could be used in this case to build the network. But given its existing Frame Relay network, the company elected to use VoFR technologies. The company opts for an integrated access solution for VoFR functionality to bypass the PSTN for its interoffice calls.

Table 13-10 shows the estimated on-net voice and fax traffic volume and PSTN expense for the existing network.

Table 13-10 *Case 2 PSTN Volume and Expenses*

Location	Number of People	Average Minutes per Person per Day	On-Net % to HQ	Workdays per Month	Total Minutes per Person per Month	Total Minutes per Office per Month	Cost per Minute*	Monthly Cost per Office
Stockholm	45	90	20	21.67	390	17,553	$0.39	$6,846
Milan	55	90	20	21.67	390	21,453	$0.34	$7,294
London	50	90	20	21.67	390	19,503	$0.23	$4,486
Madrid	40	90	20	21.67	390	15,602	$0.39	$6,085
Paris	70	90	20	21.67	390	27,304	$0.26	$7,099
Munich	70	90	20	21.67	390	27,304	$0.26	$7,099
Geneva	65	90	20	21.67	390	25,354	$0.35	$8,874
Dublin	45	90	20	21.67	390	17,553	$0.36	$6,319
TOTAL						**171,626**		**$54,101**

* This is the average cost per minute and assumes that 50 percent of calls are to headquarters and 50 percent from headquarters. Cost of a voice call based on carrier quote, off-net pricing with discount.

Reviewing the Data Network

The exact configuration of the data network and the ongoing monthly expenses, or run-rate, are shown in Table 13-11. Note there is a single PVC between each branch and headquarters.

Table 13-11 *Case 2 Current Data Network Configuration*

Location	Purpose	Access Line Speed	Initial Port Speed	Initial PVC CIR	Frame Relay Charges*
Brussels	Headquarters	E1	768	—	$2,800
Stockholm	Branch	E1	128	64	$2,690
Milan	Branch	E1	128	64	$3,010
London	Branch	E1	128	64	$3,010
Madrid	Branch	E1	128	64	$3,170
Paris	Branch	E1	128	64	$3,010

continues

Table 13-11 *Case 2 Current Data Network Configuration (Continued)*

Location	Purpose	Access Line Speed	Initial Port Speed	Initial PVC CIR	Frame Relay Charges*
Munich	Branch	E1	128	64	$3,010
Geneva	Branch	E1	128	64	$3,110
Dublin	Branch	E1	128	64	$3,110
TOTAL					**$26,920**

* The port cost and the PVC cost are included in the Frame Relay Charges column.

Case 2: Establishing Technologies Guidelines

The first step is to determine the additional bandwidth required on the data network to support the voice and fax traffic. Determining the required bandwidth can be done in one of two ways:

- The best way is to collect traffic information from both the key system or PBX and the router, and graphically add the voice and data traffic to see how often the combined voice and data traffic exceeds the current available bandwidth. However, this kind of traffic information is often unavailable or difficult to obtain.

- If this information is not available, the easiest method to determine the proper upgrade is to estimate whether (and how much) extra bandwidth is required, provision that, and then add the voice traffic to the data stream while tracking two measures of performance: user-reported voice quality and data latency. If either performance measure appears to be suffering, then more bandwidth should be added.

Often, data and voice traffic peak at different times during the day. Consequently, the data network will frequently benefit from the added bandwidth.

The network was redesigned considering the following assumptions:

- The total voice and fax call volume per person equals about 90 minutes per day.

- The call volume between headquarters and branch personnel represents about 20 percent of the total call volume, or about 15 minutes. Plus, a peak-load-hour loading factor of 17 percent is appropriate.

- The Cisco voice compression module uses 8 kbps plus 3-kbps overhead (11 kbps per voice channel); thus, each 64-kbps trunk supports five voice channels.

- The branch PBXs require additional trunk modules.

Trunking Analysis

The amount of voice and fax traffic at each branch office that would optimally be diverted from the PSTN to the multiservice network is determined using the assumptions and the appropriate traffic engineering tables, given the desired P (Blocking Probability) grade of service. This firm chose a P.05 grade of service. Table 13-12 summarizes the calculations and provides the number of PBX trunks that would be required at each site.

Table 13-12 *Case 2 Trunking Analysis*

Number of Users	Number of Sites	Hours/ Day on Phone	Minutes /Day	Busy Hour (17%)	Minutes / Busy Hour	% Traffic to HQ	Total Erlangs * to HQ	Required Lines for 0.05 Blocking Probability
40	1	1.5	3600	0.17	612	20	2.04	6
45	2	1.5	4050	0.17	689	20	2.30	6
50	1	1.5	4500	0.17	765	20	2.55	6
55	1	1.5	4950	0.17	842	20	2.81	7
60	0	0	—	0	—	0	0	0
65	1	1.5	5850	0.17	995	20	3.32	8
70	2	1.5	6300	0.17	1,071	20	3.57	8
						Total Erlangs to HQ	**22.44**	**31**

* An erlang is a unit of traffic measurement for telecommunications. An erlang is used to represent the continuous use of one voice path. Erlangs are used to describe the total traffic volume of one hour. For more information, visit www.erlang.com/whatis.html.

Based on this trunking analysis, the company determined the additional bandwidth required on the Frame Relay network to support the voice and fax traffic. From the assumptions, each 64 kbps of bandwidth can provide a minimum of five voice channels.

Case 2: Planning Capacity

The company opted for a multiservice integrated access solution. It chose an MC3810 access router to transport interoffice voice traffic across the Frame Relay network.

Figure 13-13 shows the Frame Relay topology, and you can see that the network was redesigned to use a single PVC configured to carry both data and voice traffic. The network was redesigned to keep expenses down because public Frame Relay carriers typically charge on a per-PVC basis. To successfully accomplish the redesign, the MC3810 access router has been configured to prioritize voice and data traffic.

Figure 13-13   *Pan-European Firm Redesigned Network Topology*

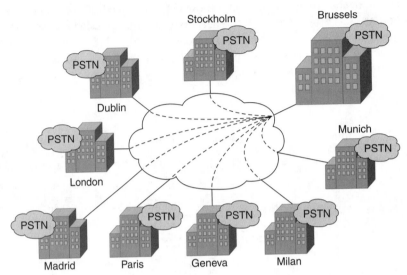

All circuit and port speeds were increased, as was the PVC CIR. An additional headquarters access line was installed to support the added traffic.

Note that voice traffic does not pass through the router portion of the MC3810 access router, but instead through the switch. Consequently, in this case, the voice traffic is not encapsulated in IP. Of course, the data traffic does go through the router because it is IP traffic.

Regional Offices Network Redesign

At the regional offices, the lines going from the key system to the MC3810 are the PBX trunks that are moved from the VPN to the MC3810 to accommodate the voice traffic. The MC3810 is configured for VoFR using bundled DLCIs and fax relay.

Figure 13-14 shows the redesign of each of the regional offices. The PBX least-cost routing tables are configured to preferentially select MC3810 trunks for on-net voice and fax.

Meanwhile, the PSTN continues to carry local calls and back up voice traffic on the Frame Relay network. On-net fax traffic is also routed across the Frame Relay network.

As a conservative measure, to accommodate added voice traffic, the Frame Relay network access line speeds were increased, as were the CIRs. These expenses were included in the financial model along with estimates for additional PBX or key system modules that might be required.

An estimated installation charge for PBX, router, and so forth also was included to cover that expense.

Figure 13-14 *Pan-European Firm Regional Office Redesign*

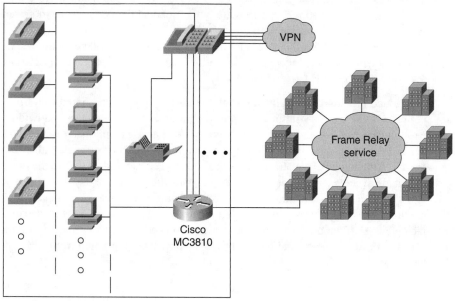

Regional offices: 40–70 people each

Network Upgrade

Recall that each branch office was originally linked to the Frame Relay network via a 128-kbps Frame Relay port (using an E1 access line), with the PVC CIR set to 64 kbps. The original link was sufficient to handle the original data traffic. However, it will not be enough to carry the added compressed voice traffic. The bandwidth at each location also must be upgraded.

The company decided that because it was redesigning its voice/data network, this would be an ideal time to upgrade its network at all regional and branch offices.

Upgrade Considerations and Strategy

The company will make a major change in its voice and data network, but rather than conduct traffic studies to determine the minimum upgrades each location would require, it was felt that upgrading all locations to the same high bandwidth level would provide a conservative approach and leave room for growth in either voice or data traffic.

The upgrade is based on the largest branch offices in the company. The largest branch offices, containing 70 people, were reengineered to direct a maximum of 8 voice lines of traffic over the Frame Relay network. At its maximum, this data stream could reach 88 kbps (because each voice line can load a maximum of 11 kbps). The branch port speeds were doubled from 128 kbps to 256 kbps, far more than the maximum added voice load, to provide improved performance (decrease buffer delay at the port) for the more sensitive voice traffic, and to leave room for growth. The PVC CIR was increased from 64 to 128 kbps to ensure that delay due to vendor network congestion would be minimized (although as always with a Frame Relay network, close watch must be kept on latency measurements to ensure that the PVC CIR is set properly).

This added Frame Relay bandwidth is also certain to improve current data traffic performance. The voice traffic, even at its maximum, still takes up less bandwidth than the new doubled port speed delivers. And voice traffic, being exceptionally variable over time, leaves the new doubled Frame Relay pipe open for more and faster data traffic for most of the day. Although the voice traffic at its peak adds an extra 88-kbps load to the Frame Relay circuit, on average it adds less than 30 kbps of data throughout the day. The rest of the added 128-kbps bandwidth is available for improved data performance. If more savings are required, the smaller branch locations' circuit and port speed could be upgraded in a smaller increment. Half the branches under consideration needed only 5 or 6 voice circuits to be added to the data stream; adding 64 kbps to the port size (upgrading from 128 kbps to 192 kbps) would have been more than sufficient.

As long as the business case could support it, however, it was preferred to upgrade the network once and then reduce bandwidth, and expenses, over time as experience is gained with the new traffic patterns.

Headquarters Network Redesign

At headquarters, tandem switching is configured on the Cisco MC3810 to allow voice traffic from one office to transit through, and terminate at a different regional office's PBX. Switching through the router eliminates the need to tandem through the PBX. Figure 13-15 shows the redesign of the HQ network.

Upgraded Network and Expenses

The company had the MC3810 installed in the branches and the headquarters. The port speeds and the CIR of the PVCs were increased to accommodate the voice traffic. Adding a separate PVC for the voice was unnecessary because the MC3810 prioritizes the traffic on a single PVC, thus avoiding the cost of another PVC.

Each branch office connected the appropriate number of PBX trunks to the Cisco MC3810. The PBX can direct approximately 95 percent of traffic destined for headquarters preferentially to one of the trunks connected to the Cisco MC3810. The remaining overflow traffic, an estimated 5 percent, is directed to the PSTN.

Figure 13-15 *Pan-European Firm Headquarters Network Redesign*

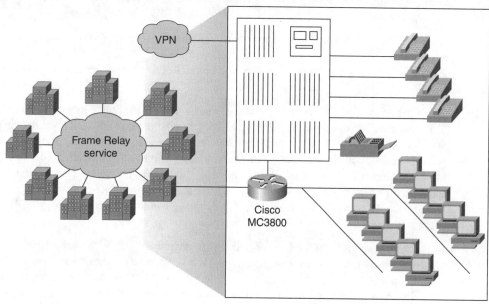

Regional offices: 40–70 people each

By using G.729 encoding, the MC3810 can compress each 64-kbps voice channel to a data stream of approximately 11 kbps (8 kbps or less plus 3-kbps overhead) and forward the compressed voice traffic over the Frame Relay network.

NOTE   The 11 kbps is a conservative measure because it does not include the benefits of the silence-suppression techniques implemented in the MC3810. The MC3810 can support up to 24 channels of 8-kbps compressed voice that can be transported over public or private Frame Relay, ATM, or leased-line networks.

Figure 13-15 shows the highlights of the upgraded network using an MC3810 for Frame Relay. There is an additional trunking option called *multiflex trunking*.

Option: A Second PVC

The MC3810 was configured to support tandem switching of the branch-to-branch voice traffic. Tandem switching relieves the central site PBX of this task and avoids the PBX recompression cycle to maintain high-quality voice. This savings was not included in the financial analysis.

Some network services are not available in all locations. For those areas of the world preparing to use ATM services, the Cisco MC3810 can be modified via a software change to support ATM.

A number of different network designs were available to this firm. One option would be to add a second PVC to transport the voice traffic. However, the second PVC option would also result in higher expenses because public Frame Relay providers commonly charge for each PVC.

A less expensive method would be to utilize the same PVC for both data and voice traffic. The dual-use PVC option means sharing a properly engineered and sized PVC between voice and data. But PVC sharing can be done only if the access device supports the prioritization of real-time traffic, voice, and video, over non-real-time traffic. This design is available when the firm uses an MC3810 because this access router can prioritize the delay-sensitive traffic (that is, voice) while ensuring that the delay-insensitive traffic (that is, data) is also serviced properly. Prioritizing real-time traffic would avoid to some degree the potential contention between the two different traffic types.

In addition, Cisco IOS technology provides voice switching and advanced call management. For example, the headquarters MC3810 can be configured to switch calls between branches. Although the call volume may be low, the per-minute cost in this case is very high. By off-loading this tandem function from the PBX, the delay budget is reduced and the quality is maintained by avoiding the compression/decompression cycle required by the PBX. This additional savings was not included in this case study.

The final upgrade and the cost to upgrade are shown in Table 13-13.

Futures
The MC3810 can support E1/T1 ATM services via a software change as these services become available. Plus, the CBR and VBR support in the unit prepares this firm for video conferencing when it becomes required.

Table 13-13 *Case 2 Upgraded Equipment and Costs*

Location	Purpose	Access Line Speed	Initial Port Speed	Initial PVC CIR	Frame Relay Charges	Upgraded Port Speed	Upgraded PVC CIR	Frame Relay Charges	Cost to Upgrade
Brussels	HQ	E1	768		$2,800	2×1024		$14,300	$11,500
Stockholm	Branch	E1	128	64	$2,690	256	128	$4,958	$2,268
Milan	Branch	E1	128	64	$3,010	256	128	$6,238	$3,228
London	Branch	E1	128	64	$3,010	256	128	$4,958	$1,948
Madrid	Branch	E1	128	64	$3,170	256	128	$6,238	$3,068
Paris	Branch	E1	128	64	$3,010	256	128	$4,958	$1,948
Munich	Branch	E1	128	64	$3,010	256	128	$4,958	$1,948
Geneva	Branch	E1	128	64	$3,110	256	128	$6,238	$3,128
Dublin	Branch	E1	128	64	$3,110	256	128	$6,238	$3,128
TOTAL					**$26,920**			**$59,084**	**$32,164**

Case 2: Financial Analysis

The capital costs for the branches and the headquarters are shown in Table 13-14. The required additional bandwidth indicated in Table 13-13 costs $32,164 per month.

Table 13-14 *Case 2 Capital Costs*

Equipment/Location	Number of Locations	Cost per Location	Total
Cisco MC3810 access routers	1	$72,000	$72,000
Branch PBX trunk modules	8	$5,400	$43,200
Headquarters PBX trunk modules	1	$12,300	$12,300
Overall capital cost			**$127,500**

To calculate the payback period, divide the total capital cost by the total monthly savings. For this network design, you must first compute monthly savings. Figure 13-16 shows the equation for computing monthly savings.

Figure 13-16 *Equation for Computing Payback Period*

(Original monthly PSTN cost) x (.95) = Monthly savings

(Monthly savings) − (Monthly upgrade cost for added bandwidth)
= **Net total monthly savings**

$$\frac{\text{Total capital cost}}{\text{Net total monthly savings}} = \textbf{Payback period in months}$$

Table 13-15 illustrates the net monthly savings. This firm saves more than $230,000 per year by moving on-net voice traffic onto the Frame Relay network.

Table 13-15 *Case 2 Savings and Payback*

Savings/Costs/Payback	Amount
Monthly PSTN voice expenses	$54,101
Multiply by 95% (P.05)	0.95
Monthly PSTN voice savings	$51,396
Required added bandwidth	−$32,164
Net total monthly savings	**$19,232**
Net annual savings	**$230,787**
Total capital costs	$127,500

Table 13-15   *Case 2 Savings and Payback (Continued)*

Savings/Costs/Payback	Amount
Installation	$20,000
Total capital costs	**$147,500**
Payback period (months)	7.7

The payback period is less than eight months. Figure 13-17 graphically shows this payback period and summarizes the important numbers.

Figure 13-17   *Financial Analysis for the Pan-European Firm*

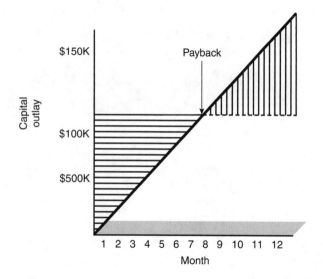

This firm reduced its annual networking expenses by over $230,000 by redesigning its network to carry voice. The company paid for the equipment in less than eight months.

The redesign costs include additional Frame Relay service expenses required to transport voice traffic.

An organization's voice cost per minute is a common expense measurement. Table 13-10 provides the cost details required to calculate the original cost per minute. The total on-net minutes per month for all offices equals 171,626 minutes and the total monthly cost is $54,101. Dividing total cost by total minutes yields an average cost per minute of $0.32 for all calls between headquarters and the branches. To carry 95 percent of this traffic over the data network, that network must be upgraded at a cost of $32,164 per month. Ninety-five percent of 171,626 equals 163,044 minutes. Dividing the upgrade costs by this amount yields an average cost per minute of $0.20. In this case, the average cost per minute was reduced from $0.32 to $0.20, a 38 percent reduction.

Case 2: Summary

In this case, annual savings of over $230,000 were realized while maintaining voice quality.

The branch-to-branch calling volume was not figured in. Though the traffic volume is likely to be small, the savings could be quite large, given the distance between the offices.

These savings could easily be realized by utilizing the tandem functions of the MC3810.

Case 3: Multinational Firm

This relatively small yet fast-growing organization has seven offices. Besides the San Jose headquarters, it has offices in London, Frankfurt, Milan, Hong Kong, Tokyo, Chicago, and New York. Figure 13-18 depicts the connections for the multinational firm.

Figure 13-18   *The Case 3 Multinational Firm Network*

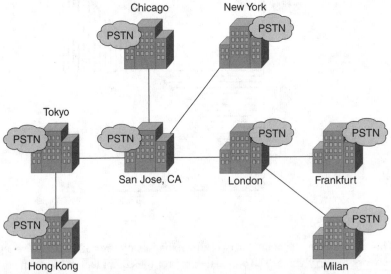

For data communications, the firm has a routed network using private lines. An eight-node data network, leased by the firm, utilizes routers and is hubbed out of its San Jose headquarters. A number of branches connect through other branches to reach the headquarters location. This arrangement was set up to hold down leased-line costs.

Key systems and small PBXs connected over the PSTN provide voice services. Again, it is assumed the firm has a VPN contract at about a 15 percent discount off list prices from standard PSTN rates.

Case 3: Auditing the Original Network Design

The branches use key systems or small PBXs to serve their phone needs, and a separate router for their leased-line IP network. Each branch office has 15 to 45 people. Figure 13-19 shows the current design.

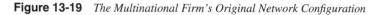

Figure 13-19   *The Multinational Firm's Original Network Configuration*

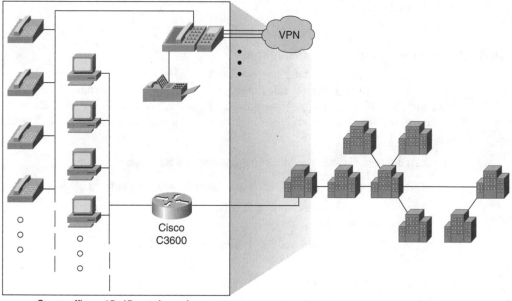

Seven offices: 15–45 people each

The majority of calls are between the branch staff members and local customers. Branch employees talk on the phone and use the fax an average of 2.5 hours per day. Only about 20 percent of this traffic is back and forth with headquarters.

Although the branch-to-headquarters traffic volume is small, it is also the most expensive on a per-minute basis because it is billed at international rates. Consequently, for international long-distance services, this firm pays approximately $32,000 per month, or about $390,000 annually.

Case 3: Setting Network Objectives

The objective of the multinational firm is to reduce long-distance toll charges from branches to headquarters. The company does not plan to make the transition to Frame Relay but would like to retain that option. It is also considering aggregating its local dial traffic for telecommuters, especially in New York and Hong Kong, where there is a great demand for telecommuting.

Case 3: Reviewing Technologies and Services

This firm has chosen to implement VoIP to leverage its current infrastructure and provide quality voice connections. The data network may be capable of handling the increased traffic, but to be sure, we have increased the bandwidth of most circuits. The increased bandwidth required on the data network costs much less than the long-distance charges.

Table 13-16 shows the current on-net voice and fax traffic volume and expense from the branch offices.

Case 3: Establishing Technologies Guidelines

It is essential that redesign of the data network to support the added voice traffic be accomplished without adversely affecting performance. The plan is to have the PSTN cost reductions pay for the redesign. Though any of the packet voice technologies could be used to build this multiservice network, given the firm's infrastructure and expertise in IP, a VoIP network is chosen.

In redesigning the network the following assumptions were used:

- There are approximately 15 people per small branch, 45 per large branch.
- The bi-directional voice and fax call volume totals about 2.5 hours per person per day per branch.
- About 20 percent of the total call volume is between headquarters and each branch location.
- A peak load hour loading factor of 17 percent is appropriate.
- The Cisco voice compression module uses 8 kbps, plus 1-kbps overhead per call. It was assumed that a 64-kbps trunk circuit supports five calls, rather than seven. This is a conservative estimate.
- In the small branches, one key system trunk module would be required, whereas two cards would be necessary for the large branches.

Using the assumptions and the following calculations, the amount of voice and fax traffic at each branch office that would optimally be diverted from the PSTN to the multiservice network is as follows:

- 2.5 hours call volume per user per day × 15 users = 37.5 hours daily call volume per office
- 37.5 hours × 60 minutes per hour = 2,250 minutes per day
- 2,250 minutes × 17 percent (peak load hour) = 382.5 minutes per peak load hour
- 382.5 minutes per peak load hour × 1 erlang/60 minutes per peak load hour = 6.375 erlangs

6.375 erlangs × 20 percent of traffic to headquarters = 1.275 erlangs volume proposed

Table 13-16 *Case 3 PSTN Voice and Fax Volume and Expenses*

Location	Purpose	No. of People	Avg. Min. per Pers. per Day	On-Net % to HQ	Workdays per Month	Tot. Min. per Pers. per Mo.	Tot. Min. per Office per Mo.	Cost per Minute*	Monthly Cost per Office
Frankfurt	Branch	15	150	20	21.67	650	9,752	$0.54	$5,266
Milan	Branch	15	150	20	21.67	650	9,752	$0.48	$4,681
London	Branch	45	150	20	21.67	650	29,255	$0.29	$8,484
New York	Branch	45	150	20	21.67	650	29,255	$0.07	$2,048
Chicago	Branch	15	150	20	21.67	650	9,752	$0.07	$683
Tokyo	Branch	15	150	20	21.67	650	9,752	$0.52	$5,071
Hong Kong	Branch	15	150	20	21.67	650	9,752	$0.63	$6,095
TOTAL							107,267		$32,326

*This is the average cost per minute and assumes that 50 percent of calls are to headquarters and 50 percent are from headquarters.

Considering these assumptions, we will use a conservative redesign. We need to determine the additional bandwidth required on the data network to support the voice and fax traffic. As indicated in Case 2, there are two approaches to determining bandwidth requirements. The best way is to collect traffic information from both the key system or PBX and the router, and then graphically add the voice and data traffic. Collecting traffic information enables one to see how often the combined voice and data traffic would exceed the available bandwidth. But this kind of traffic information is often unavailable. If this information is not available, the easiest method to establish the proper upgrade would be to estimate whether (and how much) extra bandwidth would be required and provision that amount. Then the voice traffic can be added to the data stream while tracking two measures of performance: user-reported voice quality and data latency. If either performance measure appears to be insufficient, then bandwidth should be added. Data and voice traffic frequently peak at different times in the day. The data network will frequently benefit from the added bandwidth.

Trunking Analysis

To determine the appropriate number of trunks required to carry the traffic, we next consult traffic engineering tables. This firm chose a P.05 grade of service. Shown in Table 13-17 are the applicable sections of the erlang C tables.

Table 13-17   *Case 3 Erlang C Table*

Blocking Probability (Grade of Service)	Small Branch Traffic to HQ (1.275 Erlangs)	Large Branch Traffic to HQ (3.825 Erlangs)
P = 0.01	5 trunks	10 trunks
P = 0.05	4 trunks	8 trunks
P = 0.10	3 trunks	7 trunks
P = 0.20	3 trunks	6 trunks

Using the calculated erlangs and the information in Table 13-18, it turns out that four trunks are required in the smaller offices and eight trunks in the larger ones. Table 13-18 summarizes the calculations and trunking, and it gives the figures for the larger branch office.

Table 13-18 *Case 3 Trunking Analysis*

Site	No. of Users	Call Volume per Day	Minutes per Day	Busy Hour (17%)	Minutes per Busy Hour	% of Traffic to HQ	Total Erlangs to HQ	Required Trunks for 0.05 Blocking Probability
Large	45	2.5	6750	0.17	1147.5	20	3.825	8
Small	15	2.5	2250	0.17	382.5	20	1.275	4
						2 Large	7.65	
						5 Small	6.375	
					TOTAL:	Erlangs for HQ	14.025	23

Figure 13-20 shows the redesign at the smaller branches. Here a Cisco 3620 is installed and connected to the key system. The central office trunks to the VPN remain for local and off-net voice traffic, but the quantity could be reduced. This reduction is not included in the financial model and represents additional savings.

Figure 13-20 *Multinational Firm Smaller Office Redesign*

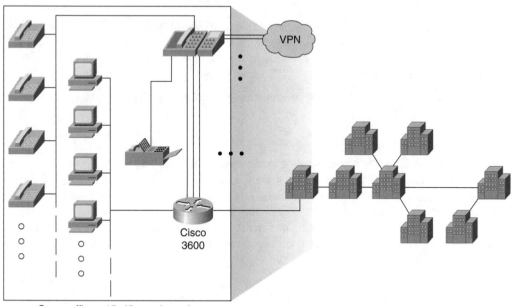

Seven offices: 15–45 people each

If required, the necessary key system interfaces are expanded. Just in case, we have added a key system interface module to each office and included it in the financials.

The key systems have been reprogrammed to include the router in the least-cost routing tables, thereby routing the calls to the voice-enabled Cisco 3600, when the destination is an interoffice number. Because this firm is leveraging its existing infrastructure, it can actually call any other office as well, but this traffic is not accounted for in this model.

At the two larger offices, a Cisco 3640 is installed and connected to the small PBX.

Case 3: Planning Capacity

Bandwidth was increased to handle the additional voice traffic on the private-line network.

To support 4 simultaneous voice calls, only 36 kbps of additional bandwidth is needed. But to be conservative, bandwidth on most circuits was increased from 64 kbps to 128 kbps. The San Jose-to-London circuit was increased further to accommodate London's own voice traffic as well as that transiting through to San Jose. The topology in this network has not changed, but the bandwidth has been increased.

The VoIP section of the design steps noted that certain mechanisms are available to ensure voice quality. These mechanisms, such as RSVP and queuing, have been configured in the Cisco routers.

Because this is a routed network, any office can now call any other office and the traffic will be routed over the IP network. However, the delay budget should first be calculated to ensure that the quality is within the firm's requirements. For calls that do not meet the requirement, the key system or PBX would be configured to use only the VPN.

Note that on-net fax traffic is also routed across the IP network.

Customer Premises Equipment

A Cisco 3620 router was installed in the smaller branches and a Cisco 3640 router was installed in the two larger branches. Each smaller branch office connected four key system FXO trunks to the Cisco 3620 router (eight trunks to the Cisco 3640 in the larger branch locations). The reprogrammed key system would then direct approximately 95 percent of traffic destined for headquarters preferentially to one of the trunks connected to the Cisco router. The remaining overflow on-net voice traffic, an estimated 5 percent, is directed to the PSTN.

The leased lines terminate at the San Jose headquarters, where the voice channels are decompressed and routed to the headquarters PBX. Because 23 channels have been removed from the PSTN and are now sent over the router network, the San Jose headquarters can remove one of the PSTN T1 access lines.

The Cisco 3600 enables high performance and low latency, which result in very high voice quality. This firm is planning a possible migration to Frame Relay and support for telecommuting. In Hong Kong and New York, where commuting to work can be very difficult, the Cisco 3600 will eventually be used as an access server for employees telecommuting. If the firm decides to make the transition to Frame Relay, the Cisco 3600 can support that protocol.

It should be noted that not all key systems provide automatic route selection for preferential voice traffic switching. If you are considering this type of configuration, contact the key system or PBX vendor to ensure this capability exists in the equipment.

Network Redesign Incremental Expense

The additional bandwidth in the multinational firm's redesigned network is not free. Table 13-19 furnishes the incremental expense details.

Table 13-19 *Case 3 Incremental Expense*

San Jose to:	Original 64 kbps (Monthly Cost)	Redesigned 128 kbps (Monthly Cost)	Redesigned 192 kbps (Monthly Cost)	Incremental Cost (Monthly Cost)
Tokyo	$8,700	$13,400		$4,700
Tokyo to Hong Kong	$5,500			—
Chicago	$1,250	$2,050		$800
New York	$1,400		$3,100	$1,700
London	$6,400		$13,400	$7,000
London to Milan	$5,100	$8,150		$3,050
London to Frankfurt	$4,250			—
TOTAL				**$17,250**

Functional Highlights

After the redesign, the firm's network is leveraged to take full advantage of the voice/data integration and is flexible enough to withstand the inevitable changes of internetworking. The following shows the functional highlights of the redesigned network:

- This design has extended the firm's IP network, thereby leveraging the expertise it has already built within the company. The compression algorithm used here was CS-ACELP, which, including overhead, utilizes only 9 kbps per call.

- The Cisco 3600 itself is modular and provides a migration path from private lines to Frame Relay, one of the firm's potential requirements.

- The VoIP is standards-based to allow future integration with H.323 applications such as video delivery and Web-based applications such as Microsoft's NetMeeting.

Case 3: Financial Analysis

Table 13-20 shows the capital costs for the branches and the headquarters in Case 3. The required additional bandwidth indicated in the preceding section costs $22,050 per month.

Table 13-20 *Case 3 Capital Costs*

Equipment/Location	Number of Locations	Cost per Location	Total
Cisco 3620 and 3640 routers			$84,000
Branch Key system modules	9	$700	$6,300
Headquarters PBX trunk modules	1	$5,418	$5,418
Overall capital cost			**$95,718**

Comparing the savings to these additional expenses, the net monthly savings are illustrated in Table 13-21. This firm saves nearly $175,000 per year by moving its internal voice traffic onto its router backbone. Using the equation in Figure 13-16, we can calculate that the payback period is less than eight months.

Table 13-21 *Case 3 Savings and Payback*

Expenses/Savings/Payback	Amount
Monthly PSTN voice expenses	$32,326
Multiply by 95 percent (P0.05)	0.95
Monthly PSTN voice savings	$30,710
E1 removed from HQ to PSTN	$950
Required added WAN links	−$17,250
Net total monthly savings	$14,410
Net total annual savings	$172,919
Capital costs	$95,718
Installation (estimate)	$15,000
Total capital costs	**$110,718**
Payback period (months)	**7.7**

The network redesign saves over $173,000. The payback period of eight months shown in Figure 13-21 is conservative given the installation of additional private-line bandwidth. After new traffic patterns are established, it may be possible to reduce bandwidth.

Figure 13-21   *Financial Analysis for the Multinational Firm*

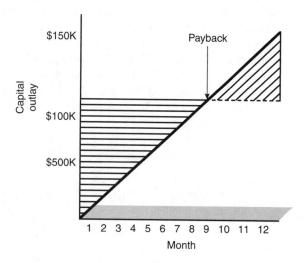

Table 13-21 provides the details required to calculate the original cost per minute. The total on-net minutes per month for all offices equals 107,267 minutes and the total monthly cost is $32,326. Dividing $32,326 by 107,267 yields an average cost per minute of $0.30 for all calls between headquarters and the branches. To carry 95 percent of this traffic over the data network, that network must be upgraded at a cost of $17,250 per month. Ninety-five percent of 107,267 equals 101,904 minutes. Dividing the monthly upgrade costs by the carried minutes yields an average cost per minute of $0.17. In this case, the firm was able to reduce its average cost per minute from $0.30 to $0.17, a 43 percent reduction.

Case 3: Summary

The multinational company in Case 3 was able to save over $173,000 by adjusting bandwidth and extending its IP network. By using the Java-based manager, configuration and deployment were simplified. By using the Cisco 3600 family of routers, the company was able to anchor its network in modular technology to allow for growth and further development.

Summary

This chapter provides background on the aspects of voice and data networking that must be considered when designing a multiservice integrated voice data network. It covers the available technologies—VoIP, VoFR, and VoATM—and provides a framework to help analyze the optimum solution. The chapter outlines the six-step process of designing an integrated voice and data network, followed by case studies demonstrating this process. The case studies show examples, using actual scenarios and costs, of how the technologies could be implemented and the benefits obtained. When upgrading networks, you need to use the six-step process to analyze traffic patterns, account for quality requirements, calculate delay budget, and decide upon the right design for your organization's individual needs. The solution you choose will be based on current and future needs as well as the existing environment and expertise.

PART V

Appendixes

Answers to Review Questions

Chapter 1

1 Name the three components of a leased-line network.

Answer: Two tail-circuits, IXC circuits, and LEC circuits.

2 Name the two types of leased-line networks.

Answer: Point-to-point and multi-drop networks.

3 Name the three layers of the Cisco network design model and briefly describe their functions.

Answer: Access layer—the connection point for end-user devices on the network; distribution layer—the aggregation point for devices on the network, functions include policy, security, and summarization; core layer—the backbone area of the network, used strictly for high-speed switching between one distribution area and another.

4 What are the chief benefits of implementing an AVVID network?

Answer: An enterprise can reduce costs and increase productivity through the use of a single network for all of their communication needs.

5 What is a commonly implemented ITU-T standard used for voice compression?

Answer: G.729 and G.729a are commonly implemented ITU-T standards used for compressing voice data.

6 What are two business markets that can benefit from implementing Cisco's AVVID technology?

Answer: Both the enterprise market and the service provider can benefit from implementing AVVID in their networks to decrease costs and increase productivity.

Chapter 2

1 What is the name for the normal state of a telephone that is hung up?

Answer: On-hook.

2 What component in the telephone set converts from 2-wire to 4-wire, and what other function does it serve?

Answer: Hybrid; it also provides feedback or local echo to the earpiece.

3 Name the three types of signaling used to establish a call and their general purpose.

Answer: Supervisory, used for setup and teardown of a connection; address, used to determine the method of encoding the digits dialed; informational, used to provide call progress and alert indicators to the telephone system users.

4 List the three types of E&M start signaling.

Answer: Wink start, immediate start, and delay start.

5 What is the most common E&M tie-line type used in North America and what are its distinguishing characteristics?

Answer: Type I, uses only two leads and a common signaling ground.

6 What is the most common E&M tie-line type used outside of North America and what are its distinguishing characteristics?

Answer: Type V, uses only two leads (E&M) and the signal ground (SG) lead for a common signaling ground.

7 What are the two types of address signaling covered in this chapter?

Answer: Pulse and DTMF.

8 What device solves N^2 problem?

Answer: A switch.

9 What are the two trunk types one would order to provide dedicated lines for incoming calls and outgoing calls?

Answer: DID and DOD trunks.

10 What are two methods of dealing with echo?

Answer: Echo suppression and echo cancellation.

Chapter 3

1 What is the rate at which analog voice samples are taken in G.711 PCM coding?

Answer: 8000 times per second.

2 What are two examples of hybrid coding?

Answer: CS-ACELP and LD-CELP.

3 How does the ITU-T express voice quality measurement?

Answer: Mean Opinion Score (MOS).

4 How many contiguous frames are contained in a Superframe?

Answer: 12.

5 What T1 line-coding method guarantees that Ones Density requirements will be met?

Answer: B8ZS.

6 What is the D4 framing pattern used in T1 Superframes?

Answer: 1000 1101 1100.

7 What is the framing format of an E1 circuit called?

Answer: Multiframe.

8 What are the two line-coding methods used for E1?

Answer: AMI and HDB3.

9 How long does one bit time last in T1?

Answer: 648 Nanoseconds.

10 What is the name of the external signaling network used in the telephony network?

Answer: SS7.

Chapter 4

1 Which Cisco product provides an end-to-end integrated multiservice solution in a single fixed configuration chassis?

Answer: The MC3810 (multiservice concentrator).

2 True or False: The 3660 router supports analog and digital voice ports.

Answer: True: The 3660 supports up to 24 analog voice interfaces or 360 digital voice interfaces.

3 True or False: The 3640 can perform VoIP/VoFR/VoATM functions.

Answer: True: All of the 3600 series devices can perform these functions.

4 List the three components of the digital T1/E1 packet trunk module.

Answer: Multiflex trunk (MFT VWIC), high-density voice network module (NM-HDV), and packet voice DSP module (PVDM).

5 Which analog interface card allows connection to a central office of the PSTN?

A) FXS
B) FXO
C) E&M

Answer: (B) The FXO interface is a interface that allows an analog connection to be directed at a CO of the PSTN.

Chapter 5

1 What is the connection type that allows for calls between two devices connected to the same router?

Answer: Local.

2 What is a connection that automatically dials when a phone is lifted off-hook?

Answer: PLAR.

3 Which interface allows for what connections to the PSTN?

Answer: FXO (foreign exchange office).

4 Which interface connects directly to a phone handset and provides dial tone, line power, and ring voltage?

Answer: FXS (foreign exchange switch).

5 Which of the voice-over technologies do analysts predict will have long-term growth potential?

Answer: VoIP.

6 The H.323 standard provides a foundation for what?

Answer: Standardized voice, video, and data communications over an IP network.

7 Hoot-and-holler allows customers to save money by eliminating what, using VoIP or VoFR networks?

Answer: Expensive leased lines.

8 The use of small, fixed-size frames in VoATM reduces what in voice networks?

Answer: Delay queuing and delay variations.

9 Which standard described in this chapter is designed to operate on various vendor equipment?

Answer: H.323.

10 What application allows users to get all of their messaging from a single point?

Answer: Unified messaging.

Chapter 6

1 Which module is required in the 3600 and 2600 series routers to provide a slot for VICs?

Answer: The voice network module (VNM).

2 On the 2600, how many voice ports can be installed in a single-network module slot?

Answer: Four ports (two WICs, with two ports each).

3 What are the three types of WICs that can be installed in the VNM?

Answer: FSX, FXO, and E&M.

4 True or False: All 3600 series routers have a console connection on the front panel.

Answer: False: the 3660 console connection is on the rear.

5 What do you check on the 2600/3600 to verify proper installation of a VIC module?

Answer: Check that the LED light corresponding to the VNM slot you installed is on.

6 MC3810 design chassis were modified to accommodate which module?

Answer: The video dial module (VDM).

7 If the MC3810 shuts down after operating for some period of time, what is a possible cause?

Answer: The device is overheating.

8 When installing voice modules in the AS5300 at least one feature card must be what?

Answer: One feature card slot must have either a quad T1/PRI or a quad E1/PRI installed for WAN and telephony device connections.

9 What does MICA stand for?

Answer: Modem ISDN channel aggregation.

10 True or False: The AS5300 uses the same cables as the 2600/3600 routers.

Answer: True.

Chapter 7

1 Which voice port standard allows connectivity to a standard PBX interface or the PSTN?

Answer: The FXO standard.

2 What is a call leg?

Answer: A call leg is a logical connection between the router and an endpoint.

3 Which global configuration command allows you to configure properties on an FXS port?

Answer: The voice-port command.

4 What is the European codec standard for ITU-T Pulse Code Modulation?

Answer: a-law.

5 True or False: Voice transmissions are like data transmissions and are not seriously affected by delay of packet delivery.

Answer: False.

6 How would you configure the router to expand the dialed digits 55512 to include a 1 and the area code 859?

Answer: num-exp 55512 185955512.

7 What is the first thing you should check when troubleshooting a voice port?

Answer: Check for dial tone.

8 What powerful troubleshooting tool can severely impact a router's performance?

Answer: The debug commands.

Chapter 8

1   What kind of Frame Relay network minimizes transit hops and maximizes the ability to establish different qualities of service?

Answer: A full mesh configuration provides the shortest distance between any two sites and the greatest flexibility with regard to establishing different QoS levels between any two sites.

2   Name the two general categories of delay and three specific types of delay within each.

Answer: Fixed delay and variable delay are two general categories of delay. Fixed delay includes propagation delay, serialization delay, and processing delay. Variable delay includes de-jitter buffers, queuing delay, and WAN or service provider delay.

3   What is the standards-based mechanism for providing fragmentation of data over Frame Relay?

Answer: The Frame Relay Forum's FRF.12 provides for fragmentation of data over Frame Relay.

4   What is the standards-based mechanism for providing fragmentation of data in a PVC used for VoFR?

Answer: The Frame Relay Forum's FRF.11 Annex C provides for fragmentation of data in a PVC used for VoFR.

5   What command is used to configure PQ-WFQ?

Answer: The voice bandwidth command in the map-class frame-relay class command group is used to provide a priority queue for voice traffic.

6   What is the preferred queuing method for VoFR?

Answer: The preferred queuing mechanism is priority queuing-weighted fair queuing (PQ-WFQ).

7   How do you configure the CIR for traffic shaping?

Answer: The frame-relay cir command in the map-class frame-relay class command group is used to provide a priority queue for voice traffic.

8   How do you configure per-virtual circuit queuing and traffic shaping for all PVCs on a Frame Relay interface?

Answer: You configure per-virtual circuit queuing and traffic shaping for all PVCs on a Frame Relay interface through the use of the frame-relay traffic-shaping command on the interface.

9 What is the maximum one-way delay acceptable when designing a VoFR network?

Answer: The maximum one-way delay acceptable when designing a VoFR network is 150–200 ms.

10 The overhead in a VoFR frame is how many bytes?

Answer: There are 6 to 8 bytes of overhead in a VoFR frame.

Chapter 9

1 What are the two general categories of ATM service classes and what are they typically used for?

Answer: Real-time and non-real-time. Real-time is typically used for delay-sensitive traffic like video and voice; non-real-time is typically used for data traffic.

2 What traffic service class is generally recommended for voice traffic and why?

Answer: Variable bit-rate-real-time, because it provides QoS to insure that delay does not exceed required limits.

3 Describe an ATM cell, including header bytes and payload bytes.

Answer: There are 53 bytes in an ATM cell, 5 in the header and 48 in the payload.

4 Describe the formula for calculating the bandwidth required for a particular codec.

Answer: The formula used is: ATM bandwidth = ([cells per sample × 53] × 8) / sample rate in milliseconds (ms).

5 What are the guidelines for calculating the ATM PVC traffic-shaping parameters for a VC used to carry voice calls?

Answer: Peak value = 16 × number of calls × 2, Average value = 16 × number of calls, Burst value = 4 × number of calls.

6 What ATM encapsulation types are used for voice and data?

Answer: encapsulation aal5mux voice is used for voice, and encapsulation all5snap is typically used for data.

7 What is the overhead factor used in calculating the total bandwidth necessary for data PVCs?

Answer: The factor used is 1.13 times the application's required bandwidth.

8 On the Cisco MC3810, what must be configured in order to permit traffic on the ATM0 interface?

Answer: The controller t1 0 must be configured with the mode atm command.

9 How many channels are used for ATM on a T1?

Answer: All 24 channels on a T1 are used for ATM.

10 Which **show** command will display the ATM configuration for each PVC?

Answer: The show atm pvc command will display the ATM configuration for each PVC, including encapsulation, service category, peak rate, average rate, burst cells, and status.

Chapter 10

1 How many bytes does the RTP header add to the UDP header?

Answer: 12 bytes.

2 What is the ITU-T standard that specifies recommended one-way voice delay?

Answer: ITU-T G.114.

3 In configuring VoIP networks, what command is used to enable fragmentation, but fragments all frames leaving the interface?

Answer: mtu *size.*

4 What ISDN signaling protocol does the ITU-T H.323 protocol use for connection requests?

Answer: Q.931.

5 What IETF RFC does RSVP follow?

Answer: RFC 2205.

6 Which show command displays the dial-peer information in a summarized table?

Answer: show dial-peer voice summary.

7 RTP header compression can reduce the IP/UDP/RTP header from _____ bytes to about 2 or 4.

Answer: 40.

8 Which queuing method uses four queues, called High, Medium, Normal, and Low?

Answer: Priority Queuing.

9 By default, a VoIP packet contains how many bytes of payload?

Answer: 20.

10 At Layer 2, how many bytes of overhead in VoIP applications does Ethernet use?

Answer: 14.

Chapter 11

1 What two Cisco IP Telephony components are used to connect to the PSTN?

Answer: A Cisco IOS Gateway and a digital trunk gateway can be used to connect to the PSTN.

2 What are three skinny station protocol message types?

Answer: Call control, message control, registration and management.

3 What are two functions of a Cisco IP Phone?

Answer: The Cisco IP Phone functions as a codec and DHCP client.

4 Which standard protocol does the Cisco IP Phone use to communicate with the Cisco CallManager?

Answer: The Skinny Station Protocol, as published by Cisco Systems, Inc., is the de facto standard used by the Cisco IP Phone.

5 What protocols are used to carry Voice over IP traffic?

Answer: IP, UDP, and RTP are used to carry Voice over IP.

6 What are the two Media Convergence Servers covered in this book?

Answer: The MCS-7822 and the MCS-7835.

7 List three voice-access gateways.

Answer: The DT-24+/DE-30+, the VG-200, and the Catalyst 6000 Family Voice T1 Module.

8 Which IP Phone(s) discussed in the text has an Ethernet switch integrated in it?

Answer: Both the Cisco IP Phone 7910+SW and the Cisco IP Phone 7960 have 10/100 Ethernet switches integral to the phone.

9 How much bandwidth does a G.729 call consume?

Answer: A G.729 call consumes 24 kbps of bandwidth.

10 What two types of conferences are supported in the CallManager architecture?

Answer: The CallManager software supports both Ad-Hoc and Meet-Me conferences using software conference bridging or hardware DSP resources.

Chapter 12

1 What are the four primary differences between a PBX and a key system?

Answer: PBXs are primarily digital while key systems can be digital or analog; PBXs are similar in architecture to a CO switch—a key system is not a switch; PBXs are typically deployed in large companies or sites—key systems are typically used for a branch office or small company; on a PBX you typically dial "9" to access an outside line—on a key system you press a button to access an outside line.

2 What are the major components of the PSTN in the North American market and who provides them?

Answer: Interexchange carriers, or long-distance companies, typically provide long-distance service. Local exchange carriers typically provide local dial tone. Public telephone operators (PTO or PTT) are typically government monopolies that provide all telephone access in their country.

3 What are the four types of switching-network designs for a PBX?

Answer: Centralized control, distributed control, dispersed control, and adjunct control PBX switch designs.

4 What types of circuit cards are typically installed in a PBX and what are their functions?

Answer: System circuit cards for processor controllers, memory, switch network, and maintenance cards. Service circuit cards for touch-tone receivers, conference, input/output, cabinet links, and host and client/server CTI links. Line cards for providing analog, digital, ISDN BRI and IP services. Trunk cards for providing analog and digital interfaces.

5 What is busy-hour call handling?

Answer: Busy Hour Call Attempts (BHCA) is a means of measuring the peak load of a telephone switch such as a PBX. The PBX should be designed to meet the peak-hour needs as far as outbound trunks are concerned so that no one receives a busy signal when trying to seize an outside line.

6 What are the main distribution frame (MDF) and intermediate distribution frame (IDF) used for?

Answer: The MDF connects telephone trunks from the local CO to the premises wiring. The IDF connects the MDF to the individual telephone wires.

7 Where is call processing handled in the client/server PBX design?

Answer: Call processing is handled by application processors, not the PBX.

8 What functions do line cards provide in a PBX?

Answer: Line cards provide analog services, digital services, and ISDN BRI interfaces to line-side devices such as phones.

9 What are the five categories of signaling in a telecommunications network?

Answer: Supervisory signaling, information signaling, address signaling, control signaling, and alert signaling.

10 What is the purpose of the PBX backplane?

Answer: The PBX backplane ties the lines, trunks, and central control circuits to the switching fabric and buses.

APPENDIX B

Further Reading

There are several good books on routing, switching, telephony, and voice-over technologies. This list is not meant to be exhaustive, but to provide a starting point. These books provide either a good narrative of the technology involved or serve as an excellent reference. As in any fast-changing technical field, there will be new books published that supersede these.

Traditional Telephony

Bates, Regis J., *Introduction to T1/T3 Networking,* Boston: Artech House, 1992.

Bell Telephone Laboratories, Inc., *Engineering and Operations in the Bell System,* Indianapolis: Western Electric Company, Inc., 1977.

Bellamy, John. *Digital Telephony,* 3rd ed., New York: John Wiley & Sons, Inc., 2000.

Bhatnagar, P. K., *Engineering Networks for Synchronization, CCS 7, and ISDN: Standards, Protocols, Planning, and Testing,* New York: Institute of Electrical and Electronics Engineers, Inc., 1997.

Bigelow, Stephen J., *Understanding Telephone Electronics,* 3rd ed., Indianapolis: SAMS, 1991.

Flanagan, William A., *T-1 Networking: How to Buy, Install and Use T-1 from Desktop to DS-3,* 5th ed., New York: Telecom Books, 1997.

Freeman, Roger L., *Telecommunication System Engineering,* 3rd ed., New York: John Wiley & Sons, Inc., 1996.

Mazda, Fraidoon, ed. *Telecommunications Engineer's Reference Book,* Oxford: Butterworth-Heinemann, 1996.

Pecar, Joseph A., Roger J. O'Connor, and David A. Garbin, *The McGraw-Hill Telecommunications Factbook,* New York: McGraw-Hill, Inc., 1993.

Reeve, Whitham D., *Subscriber Loop Signaling and Transmission Handbook: Digital,* New York: Institute of Electrical and Electronics Engineers, Inc., 1995.

Reeve, Whitham D., *Subscriber Loop Signaling and Transmission Handbook: Analog,* New York: Institute of Electrical and Electronics Engineers, Inc., 1992.

van Bose, John G., *Signaling in Telecommunications Networks,* New York: John Wiley & Sons, Inc., 1998.

Voice-over Technologies

Black, Uyless, *Voice Over IP*, Upper Saddle River, NJ: Prentice Hall PTR, 2000.

Davidson, Jonathon, and James Peters, *Voice over IP Fundamentals,* Indianapolis: Cisco Press, 2000.

Douskalis, Bill, *IP Telephony: The Integration of Robust VoIP Services,* Upper Saddle River, NJ: Prentice Hall PTR, 2000.

Flanagan, William A., *Voice Over Frame Relay,* New York: Telecom Books, 1997.

Minoli, Daniel, and Emma Minoli, *Delivering Voice over Frame Relay and ATM,* New York: John Wiley & Sons, Inc., 1998.

Minoli, Daniel, and Emma Minoli, *Delivering Voice over IP Networks,* New York: John Wiley & Sons, Inc., 1998.

Network Design, Routing, and Switching

Armitage, Grenville, *Quality of Service in IP Networks,* Indianapolis: Macmillan Technical Publishing, 2000.

Birkner, Matthew H., *Cisco Internetwork Design*, Indianapolis: Cisco Press, 1999.

Chappell, Laura, *Advanced Cisco Router Configuration*, Indianapolis: Cisco Press, 1998.

Cisco IOS Dial Solutions, Indianapolis: Cisco Press, 1998.

Clark, Kennedy, and Kevin Hamilton, *Cisco LAN Switching,* Indianapolis: Cisco Press, 1999.

Doyle, Jeff, *Routing TCP/IP.* Vol. 1, Indianapolis: Cisco Press, 1998.

Oppenheimer, Priscilla, *Top-Down Network Design*, Indianapolis: Cisco Press, 1998.

Paquet, Catherine, *Building Cisco Remote Access Networks,* Indianapolis: Cisco Press, 1999.

Teare, Diane, *Designing Cisco Networks*, Indianapolis: Cisco Press, 1999.

Useful Web Sites

www.ietf.org—The Internet Engineering Task Force, where you can find drafts and RFCs discussing new ideas for transporting multimedia over the packet network.

www.itu.ch—The International Telecommunications Union, where you can buy and download many of the standards that define the infrastructure of both the old and new voice networks.

www.cisco.com—Cisco Systems, Inc., where you can find information about Cisco's product line, white papers on implementation, and technical documentation.

www.ieee.org—The International Association of Electrical and Electronics Engineers, where you can find several LAN/MAN standards.

www.ansi.org—The American National Standards Institute, where you can find information on U.S. standards.

www.eia.org—The Electronics Industry Association/ Telecommunications Industry Association, where you can find information on EIA/TIA standards.

c5.hakker.com/signalling—A site that lists information on telephony signaling, including R1 and R2 signaling.

ciscoinanutshell.com—A site that provides links to voice and video as well as several other Cisco Systems, Inc.–related sites.

Symbols

μ-law, 59

Numerics

10-high-day (10HD) busy period, 416
1750 series routers, 91
2600 series routers, 92–94
    configuring, 147–148
    digital T1/E1 packet voice trunk, 103–104
    PVDM, 103
    VICs, 102
    voice/fax network module, 101–102
3600 series routers, 95
    Cisco 3620 modular access routers, 95
    Cisco 3640 modular access routers, 95
    Cisco 3660 modular access routers, 95
    configuring, 148, 150–151
    digital T1/E1 packet voice trunk, 103–104
    hardware, verifying installation, 156
    PVDM, 103
    VICs, 102
        port addressing, 155
        selecting, 154–155
    VNMs
        assembly, 155
        selecting, 152
    voice/fax network module, 101–102
7200 series routers, 112–114
7500 series routers, 114, 117

A

A bit, 73
ABBH (average bounding busy hour), 416
ABCD bit pattern, 74
    robbed bit format, 76
    trunk configuration, 237–239
ABR (available bit rate), 308
access codes, 124
access gateways, 375

access layer, 9
    in new-world networks, 18
access numbers, pilot number, 374
access servers, AS5300, 107
    DSP module, 170
    DSP module, selecting, 172
    feature card module, 170
    feature cards, selecting, 171–172
    MICA (modem ISDN channel
      aggregation), 170
    setting up, 168
ACD, 421–422
adaptive differential pulse code modulation
  (ADPCM), 431
adding string description to voice ports, 185
address signaling, 36, 417
Ad-Hoc conferences, 391
adjunct control PBXs, 406
adjusting
    voice port timing parameters, 212–213
    voice quality, 203
admission control, 349
    CCM, 381–383
ADPCM (adaptive differential pulse code
  modulation), 431
advanced router systems (ARSs), Cisco 7500 series,
  114, 117
A-law, 59
alert signaling, 418
alerting indicators, 38
allocation of bandwidth in voice networks, 381–383
analog signaling
    coding, 61
    delay-start, 35
    digitization
        compression, 61–64
        encoding process, 60–61
        quantization process, 58–60
        sampling, 56–57
    immediate-start, 36
    supervisory, 28
    wink-start, 34–35
analog technologies
    access gateways, 375
    E&M signaling

B

C

D

F

W-Z

CCIE Professional Development

Cisco LAN Switching

Kennedy Clark, CCIE; Kevin Hamilton, CCIE

1-57870-094-9 • AVAILABLE NOW

This volume provides an in-depth analysis of Cisco LAN switching technologies, architectures, and deployments, including unique coverage of Catalyst network design essentials. Network designs and configuration examples are incorporated throughout to demonstrate the principles and enable easy translation of the material into practice in production networks.

Advanced IP Network Design

Alvaro Retana, CCIE; Don Slice, CCIE; and Russ White, CCIE

1-57870-097-3 • AVAILABLE NOW

Network engineers and managers can use these case studies, which highlight various network design goals, to explore issues including protocol choice, network stability, and growth. This book also includes theoretical discussion on advanced design topics.

Large-Scale IP Network Solutions

Khalid Raza, CCIE; and Mark Turner

1-57870-084-1 • AVAILABLE NOW

Network engineers can find solutions as their IP networks grow in size and complexity. Examine all the major IP protocols in-depth and learn about scalability, migration planning, network management, and security for large-scale networks.

Routing TCP/IP, Volume I

Jeff Doyle, CCIE

1-57870-041-8 • AVAILABLE NOW

This book takes the reader from a basic understanding of routers and routing protocols through a detailed examination of each of the IP interior routing protocols. Learn techniques for designing networks that maximize the efficiency of the protocol being used. Exercises and review questions provide core study for the CCIE Routing and Switching exam.

CISCO SYSTEMS

CISCO PRESS

www.ciscopress.com

Cisco Career Certifications

Cisco CCNA Exam #640-507 Certification Guide
Wendell Odom, CCIE

0-7357-0971-8 • AVAILABLE IN APRIL

Although it's only the first step in Cisco Career Certification, the Cisco Certified Network Associate (CCNA) exam is a difficult test. Your first attempt at becoming Cisco certified requires a lot of study and confidence in your networking knowledge. When you're ready to test your skills, complete your knowledge of the exam topics, and prepare for exam day, you need the preparation tools found in *Cisco CCNA Exam #640-507 Certification Guide* from Cisco Press.

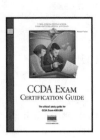

CCDA Exam Certification Guide
Anthony Bruno, CCIE & Jacqueline Kim

0-7357-0074-5 • AVAILABLE NOW

CCDA Exam Certification Guide is a comprehensive study tool for DCN Exam #640-441. Written by a CCIE and a CCDA, and reviewed by Cisco technical experts, *CCDA Exam Certification Guide* will help you understand and master the exam objectives. In this solid review on the design areas of the DCN exam, you'll learn to design a network that meets a customer's requirements for perfomance, security, capacity, and scalability.

Interconnecting Cisco Network Devices
Edited by Steve McQuerry

1-57870-111-2 • AVAILABLE NOW

Based on the Cisco course taught worldwide, *Interconnecting Cisco Network Devices* teaches you how to configure Cisco switches and routers in multi-protocol internetworks. ICND is the primary course recommended by Cisco Systems for CCNA #640-507 preparation. If you are pursuing CCNA certification, this book is an excellent starting point for your study.

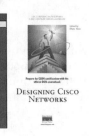

Designing Cisco Networks
Edited by Diane Teare

1-57870-105-8 • AVAILABLE NOW

Based on the Cisco Systems instructor-led and self-study course available worldwide, *Designing Cisco Networks* will help you understand how to analyze and solve existing network problems while building a framework that supports the functionality, performance, and scalability required from any given environment. Self-assessment through exercises and chapter-ending tests starts you down the path for attaining your CCDA certification.

CISCO SYSTEMS
CISCO PRESS

www.ciscopress.com

Cisco Press Solutions

Enhanced IP Services for Cisco Networks
Donald C. Lee, CCIE

1-57870-106-6 • AVAILABLE NOW

This is a guide to improving your network's capabilities by understanding the new enabling and advanced Cisco IOS services that build more scalable, intelligent, and secure networks. Learn the technical details necessary to deploy Quality of Service, VPN technologies, IPsec, the IOS firewall and IOS Intrusion Detection. These services will allow you to extend the network to new frontiers securely, protect your network from attacks, and increase the sophistication of network services.

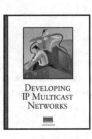

Developing IP Multicast Networks, Volume I
Beau Williamson, CCIE

1-57870-077-9 • AVAILABLE NOW

This book provides a solid foundation of IP multicast concepts and explains how to design and deploy the networks that will support appplications such as audio and video conferencing, distance-learning, and data replication. Includes an in-depth discussion of the PIM protocol used in Cisco routers and detailed coverage of the rules that control the creation and maintenance of Cisco mroute state entries.

Designing Network Security
Merike Kaeo

1-57870-043-4 • AVAILABLE NOW

Designing Network Security is a practical guide designed to help you understand the fundamentals of securing your corporate infrastructure. This book takes a comprehensive look at underlying security technologies, the process of creating a security policy, and the practical requirements necessary to implement a corporate security policy.

CISCO SYSTEMS

CISCO PRESS

www.ciscopress.com

Cisco Press Solutions

EIGRP Network Design Solutions

Ivan Pepelnjak, CCIE

1-57870-165-1 • AVAILABLE NOW

EIGRP Network Design Solutions uses case studies and real-world configuration examples to help you gain an in-depth understanding of the issues involved in designing, deploying, and managing EIGRP-based networks. This book details proper designs that can be used to build large and scalable EIGRP-based networks and documents possible ways each EIGRP feature can be used in network design, implmentation, troubleshooting, and monitoring.

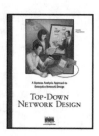

Top-Down Network Design

Priscilla Oppenheimer

1-57870-069-8 • AVAILABLE NOW

Building reliable, secure, and manageable networks is every network professional's goal. This practical guide teaches you a systematic method for network design that can be applied to campus LANs, remote-access networks, WAN links, and large-scale internetworks. Learn how to analyze business and technical requirements, examine traffic flow and Quality of Service requirements, and select protocols and technologies based on performance goals.

Cisco IOS Releases: The Complete Reference

Mack M. Coulibaly

1-57870-179-1 • AVAILABLE NOW

Cisco IOS Releases: The Complete Reference is the first comprehensive guide to the more than three dozen types of Cisco IOS releases being used today on enterprise and service provider networks. It details the release process and its numbering and naming conventions, as well as when, where, and how to use the various releases. A complete map of Cisco IOS software releases and their relationships to one another, in addition to insights into decoding information contained within the software, make this book an indispensable resource for any network professional.

Cisco Press Solutions

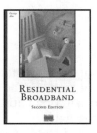

Residential Broadband, Second Edition

George Abe

1-57870-177-5 • AVAILABLE NOW

This book will answer basic questions of residential broadband networks such as: Why do we need high speed networks at home? How will high speed residential services be delivered to the home? How do regulatory or commercial factors affect this technology? Explore such networking topics as xDSL, cable, and wireless.

Internetworking Technologies Handbook, Second Edition

Kevin Downes, CCIE, Merilee Ford, H. Kim Lew, Steve Spanier, Tim Stevenson

1-57870-102-3 • AVAILABLE NOW

This comprehensive reference provides a foundation for understanding and implementing contemporary internetworking technologies, providing you with the necessary information needed to make rational networking decisions. Master terms, concepts, technologies, and devices that are used in the internetworking industry today. You also learn how to incorporate networking technologies into a LAN/WAN environment, as well as how to apply the OSI reference model to categorize protocols, technologies, and devices.

OpenCable Architecture

Michael Adams

1-57870-135-X • AVAILABLE NOW

Whether you're a television, data communications, or telecommunications professional, or simply an interested business person, this book will help you understand the technical and business issues surrounding interactive television services. It will also provide you with an inside look at the combined efforts of the cable, data, and consumer electronics industries' efforts to develop those new services.

www.ciscopress.com

Cisco Press Fundamentals

IP Routing Primer
Robert Wright, CCIE

1-57870-108-2 • **AVAILABLE NOW**

Learn how IP routing behaves in a Cisco router environment. In addition to teaching the core fundamentals, this book enhances your ability to troubleshoot IP routing problems yourself, often eliminating the need to call for additional technical support. The information is presented in an approachable, workbook-type format with dozens of detailed illustrations and real-life scenarios integrated throughout.

Cisco Router Configuration
Allan Leinwand, Bruce Pinsky, Mark Culpepper

1-57870-022-1 • **AVAILABLE NOW**

An example-oriented and chronological approach helps you implement and administer your internetworking devices. Starting with the configuration devices "out of the box;" this book moves to configuring Cisco IOS for the three most popular networking protocols today: TCP/IP, AppleTalk, and Novell Interwork Packet Exchange (IPX). You also learn basic administrative and management configuration, including access control with TACACS+ and RADIUS, network management with SNMP, logging of messages, and time control with NTP.

IP Routing Fundamentals
Mark A. Sportack

1-57870-071-x • **AVAILABLE NOW**

This comprehensive guide provides essential background information on routing in IP networks for network professionals who are deploying and maintaining LANs and WANs daily. Explore the mechanics of routers, routing protocols, network interfaces, and operating systems.

Cisco Press Fundamentals

Internetworking Routing Architectures
Bassam Halabi

1-56205-652-2 • **AVAILABLE NOW**

Explore the ins and outs of interdomain routing network designs. Learn to integrate your network into the global Internet, become an expert in data routing manipulation, build large-scale autonomous systems, and configure the required policies using the Cisco IOS language.

For the latest on Cisco Press resources and Certification and

Training guides, or for information on publishing opportunities, visit

www.ciscopress.com

Cisco Press

ciscopress . com

Committed to being your long-term learning resource while you grow as a Cisco Networking Professional

Help Cisco Press **stay connected** to the issues and challenges you face on a daily basis by registering your product and filling out our brief survey. Complete and mail this form, or better yet ...

Register online and enter to win a **FREE** book!

Jump to **www.ciscopress.com/register** and register your product online. Each complete entry will be eligible for our monthly drawing to win a FREE book of the winner's choice from the Cisco Press library.

May we contact you via e-mail with information about **new releases, special promotions**, and **customer benefits?**

❒ Yes   ❒ No

E-mail address _____

Name _____

Address _____

City _____ State/Province _____

Country_____ Zip/Post code _____

Where did you buy this product?

❒ Bookstore   ❒ Computer store/Electronics store   ❒ Direct from Cisco Systems
❒ Online retailer   ❒ Direct from Cisco Press   ❒ Office supply store
❒ Mail order   ❒ Class/Seminar   ❒ Discount store
❒ Other_____

When did you buy this product? _____ Month _____ Year

What price did you pay for this product?

❒ Full retail price   ❒ Discounted price   ❒ Gift

Was this purchase reimbursed as a company expense?

❒ Yes   ❒ No

How did you learn about this product?

❒ Friend   ❒ Store personnel   ❒ In-store ad   ❒ cisco.com
❒ Cisco Press catalog ❒ Postcard in the mail   ❒ Saw it on the shelf ❒ ciscopress.com
❒ Other catalog   ❒ Magazine ad   ❒ Article or review
❒ School   ❒ Professional organization ❒ Used other products
❒ Other_____

What will this product be used for?

❒ Business use   ❒ School/Education
❒ Certification training   ❒ Professional development/Career growth
❒ Other_____

How many years have you been employed in a computer-related industry?

❒ less than 2 years   ❒ 2–5 years   ❒ more than 5 years

Have you purchased a Cisco Press product before?

❒ Yes   ❒ No

Cisco **Press**

ciscopress.com

How many computer technology books do you own?
❏ 1   ❏ 2–7   ❏ more than 7

Which best describes your job function? (check all that apply)
❏ Corporate Management   ❏ Systems Engineering   ❏ IS Management   ❏ Cisco Networking
❏ Network Design   ❏ Network Support   ❏ Webmaster   Academy Program
❏ Marketing/Sales   ❏ Consultant   ❏ Student   Instuctor
❏ Professor/Teacher   ❏ Other _____

Do you hold any computer certifications? (check all that apply)
❏ MCSE   ❏ CCNA   ❏ CCDA
❏ CCNP   ❏ CCDP   ❏ CCIE   ❏ Other _____

Are you currently pursuing a certification? (check all that apply)
❏ MCSE   ❏ CCNA   ❏ CCDA
❏ CCNP   ❏ CCDP   ❏ CCIE   ❏ Other _____

On what topics would you like to see more coverage?
_____
_____

Do you have any additional comments or suggestions?
_____
_____

Thank you for completing this survey and registration. Please fold here, seal, and mail to Cisco Press.
Cisco Voice over Frame Relay, ATM, and IP (1-57870-227-5)

Indianapolis, IN 46278-8046
P.O. Box #781046
Customer Registration—CP050227
Cisco Press

ciscopress.com
Indianapolis, IN 46290
201 West 103rd Street
Cisco Press